Embedded, Cyber-Physical, and IoT Systems

Shuvra S. Bhattacharyya • Miodrag Potkonjak
Senem Velipasalar
Editors

Embedded, Cyber-Physical, and IoT Systems

Essays Dedicated to Marilyn Wolf on the Occasion of Her 60th Birthday

Springer

Editors
Shuvra S. Bhattacharyya
Dept. of ECE and UMIACS
University of Maryland
College Park
MD, USA

Miodrag Potkonjak
Dept. of Computer Science
University of California
Los Angeles
CA, USA

Senem Velipasalar
Dept. of EECS
Syracuse University
Syracuse NY, USA

ISBN 978-3-030-16951-0 ISBN 978-3-030-16949-7 (eBook)
https://doi.org/10.1007/978-3-030-16949-7

This Springer imprint is published by the registered company Springer Nature Switzerland AG.
The registered company address is: Gewerbestrasse 11, 6330 Cham, Switzerland

Foreword

The search for efficient means of computing and automation has permeated modern society starting from the time of the industrial revolution. Progress has been driven by visionaries who could quickly grasp the entire flow of the computing chain, from data acquisition to information elaboration, to visualization, and to actuation. Such visionaries also provided solutions matching the technology at hand. In this way, computing morphed through the years, from large mainframes to portable devices. Automation became ubiquitous through the use of embedded computing nodes.

Within computing systems, the balance of hardware and software and their concurrent design—dubbed hw/sw co-design—has been an important enabling methodology. Marilyn Wolf has been a pioneer in analyzing co-design tools and methods as well as in creating procedures for optimal co-design in various metrics. As a founding parent of co-design in electronic design automation, her seminal work brought consciousness to the research community of the multitude of issues in co-design and of the synergy of circuit, architecture, and algorithm techniques in addressing the problem.

Later on, as the field of embedded systems emerged as the fastest growing segment of the computing industry, Marilyn Wolf's leadership in research and education became prominent. Her books allowed many of us to understand the various abstractions of distributed computing and control and leverage them in designing tools for embedded systems analysis and synthesis. Her textbook, *Computers as Components*, was seminal as it allowed engineers to put computing elements in the perspective of large systems.

Next, as distributed systems became elaborate means of processing the information and as such information became semantically richer, Marilyn Wolf contributed to the field by proposing and perfecting visual acquisition capturing and processing systems. Distributed image processing is a key constituent of security and defense, and it is used also to monitor industrial production and the environment. Her contribution is present in many tangible aspects of modern society.

Overall, Marilyn contributed to the advancement of electrical engineering and computer science, both in academia and in industry. Her contribution to education and dissemination of technology are well known throughout the world. This book serves as a reference point for scientists and engineers to understand the complex field of computing and its evolution through her outstanding contributions.

EPFL, Lausanne, Switzerland Giovanni De Micheli
2019

Preface

This Festschrift volume is published in honor of Marilyn Wolf, on the occasion of her 60th birthday. The book covers various topics in Embedded, Cyber-Physical, and Internet of Things (IoT) Systems, with emphasis on topics related to Smart Cameras, Hardware/Software Co-Design, and Multimedia Applications. Marilyn has made pioneering contributions of great impact in all of these areas. Embedded systems are everywhere; cyber-physical systems enable monitoring and control of complex physical processes with computers; and IoT technology is of increasing relevance in major application areas, including factory automation, and smart cities. Smart cameras and multimedia technologies introduce novel opportunities and challenges in embedded, cyber-physical, and IoT applications. Advanced hardware/software co-design methodologies provide valuable concepts and tools for addressing these challenges. The purpose of the book is to provide a collection of timely articles that cover important topics in the aforementioned areas, which represent major themes in Marilyn's career.

Marilyn Wolf

Attendees at the Workshop on Embedded Computing, which was held in honor of Professor Wolf on September 21, 2018: Jacques Florence, Graham Hellestrand, Chia-Han Lee, Weichen Liu, Burak Ozer, Nikshep Patil, Dimitrios Serpanos, Umer Tariq, Marilyn Wolf, Yuan Xie, Jiang Xu, Jishen Zhao

Marilyn Wolf is the Georgia Research Alliance Eminent Scholar and Rhesa "Ray" S. Farmer, Jr., Distinguished Chair in Embedded Computing Systems in the School of Electrical and Computer Engineering at Georgia Institute of Technology. Her major distinctions and awards include the ASEE Frederick E. Terman Award (2003), IEEE Circuits and Systems Society Education Award (2006), IEEE Computer Society Golden Core Award, and Fellow of both the ACM and IEEE. She has helped to found major international conferences, including CODES (now CODES-ISSS as part of Embedded Systems Week) and MPSoC. She has written several books that are used widely throughout the world. These include *Embedded System Interfacing*; *Computers as Components*; *High-Performance Embedded Computing*; *The Physics of Computing*; *Smart Camera Design: Algorithms, Architectures, and Art*; and *Internet-of-Things (IoT) Systems*.

The diverse topics of the chapters in this Festschrift help to reflect the great breadth and depth of Marilyn's contributions in research and education. The chapters have been written by some of Marilyn's closest collaborators and colleagues.

The completion of this Festschrift follows a *Workshop on Embedded Computing*, which was held in honor of Marilyn on September 21, 2018, at the Georgia Tech campus in Atlanta, Georgia. The workshop featured technical presentations, as well as personal stories of the tremendous positive influence that Marilyn has had on the careers and lives of people who have worked with her, including those who had the fortune to study under her mentorship. The workshop also included several video greetings from scholars around the world who were unable to attend the event in person.

We would like to thank all of the authors who contributed to this Festschrift and the reviewers who provided constructive feedback during the development of the chapters. We would also like to thank Paul Drougas, Jennifer Evans, and Rachel Toy at Springer for their support.

To Marilyn, we would like to extend our heartiest congratulations on her truly outstanding contributions in research and education. We look forward to many more years of friendship and collaboration with her.

College Park, MD, USA
Los Angeles, CA, USA
Syracuse, NY, USA

Shuvra S. Bhattacharyya
Miodrag Potkonjak
Senem Velipasalar

Contents

i-Core: A Runtime-Reconfigurable Processor Platform for Cyber-Physical Systems

Marvin Damschen, Martin Rapp, Lars Bauer, and Jörg Henkel

Abstract We provide an overview of *i*-Core (invasive Core), a processor platform with a runtime-reconfigurable instruction set. *i*-Core couples a general-purpose processor core with a reconfigurable fabric that enables the configuration of application-specific hardware accelerators at runtime. This way, *i*-Core can adapt its instruction set to choose a runtime trade-off between the resources allocated and the performance achieved for application-specific acceleration. The adaptivity of *i*-Core is leveraged to advance several research areas that are addressed in this chapter. First, we provide an overview of the *i*-Core architecture. Then, we summarize our research findings in the areas of task scheduling, reliability, and hard real-time as well as multi-core systems. The focus of this summary is on our recent findings in the context of worst-case execution time guarantees.

1 Introduction

Reconfigurable computing, i.e., performing computations using a reconfigurable fabric such as field-programmable gate arrays (FPGAs), was introduced in the early 1990s [40]. Today it is an established computing paradigm in a growing number of application domains in research and industry, not only in embedded computing (e.g., signal processing [39], computer vision [30], or encryption [29]), but also in high-performance and scientific computing (e.g., financial pricing [19] or DNA-sequencing [12]), data centers (e.g., searching [35] or database queries [20]), networks (routing [33], intrusion detection [18]), and others. In these domains, applications generally comprise several compute-intensive loops, so-called *computational kernels*, that benefit greatly from implementation as application-specific hardware accelerators in terms of performance and energy efficiency. FPGAs enable the utilization of application-specific hardware accelerators without fabricating

M. Damschen (✉) · M. Rapp · L. Bauer · J. Henkel (✉)
Karlsruhe Institute of Technology, Karlsruhe, Germany
e-mail: damschen@kit.edu; martin.rapp@kit.edu; lars.bauer@kit.edu; henkel@kit.edu

S. S. Bhattacharyya et al. (eds.), *Embedded, Cyber-Physical, and IoT Systems*,
https://doi.org/10.1007/978-3-030-16949-7_1

custom chips and they provide flexibility as well as the ability for upgrades like software.

Several alternatives exist when designing reconfigurable cyber-physical systems that combine a general-purpose CPU with a reconfigurable fabric. Generally, a tighter integration reduces communication latency between CPU and reconfigurable fabric, but requires more effort in the architectural design of the system. Numerous products are available that add an FPGA as a separate chip to an existing system by attaching it to the system's peripheral bus. However, it is crucial for cyber-physical applications to minimize (1) the communication latency between CPU and reconfigurable fabric to enable acceleration of short-running kernels (e.g., in control loops) and (2) the system's power consumption as well as (3) area footprint. Therefore, the advancing trend of processor integration has resulted in *reconfigurable SoCs* that combine FPGAs and CPUs on a single chip (e.g., Xilinx Zynq or Intel (formerly Altera) SoC FPGA), which have led to the wide adoption of reconfigurable systems in cyber-physical systems, e.g., in implementations of advanced driver assistance systems in the automotive domain. In reconfigurable SoCs, CPU and FPGA are still separate processing devices that communicate over the (internal) system bus.

In this chapter, we demonstrate that an even tighter integration of CPU and FPGA than in current reconfigurable SoCs is beneficial to target non-functional properties of cyber-physical systems based on *i*-Core. *i*-Core (invasive Core [28]) is a *reconfigurable processor* that attaches an FPGA-based reconfigurable fabric directly to the CPU pipeline. In the remainder of this chapter, we first introduce the *i*-Core architecture. Then, an overview of our recent research is given, starting with task scheduling (Sect. 2) for reconfigurable processors. We then focus on two non-functional properties that are commonly found in cyber-physical systems: reliability constraints (Sect. 3) and worst-case execution time guarantees (Sect. 4). Finally, we summarize our work on reconfigurable multi-core architectures (Sect. 5).

1.1 The i-Core Architecture

This section introduces key concepts of the *i*-Core architecture as a basis for the following sections. Details, including comparisons to other reconfigurable system designs that couple a reconfigurable area with a processor core, can be found in [8]. An overview of the *i*-Core architecture is shown in Fig. 1. *i*-Core is a reconfigurable processor, i.e., it is based on a general-purpose processor (GPP) pipeline and enables the execution of runtime-reconfigurable *custom instructions* (CIs). CIs extend the processor's core instruction set architecture (cISA) by application-specific instructions that are realized using (1) microcode and (2) reconfigurable accelerators that are detailed in the following.

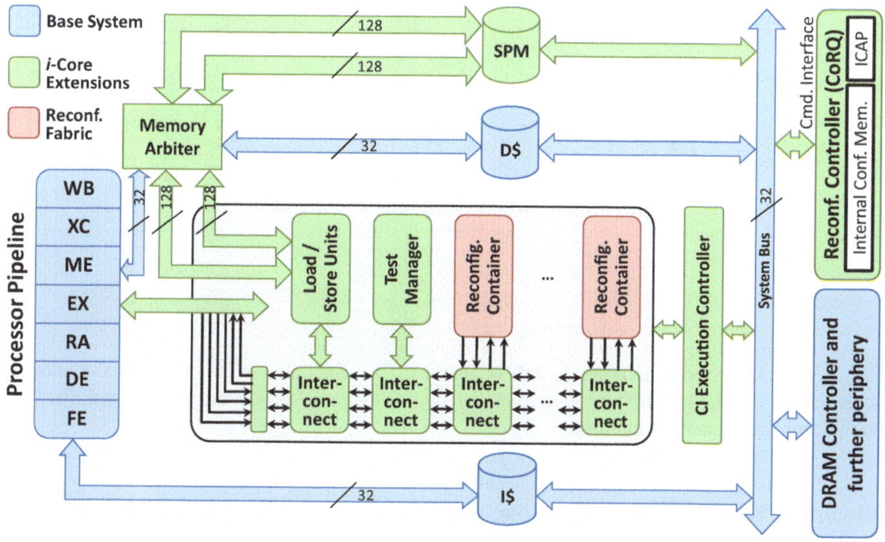

Fig. 1 The main components that the *i*-Core processor platform comprises

1.1.1 Microcoded Custom Instructions

When the processor pipeline encounters a CI in its execute (EX) stage, the pipeline stalls and initiates execution of the respective microprogram (i.e., a program written in microcode) that implements the functionality of the encountered CI on the CI execution controller. The microprogram controls all resources of the reconfigurable fabric:

- Load/store units (LSUs) enable access to the main memory (through the processor's L1 data cache (D$)) and high-bandwidth scratchpad memory (SPM) for CIs
- Reconfigurable containers (RCs), (embedded) FPGAs that provide the reconfigurable area for runtime-reconfiguration of accelerators (one accelerator per container, each of similar complexity as, e.g., floating-point multiply-accumulate or a dozen integer operations)
- Interconnects connect LSUs, RCs, and the processor's register file to a common (four word-wide segmented) bus.

Note that a single microprogram can utilize one or more accelerators. In other words, *the functionality defined by a CI is realized using one or more accelerators*. Application-specific hardware accelerators provide an important trade-off: the more area is utilized, the higher the resulting performance. At the same time, multiple accelerators compete for the constrained reconfigurable area. This trade-off is the result of instruction-level parallelism that can be exploited when more hardware resources are added to an application-specific accelerator. The main benefit of

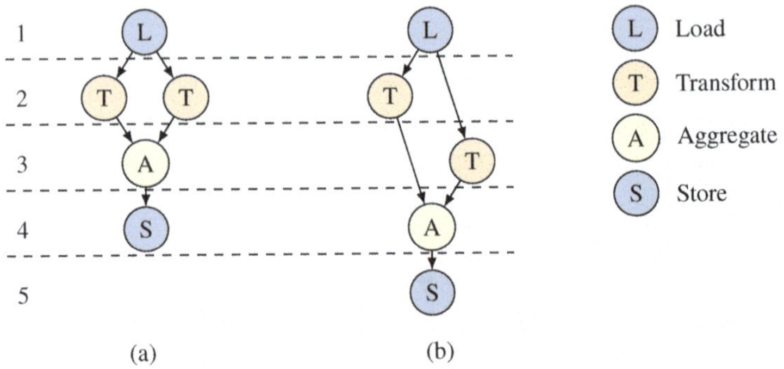

Fig. 2 CIs define computations as DFGs that can be scheduled with different amounts of accelerators, resulting in different latencies. (**a**) Two "transform" accelerators configured. (**b**) One "transform" accelerator configured

allowing CIs to utilize more than one accelerator is that this trade-off can be chosen at runtime by providing several microprograms that implement the same CI, but utilize different amounts of accelerators that each implement a part of the CIs functionality. Consequently, CIs define computations as data-flow graphs (DFG) where nodes are accelerators and load/stores. Figure 2 shows a simplified example of a CI that loads input data, performs transformations on the data, aggregates results, and finally stores them. Depending on how many "transform" accelerators are configured in the RCs at runtime, the DFG can be scheduled in four steps (see Fig. 2a) or five steps (see Fig. 2b). Each of these schedules corresponds to a microprogram for the CI execution controller, which implements the CI.[1]

The CI model employed by *i*-Core provides additional opportunities for performance increases over a model where a CI corresponds to a single accelerator:

- *Sharing*. Because the functionality of a CI is "split" in several accelerators, the computations of a single accelerator become more general. Thus, once configured accelerators create opportunities to be utilized by multiple CIs (implementing different functionality).
- *Upgrading*. Reconfiguration takes time. The *reconfiguration delay* on modern architectures is still in the range of milliseconds, depending on the configuration size. In the *i*-Core CI model, a CI can already be executed in hardware as soon as each required accelerator type is configured at least once on the reconfigurable fabric (like in Fig. 2b). The more accelerators finish reconfiguration during runtime, the more parallelism can be exploited and the lower the CI latency gets (e.g., runtime reconfiguration of an additional "transform" accelerator leads to schedule Fig. 2a instead of Fig. 2b).

[1]In the remainder of this chapter we will refer to CI implementation and CI microprogram interchangeably, depending on the focus of the respective section.

More details about such *modular* CIs, the load/store unit (see Fig. 1), and the address generation can be found in [8, 11].

1.1.2 Software Emulation

So far, we discussed how CIs are executed, assuming all accelerators required by a certain implementation are currently configured on the reconfigurable fabric. However, CIs can be *unavailable*, i.e., there exists no schedule of the CI's DFG for the accelerators that are currently configured. Two alternatives exist to handle the case that the *i*-Core attempts to execute an unavailable CI at runtime: *stalling* and *software emulation*. Figure 3 visualizes the different execution models. Software only (top) corresponds to execution of iterations of a kernel that does not utilize any CIs. *Stalling* (middle) executes CIs on the reconfigurable fabric and trying to execute an unavailable CI is an error. Therefore, CIs need to be configured before the kernel is entered and the execution is stalled until the required CIs are available. Finally, *software emulation* (bottom) triggers functionally equivalent software execution of an unavailable CI on the *i*-Core's pipeline using the base processor's cISA. It enables execution of the kernel while required CIs are still being reconfigured. Thus, progress can already be made without any CIs and as soon as reconfiguration of a CI finishes, the CI is utilized to speed up the following iteration of the kernel. While software emulation is always beneficial at runtime, it is more complex to analyze for execution time guarantees than stalling, which will be detailed in Sect. 4. Software emulation is realized using two alternative approaches in *i*-Core: an "unimplemented instruction trap" with a corresponding trap handler or CI Invocations that branch to either the CI or software emulation before trying to execute the CI. CI Invocations are beneficial for execution time guarantees and are detailed in Sect. 4.2.

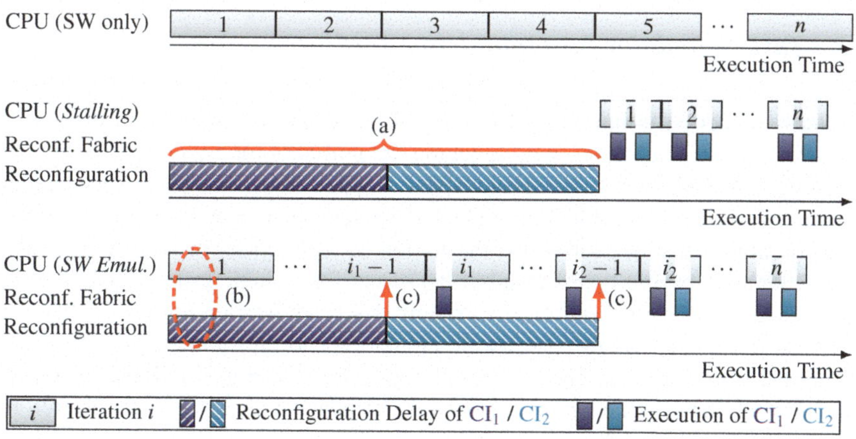

Fig. 3 Timelines of executing a kernel using software only, stalling and software emulation

This section provided an overview of the *i*-Core architecture as a background for the following sections. *i*-Core exists as a constantly evolving hardware prototype, currently it is based on the Gaisler LEON3 processor[2] and synthesizes to Xilinx Virtex-7 FPGAs. A more detailed explanation of how the architecture is realized can be found in [8, 9, 11]. Additionally, a SystemC-based cycle-accurate simulator is available [10] for early evaluation of runtime system algorithms. In the following, task scheduling on runtime-reconfigurable processors like *i*-Core is discussed.

2 Task Scheduling for Runtime-Reconfigurable Processors

While application-specific accelerators can provide significant speedup in domains such as mobile computing (e.g., 5G, cryptography) and robotics (e.g., image/audio processing, feature matching), applications in these domains are typically composed of multiple tasks. Systems for these applications are often dynamic multi-tasking systems, i.e., task arrival is not known at design time and task duration is dependent on input data (e.g., the recognized objects in a camera-based mobile robot). As long as such dynamically changing multi-tasking scenarios are not efficiently supported on reconfigurable processors, their inherent efficiency advantages are inaccessible for these demanding domains. In the following, we present how efficient multi-tasking support is enabled for reconfigurable processors by presenting two task schedulers for scenarios where it is unknown at compile time which tasks will execute at the same time (and which CIs will be required). First, we present a task scheduler that is optimized for reducing the *tardiness* (i.e., the total time by which deadlines were missed over all tasks), then a task scheduler for improving the *makespan* (i.e., the completion time of the tasks).

2.1 Task Scheduler for Optimizing Tardiness

Tardiness reduction is important for application scenarios where at least some of the tasks have soft deadlines, e.g., reducing the tardiness for a video recording and encoding task leads to a reduced number of dropped frames. In [7] we present the performance aware task scheduling (PATS) strategy that aims to reduce tardiness of all running tasks. We introduce the notion of *task efficiency* in reconfigurable processors and observe that it changes over time for a single task: A task that has just started executing a kernel (i.e., its required accelerators are not yet configured) will need to execute CIs using software emulation and thus, have a *low efficiency*. In contrast, a task that has finished reconfiguring its accelerators will perform more computations in the same amount of time and thus, have *high efficiency*. Therefore,

[2]https://www.gaisler.com/.

our scheduler favors executing tasks with high efficiency (unless it would lead to a deadline miss for a task with low efficiency). Tasks with low efficiency can perform their reconfigurations in parallel to execution of tasks with high efficiency and achieve a high efficiency until they are executed at a later time. PATS was compared with standard scheduling approaches like earliest deadline first (EDF), rate-monotonic scheduling (RMS), and the scheduler used in the Molen processor [37] on reconfigurable processors with different fabric sizes and tasksets with different deadlines. PATS achieves a 1.45× better tardiness on average (maximum 1.92×, minimum 1.14× better) compared to the other schedulers.

2.2 Task Scheduler for Optimizing Makespan

For systems without deadlines, such as in high-performance computing, an important performance metric is the makespan (i.e., the completion time) of a taskset. To improve the makespan on a reconfigurable processor, we introduced a combined task scheduling and reconfigurable area allocation approach: makespan optimization for reconfigurable processors (MORP) [24]. Using the notion of task efficiency introduced in PATS, MORP is based on the observation that task efficiency is low after an application switches from one kernel to another (as it requires reconfiguration of accelerators for the new kernel), an effect we call reconfiguration-induced cycle loss (RiCL). By reducing the RiCL of a taskset, we improve its makespan. The largest potential for RiCL reduction is in complex tasks that switch between different kernels during their execution time (e.g., video encoders). Such tasks are scheduled by our approach as *primary* tasks, which are initially assigned the full reconfigurable fabric. Using combination of offline profiling and lightweight online monitoring, the approach predicts when the primary task will switch from one kernel to another. A short time before leaving a kernel (at high efficiency), a small share of the reconfigurable area is already reallocated to a *secondary* task. While accelerators of the secondary task are reconfigured, the primary task completes its current kernel. Upon switching to its next kernel, the task efficiency of the primary task drops (as its required accelerators are not yet configured). Therefore, the system temporarily schedules the secondary task, which by this point has a higher efficiency than the primary task. The primary task reconfigures its accelerators while the secondary task is running. During its final reconfigurations (for the upcoming kernel), the primary task reacquires the reconfigurable area that was allocated to the secondary task. At this point, the MORP switches to the primary task at high efficiency. Figure 4 shows results of MORP in comparison to other schedulers for different tasksets (each containing a subset of: H.264, SUSAN, AdPCM, AES, SHA). Our approach achieves an average makespan reduction by 6.5% and 20.3%, compared to the SPT scheduler (shortest processing time, optimal for makespan minimization on a non-reconfigurable processor) and RR (round robin) scheduling, respectively. Compared to the theoretical lower bound of makespan (where RiCL is assumed to be zero for any taskset), our approach produces results that are on

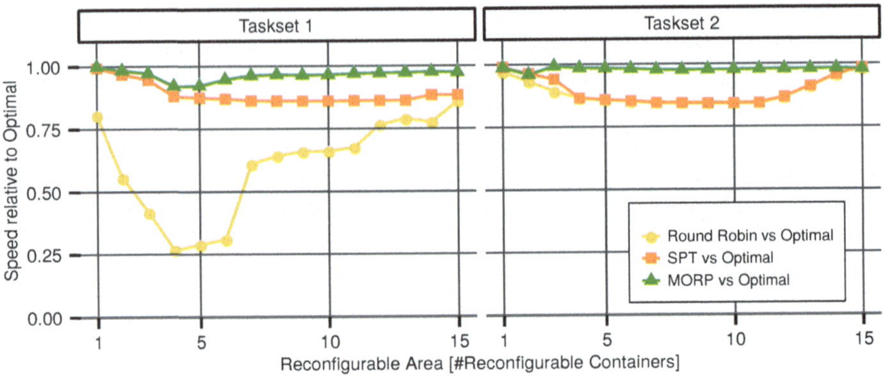

Fig. 4 Comparison of MORP to state-of-the-art scheduling approaches

average only 2.8% worse than this theoretical optimal result. The other evaluated schedulers produce schedules with makespans that are 14–20% worse than optimal.

In summary, our runtime-reconfiguration aware task schedulers can achieve 1.45× better tardiness (PATS) than comparable state-of-the-art schedulers and a makespan reduction that on average is only 2.8% worse than the theoretical lower bound. This section presented how efficient multi-tasking support is enabled on runtime-reconfigurable processors. When processors are embedded into a bigger system, like a cyber-physical system, generally non-functional requirements (e.g., real-time or reliability constraints) need to be met. Deploying runtime-reconfigurable processors in such systems provides opportunities and challenges as the following sections detail. The next section focuses on the non-functional property of reliable execution.

3 Reliable Reconfigurable Processors

Reliability concerns due to technology scaling have been a major focus of researchers and designers for several technology nodes. The reliability of reconfigurable processors is threatened not only by soft errors, but also by aging effects and latent defects. Latent defects are present in the material (unobserved) and may manifest as permanent fault during normal operation. Aging effects degrade the properties of transistors and may lead to a reduced performance (via increased transistor threshold voltage and correspondingly reduced maximal frequency) or even permanent faults (e.g., time-dependent dielectric breakdown) [27]. Soft errors instead are transient and may lead to bit flips in memory cell or logic latches [26].

In a reconfigurable processor, the non-reconfigurable components (e.g., CPU pipeline; see Fig. 1) may potentially be fabricated in an older technology node, as the system performance does not stem from high pipeline frequency, but from high parallelism of the reconfigurable accelerators. But the reconfigurable containers (RCs) should always be implemented in cutting-edge technology, to close the area/frequency gap compared to non-reconfigurable accelerators. 2.5D stacking with a silicon interposer and stacked chiplets can be used to integrate the non-reconfigurable components with the cutting-edge reconfigurable fabric. Xilinx is using 2.5D stacking since its Virtex-7 series and also Achronix promotes it as an alternative to their commercial embedded FPGAs [3]. As the RCs are manufactured in the latest technology nodes, reliability concerns are even more serious for them. However, the inherent flexibility and adaptivity due to the reconfigurability of the RCs also provides opportunities for fault mitigation. In this section, we present an overview of our comprehensive efforts in the OTERA project (online test strategies for reliable reconfigurable architectures; based on the *i*-Core) that allow to ensure functional RCs, correct reconfiguration, tolerate faults, mitigate aging, and guarantee a target reliability for observed soft-error rates.

3.1 Testing for Structural Defects and Correct Reconfiguration

In order to test the reconfigurable containers (RCs) and their accelerators for correct functionality, we added a "test manager" in a similar way as the RCs (see Fig. 1). It supports a *pre-configuration test* (PRET) and a *post-configuration test* (PORT) [6]. PRET tests the RC-internal resources for permanent faults and PORT tests whether the accelerator was configured correctly and whether it meets the target frequency. In order to test the RC-internal resources, there needs to be a way to apply test input data to RCs and to analyze their result. Therefore, PRET cannot be performed using the application-specific accelerators, but special *test configurations* (TCs, i.e., accelerators with the purpose to test RCs) need to be used.

To test the CLB array of a RC for permanent *stuck-at* faults, we developed nine TCs that test *all* CLB features, i.e., LUTs as logic, LUTs as shift registers, LUTs as distributed RAMs, multiplexers, registers, and carry-chain logic [1]. To perform a test with PRET, at first, one of the TCs needs to be configured into a RC. Afterwards, a special CI is executed that instructs the "test manager" to create and send test patterns to the RC-under-test, let it compute, and receive and analyze its computation result. The TCs use between 6 and 320 test patterns to perform their specific test. Their lowest frequency is 154 MHz, i.e., faster than the lowest frequency of the application-specific accelerators.

To evaluate the testing overhead, we use an H.264 video encoder with 9 CIs that permanently reconfigures the RCs to adapt them to the current *processing kernel* (i.e., motion estimation, encoding, or in-loop deblocking filter). The application requests application-specific reconfigurations and after every nth request, PRET *inserts* a small test. The RC that the application wanted to reconfigure anyway is reconfigured to contain one of the TCs and then its test patterns are applied. After the test, the actually requested application-specific accelerator is reconfigured into this RC. Note that the application continues executing during reconfiguring the TC

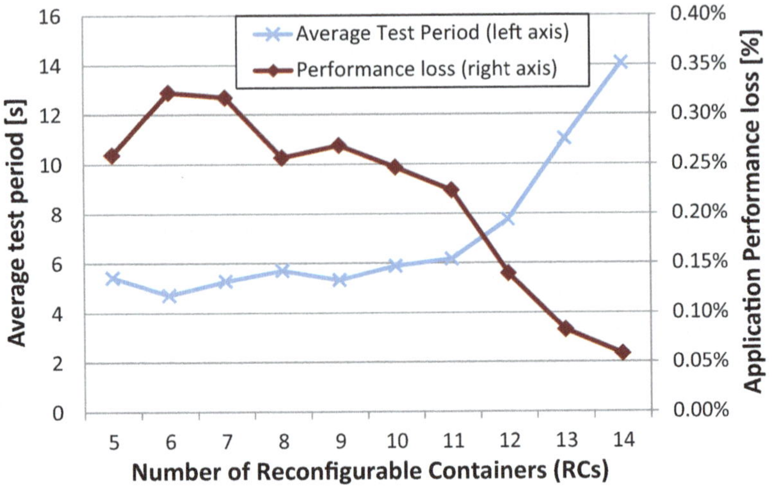

Fig. 5 Delay until all RCs are exhaustively tested by PRET vs. application performance loss due to testing

and note that many TCs are required until the entire reconfigurable fabric is tested (9 TCs per RC).

For the results shown in Fig. 5 we performed one TC after every four application-specific reconfigurations [1]. The figure analyzes the average test period and the application performance loss for reconfigurable processors with varying number of RCs. The average test period denotes the time in seconds until *all* RCs in the system are tested by *all* nine TCs each. The application performance loss denotes the additionally needed time compared to a system without any testing. Despite the very low overhead (on average 0.22%), the structural test completes very fast (on average 7.15 s). It is notable that the test period increases significantly for more than 11 RCs (while at the same time the overhead reduces). The reason is that for reconfigurable processors with so many RCs, some of the RCs are less often reconfigured by the application and thus they are tested less often and thus dominate the test period. To reduce the test period for these cases, we can use periodic testing of RCs that were not reconfigured for a longer time. For brevity, details on this extension and details on PORT (similar to PRET but without demanding extra TCs) and test scheduling are omitted and can be found in [6].

3.2 Fault Tolerance and Aging Mitigation

The pre-configuration test (PRET) described in Sect. 3.1 allows to detect permanent faults, but it provides no means to deal with them. In this section we present the idea of *diversified configurations* that (1) allows to tolerate permanent faults in RCs and that (2) can also be used to mitigate aging, i.e., their root cause. Whenever a permanent fault manifests in a CLB of a RC (detected by PRET), accelerators that use this CLB as part of their operation will no longer function correctly in all cases. However, accelerators that do not use that CLB will not be affected by the fault. The main idea of diversified configurations is to provide different implementations (i.e., place and route results) of the same accelerator that differ in their resource usage. In [47] we present the algorithm that generates a minimal set of diversified configurations for an accelerator such that for any single CLB that is found to be faulty, a diversified configuration exists that does not use that CLB. Additionally, the algorithm can generate even more diversified configuration, such that also multi-CLB-errors can be tolerated. We use the `PROHIBIT` constraint from Xilinx to ensure that the specified CLBs are not used and then let the Xilinx P&R tools freely determine the implementation of the accelerator on the remaining CLBs. In our evaluation for several applications (H.264, ADPCM, AES, JPEG), two to seven diversified configurations are sufficient to provide fault tolerance for any single-CLB fault at a minimal frequency loss of 5.6% compared to the accelerator implementations without diversification [47].

As diversified configurations differ in their usage of CLBs—and thus also in how much *stress* they induce to individual CLBs—they can also be used to mitigate aging. For instance, the degree of hot-carrier injection (HCI) aging depends (among others) on the average toggle rate of transistors. Higher toggle rate (at otherwise identical conditions) leads to faster HCI aging, which eventually increases the transistor threshold voltage. By continuously switching between different diversified configurations, the stress can be balanced over all transistors, instead of being accumulated into few transistors that would otherwise age fastest and thus affect the performance or reliability goals of the system first. Figure 6a shows the stress that an accelerator (`alu4` from the MCNC benchmark suite) induces into its RC, when it remains configured in the same RC for the entire application execution time [46]. Each square in Fig. 6a corresponds to one CLB of the RC and the color shows the average toggle rate of the transistors in this CLB. Figure 6b shows the stress when using a maximally diversified configuration of the same accelerator and Fig. 6c shows the stress when periodically reconfiguring between these two configurations. Even though the stress is more balanced, the peak stress in Fig. 6c is not reduced significantly. Figure 6d achieves a significantly reduced peak stress by switching between four diversified configurations (the minimal number to tolerate all single-CLB faults for this accelerator). The maximum HCI stress reduction when using the minimum number of configurations for single-CLB fault tolerance ranges up to 68.9%, which increases the time to failure by 222% [46].

In addition to balancing the stress within a RC by switching between diversified configurations (intra-RC stress balancing), we also developed a stress-aware placement algorithm (STRAP) that decides for all accelerators that shall be reconfigured, which diversified configuration shall be used and into which RC it shall be reconfigured (inter-RC stress balancing) [47, 49]. This is important as some accelerators may be used significantly more often than others, which could lead to some RCs being significantly more stressed than others (even though being

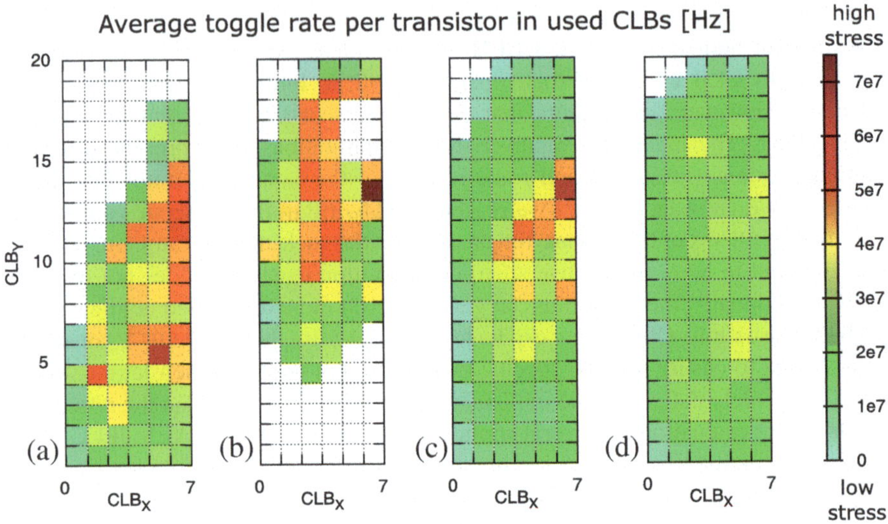

Fig. 6 The average toggle rate (stress) of the CLBs of a reconfigurable container after several executions of (**a**) an accelerator, (**b**) a maximally diversified implementation of the same accelerator, (**c**) an alternating schedule (reconfiguration) of the first two implementations, (**d**) a balanced schedule of four implementations

reasonable balanced within the RC). As the placement decision has to be made at runtime and depends on the accumulated stress due to the CI executions so far, we developed an efficient search space pruning technique to reduce the runtime overhead by calculating guaranteed lower and upper bounds of the achievable stress distribution. Details on the algorithms can be found in [47, 49]. Altogether, using our diversified configurations with our fault-tolerant and stress-balancing accelerator placement, an H.264 video encoder delivers from 1.9× up to 3.7× the performance of a baseline system in the presence of 4 to 40 faulty CLBs, while at the same time achieving up to 6.8× higher mean-time to failure (MTTF) than a baseline system for a set of benchmarks [47].

3.3 Guaranteed Reliability for Reconfigurable Processors

Besides permanent faults and aging effects, single-event upsets (SEUs) are the most-demanding challenge for reliable reconfigurable processors. SEUs manifest themselves as a temporary *bit flip* in the logic that may be captured by state-holding elements. As reconfigurable processors use SRAM-based reconfigurable fabrics to implement their RCs, they contain rather large amounts of state-holding elements and are thus especially susceptible to SEUs. Additionally, if an SEU flips a configuration bit, then this may actually alter the function of the accelerator (defined by the configuration bits) and this fault is only corrected when the RC is reconfigured again. Instead, the pipeline of reconfigurable processors (see Fig. 1) is less susceptible to SEUs, as it is not implemented using a reconfigurable fabric. It is also not critical for the overall performance (most performance-relevant parts are implemented as CIs), so it can be implemented using an older technology node, which is inherently less susceptible for SEUs. Therefore, the software implementation of a CI has the highest reliability and can be used when the reliability of a CI implementation on the reconfigurable fabric is too low.

Complex accelerators that use more resources have more *critical configuration bits* (i.e., those bits that define the functionality of the accelerators) and therefore have a higher probability that one of their critical bits is flipped due to an SEU. Based on the number of critical bits and the SEU rate,[3] we can determine the reliability of an accelerator, i.e., the probability that it produces a correct result at a certain time. The reliability starts at 100% right after the accelerator was reconfigured and it reduces exponentially over time until it is *scrubbed*[4] or reconfigured (see Fig. 7a). To increase reliability, modular redundancy can be applied to these accelerators (see Fig. 7b). This reduce the rate at which the reliability decreases, since even in case of an SEU, the system is able to correct it. Alternatively, the scrubbing frequency can be increased (see Fig. 7c), but the bandwidth to access the configuration data is limited and a higher scrubbing frequency also affects the accelerator reconfiguration time, as scrubbing and reconfiguring both need to access the same configuration port.

Only slight changes in the hardware architecture are needed to implement modular redundancy, such that any two (three) neighboring containers can be combined to a DWC (TMR) pair. In case of TMR (triple modular redundancy), the CI may complete, but the faulty RC needs to be scrubbed soon to avoid aggregating multiple faults in the TMR pair. In case of DWC, the CI has to be aborted and has to execute in software emulation instead. Future executions of the same CI are directly sent to software and both RCs need to be scrubbed, as it is not known which of them caused the fault. However, applying modular redundancy means that less accelerators are available for parallel execution, which means that the overall application execution time increases. At the same time, also the *resident time* of accelerators increases, i.e., it takes longer until they are reconfigured by the application. The longer the resident time, the higher the probability that one of the critical configuration bits gets corrupted and the lower the reliability of the accelerator. Therefore we also have to increase the scrubbing frequency with the corresponding drawbacks.

Instead of statically choosing the degree of redundancy, reconfigurable processors can change between performance (no redundancy), DWC, and TMR at runtime and that can be decided for each accelerator independently [48]. The

---

[3]Determined by a soft-error monitor, e.g., using the number of ECC errors in memory, caches, etc.

[4]The configuration bits of a RC are read back and their ECC values are checked for errors with subsequent correction, if needed.

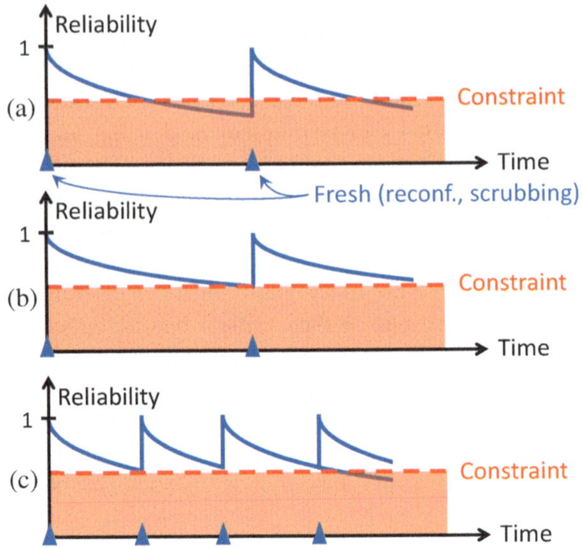

Fig. 7 The reliability of a reconfigurable container over time: (**a**) after reconfiguration or scrubbing, the reliability is known to be "1" and then decreases exponentially, potentially violating a given reliability constraint. (**b**) In order to avoid the violation, modular redundancy can be used to reduce the reliability decrease. (**c**) Alternatively (or in combination), more frequent scrubbing can be used

main challenge is to select the accelerators and their redundancy modes to satisfy application-determined reliability constraint, while at the same time maximizing the performance. We developed an algorithm that conducts the decision in multiple steps [45, 48]. The main idea is to combine all reliability impacting factors for an accelerator execution in one metric, the *effective critical bits* (ECBs), and then to distribute the ECBs among kernels and CIs. The metric is inspired by the critical bits, i.e., the configuration bits of an accelerator that define its functionality. These bits need to have the correct value in order to execute the accelerator correctly. Applying modular redundancy to an accelerator increases its reliability. That is basically the same effect as if the accelerator would have *less* critical bits, and that is exactly what we express as a reduction in ECBs of that accelerator. Similarly, increasing the scrubbing rate reduces the ECBs for all accelerators. For a CI that executes in software emulation, its ECBs basically correspond to zero, i.e., it is very reliable.

For a given target reliability and SEU rate, we calculate the number of ECBs that we can maximally tolerate [45]. Then, we partition these ECBs among the different kernels of the application, based on their resource requirements and expected execution time. In the next step, the ECBs of a kernel are partitioned among its CIs. Those CIs that have long execution time and require a large number of accelerators obtain the most budget, i.e., we assign more ECBs to complex and/or slow CIs in order to realize them in faster hardware implementations [45]. This ECB budgeting only needs to be recomputed if the target reliability or the monitored SEU rate change. Finally, we select the minimally needed scrubbing frequency and the redundancy modes for accelerators to implement the CIs such that they do not violate the given ECB budget and maximize the application performance [48]. Thereby the approach guarantees an application-specific minimum level of reliability for the CIs and the application. Since the accelerators are not over-protected, the performance of the application is maximized for the given target reliability. We evaluated the approach using an H.264 video encoder and varying SEU rates and we compared it against threshold-based methods that reconfigure all accelerators to DWC (or TMR) when exceeding a threshold (or implement the CIs in software if too few RCs are available). Compared to these threshold-based methods, out approach guarantees the same target reliability while providing 20.0% (DWC) or 42.6% (TMR) higher application performance on average. In the best case, up to 34.8% (DWC) and 68.3% (TMR) faster execution is obtained without violating the given reliability constraint.

This concludes our discussion of reliability concerns in runtime-reconfigurable processors. Orthogonal to reliability concerns, designers of cyber-physical systems need to deal with timing worst-case execution time concerns that are discussed in the following section.

4 Worst-Case Execution Time Guarantees

In contrast to general-purpose computing systems, cyber-physical systems need to meet non-functional requirements like timing constraints. Failing to meet a given deadline can lead to severe malfunctions, therefore a *timing validation* is performed to guarantee the timing constraints [44]. As part of the timing validation, a schedulability analysis is performed to guarantee that the set of tasks that should be executed on a system can be scheduled at runtime under any circumstances. The input to the schedulability analysis is the *worst-case execution time* (WCET), which needs to be known for every task from the taskset [44].

Statically determining the WCET of a task is a complex problem. Due to the undecidability halting problem, it is in general impossible to determine the precise WCET of a task or its worst-case input [34]. WCET analysis is further complicated by the fact that modern processor design focuses on reducing the average execution time: Features like pipelining, caches, and branch prediction introduce a microarchitectural state, i.e., the latency of an instruction is dependent on the execution history. Even with recent advances in research, WCET analysis lags years behind current microarchitectures with out-of-order scheduling pipelines, several hardware threads, and multiple (shared) cache layers [38]. The real challenge for a successful timing validation is to obtain *tight bounds* of the execution time, i.e., the overestimation should be as low as possible. Therefore, performance features amenable for timing analysis are requested [4, 21, 41] to face the increasing performance demands of real-time systems. Improving the WCET of a task is highly desirable as it may enable a cyber-physical system to meet previously infeasible timing constraints or make room for power optimizations.

This section introduces WCET analyses and optimization of tasks on runtime-reconfigurable processor designs like *i*-Core as one way to escape the scarcity of timing-analyzable performance features. First, this section introduces a reconfiguration controller that provides guaranteed reconfiguration delays to enable runtime reconfiguration in hard real-time systems. Afterwards, it is shown how WCET bounds can be obtained for processors with a runtime-reconfigurable instruction set. Finally, this section presents how WCET bounds can be optimized by selecting WCET-optimizing CIs.

4.1 Guaranteed Reconfiguration Delays

The most straight-forward approach of improving WCET guarantees a task using runtime reconfiguration with the constraints of timing-analyzability and reasonable implementation effort is to apply the *stalling* model [17] as shown in Fig. 3 (middle). Using the stalling model, a task that requests reconfigurations of accelerators can be analyzed for WCET guarantees with established timing analysis techniques by adding the reconfiguration delay (see Fig. 3(a)) to the WCET of the basic block that requests the reconfiguration. The assumption is that the reconfiguration delay can be determined statically, which is reasonable for the stalling approach when the CPU is stalled and its memory accesses cannot interfere with reconfiguration on main memory or a shared system bus. However, stalling is not state of the art in reconfigurable systems, because the CPU is forced to remain idle during reconfiguration and cannot, e.g., execute computations that do not depend on the accelerators being reconfigured.

An approach that enables the CPU to perform useful operations in parallel to reconfiguration is software emulation as explained in Sect. 1.1.2 and shown in Fig. 3 (bottom). Software emulation is an established technique in average-case optimizing reconfigurable systems, because it provides considerable performance improvements. For real-time systems, however, it poses new challenges:

- Figure 3(b): Static timing analysis needs to capture potential conflicts on main memory or a shared system bus when the reconfiguration process and the CPU act on memory in parallel.
- Figure 3(c): When reconfiguration and execution on CPU execute in parallel, they need to be synchronized at some point. The main question for WCET estimates is: how far did the task proceed (in the worst case) during the reconfiguration delay? In other words, from what point is it safe to assume during static timing analysis that, e.g., accelerator A is readily configured on the reconfigurable fabric and execution sped up? This question is addressed in Sect. 4.2.

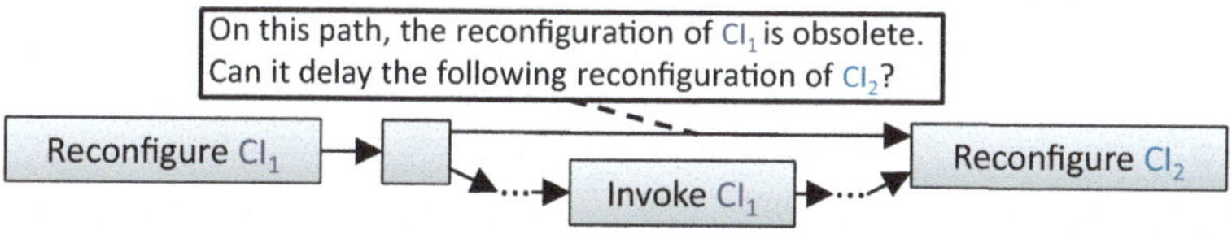

Fig. 8 Control-flow graph that shows how one reconfiguration request can delay another reconfiguration request, thus impairing timing analysis

- Figure 8: At runtime, execution could follow a faster path than the worst-case path. A new reconfiguration request could be reached, while a reconfiguration for a previous kernel is still in progress. The possibility of an already occupied reconfiguration port can lead to delays that are hard to analyze and therefore introduce pessimism in the resulting WCET bound. To be able to guarantee that the reconfiguration port is unoccupied for each reconfiguration request, it needs to be possible to abort reconfigurations.

Limiting tasks to the stalling model might seem as the favorable way to enable reconfiguration in real-time systems due to the potentially complex analysis of software emulation. When scheduling multiple real-time tasks, however, the stalling model poses similar challenges even on a uniprocessor system: when a task that requests a reconfiguration is stalled, another task can execute in parallel to the reconfiguration delay (see Fig. 3(a) and, e.g., [13]). In Sect. 4.2 it will be shown that software emulation always provides a considerable speedup at runtime, but there are cases where additional WCET overestimation compared to stalling diminishes the speedup on the WCET guarantee.

4.1.1 Enabling Runtime Reconfiguration in Real-Time Systems with CoRQ

To address the challenges of runtime reconfiguration in real-time systems, we designed a reconfiguration controller *command-based reconfiguration queue* (CoRQ[5] [16]) that enables the CPU (any CPU, not necessarily the *i*-Core) to issue sequences of reconfiguration requests, provides guaranteed reconfiguration delays, and relieves the CPU from managing accelerator availability. CoRQ informs the CPU of finished reconfigurations in a predictable way; the CPU never has to poll or be interrupted to obtain the information that an accelerator has become available (following a reconfiguration). CoRQ processes 32-bit commands and can be instantiated with an internal memory to store *bitstreams* (configuration data for the reconfigurable fabric). Commands are issued by the CPU using load/stores over the system bus (see Fig. 1). They are either executed immediately or enqueued in an internal FIFO queue (denoted as *immediate* or *queueable* commands, respectively,

[5]Open-source project available at: https://git.scc.kit.edu/CES/corq.

Table 1 CoRQ commands with cycles spent in EXE state

Command	**Im**mediate/**Qu**eueable	Latency$_{\mathrm{EXE}}$ (cycles)[a]
`clearQ`	Im, Qu	5
`abortReconf`	Im	5
`configBitsInt`[b]	Qu	$6 + \lceil B/4 \rceil$
`sendGPIO`	Qu	1

B—size of bitstream (byte)
[a]Discussed in Sect. 4.1.2
[b]Detailed in Sect. 4.1.3

```
1 clearQ        (Im: Ensure command queue is empty)
2 abortReconf   (Im: Ensure free reconfiguration port)
3 configBitsInt (Qu: Configure bitstream B1 from internal memory)
4 sendGPIO      (Qu: Store info 'CI1 available')
5 configBitsInt (Qu: Configure bitstream B2 from internal memory)
6 sendGPIO      (Qu: Store info 'CI2 available')
```

Listing 1 CoRQ commands for implementing software emulation

in the following). The immediate commands are used to control CoRQ itself (stop/resume processing enqueued commands, clear queue, reset) and abort a running reconfiguration. Queueable commands relieve the CPU from managing reconfigurations, i.e., they configure bitstreams (from internal or external memory), provide information about available CIs through a general-purpose interface (or send an interrupt to the CPU), and can even stall/unstall the CPU to implement the stalling model. CoRQ currently supports 11 commands, the subset of 4 commands that realize software emulation is shown in Table 1. In the following we illustrate how software emulation can be realized with CoRQ, realization of the stalling model is detailed in [16].

A reconfiguration of two accelerators using software emulation (executing software in parallel, see bottom timeline of Fig. 3) is realized using commands as shown in Listing 1. The CPU proceeds executing software after issuing the commands to CoRQ, thus reconfiguration is performed in parallel to execution on the CPU. To be able to guarantee the reconfiguration delay, it needs to be ensured that no earlier reconfiguration requests are still pending and occupy the reconfiguration port (see Fig. 8). Therefore, first all remaining commands are cleared and reconfiguration (if any) is aborted (Lines 1 and 2). Afterwards, a bitstream from internal memory is configured. This way, loading the bitstream does not conflict with memory accesses from the CPU to main memory. Once reconfiguration completes, `sendGPIO` is executed (Line 4) to notify the CPU that the first CI has become available (each CI only uses one accelerator in this simplified example). Then, the CI can immediately be used once it is configured (see Fig. 3 (bottom)), without waiting for the whole set of commands to have finished processing by CoRQ or executing a software handler to manage CI availability. A second CI is configured in Lines 5 and 6 of the example.

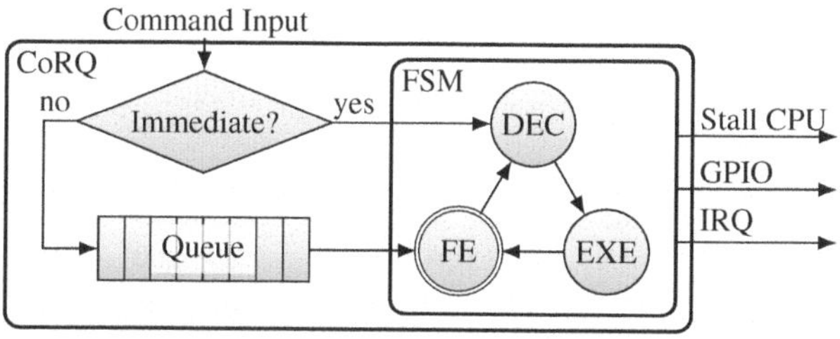

Fig. 9 High-level view of how CoRQ processes commands

The example illustrates how software emulation can be realized using CoRQ with simple sequences of commands issued by the CPU. In the following, the command execution and timing behavior of commands are detailed.

4.1.2 Command Execution

Commands are processed by CoRQ using a finite state machine (FSM) consisting of three states: fetch from queue (FE), decode (DEC), and execute (EXE) (see Fig. 9). Fetching a command takes a single cycle, the DEC state takes two cycles, and the latency of EXE depends on the command (see Table 1). Immediate commands control CoRQ itself, and thus have priority over commands from the queue. Executing either an immediate or a queueable command takes $3+\text{latency}_{\text{EXE}}$ cycles. Enqueueing a command takes 2 cycles for identifying it as queueable and writing it to the queue. Commands can simultaneously be enqueued to and fetched from the queue. The realization of this simultaneous access (with a double-ported FIFO) incurs an additional delay of 2 cycles for commands to become visible to the FSM if the FIFO was empty.

4.1.3 Guaranteed Reconfiguration Delay

CoRQ can load bitstreams from arbitrary addresses; however, accessing the system bus and a shared main memory (especially DDR) can incur memory access delays that are hard to bound for WCET guarantees. Reconfiguration delays are guaranteed when the CoRQ-internal memory is used (`configBitsInt`). The CoRQ-internal memory is implemented using SRAM (so-called block RAMs on Xilinx FPGAs) such that one word of configuration data can be fed to the reconfiguration port in each cycle. Thus, CoRQ utilizes the configuration port's full bandwidth (see Sect. 4.1.4). Additionally, the `configBitsInt` command requires five setup cycles and a single cycle at completion. Let B denote the size of the bitstream in bytes (see Table 1), then $\text{latency}_{\text{EXE}} = 6 + \lceil B/4 \rceil$ cycles. Including the latency of FE and DEC, *configuring a single bitstream from CoRQ-internal memory (`configBitsInt`) is guaranteed to take exactly* $9 + \lceil B/4 \rceil$ *cycles.*

Table 2 Resource utilization

	LUTs	Flip-flops	BRAM
LEON3 CPU (standard config.)	8144	3450	14
CoRQ	398	546	1
Internal Mem. of CoRQ (384 KB)	233	6	96
Available on VC707	303,600	607,200	1030

In the example of Sect. 4.1.1, the latency of the command sequence is simply the sum of the latencies of the queueable commands: Configuring two bitstreams using software emulation (see Listing 1 and Fig. 3 (bottom)) results in a latency of $t_{\texttt{clearQ}} + t_{\texttt{abortReconf}} + t_{\texttt{configBitsInt}} + t_{\texttt{sendGPIO}} + t_{\texttt{configBitsInt}} + t_{\texttt{sendGPIO}} = 8{+}8{+}(9{+}\lceil B_1/4\rceil){+}4{+}(9{+}\lceil B_2/4\rceil){+}4 = 42{+}\lceil B_1/4\rceil{+}\lceil B_2/4\rceil$. This latency starts once the immediate command `clearQ` reaches CoRQ and is running in parallel to the CPU that sends the commands to CoRQ. Executing previous commands always takes at least as long as the delay for enqueueing the current command, therefore enqueueing the commands does not add to the delay.

4.1.4 Results

The resource utilization of CoRQ, when added to the default Gaisler LEON3 design (GRLIB GPL 1.4.1) targeting the Xilinx VC707 board, is shown in Table 2. The design instantiates a single LEON3, uses the DDR3 on the VC707 as main memory, and runs at 100 MHz.

The reconfiguration port (ICAP in Xilinx devices) can process 4 byte each cycle at maximum 100 MHz on the VC707. Therefore, the theoretical maximum reconfiguration bandwidth is 381.47 MiB/s.[6] In the following, we reconfigure 25 partial bitstreams of $B = 57{,}248$ bytes each, which together takes a minimum of 357,800 cycles when assuming the theoretical maximum reconfiguration bandwidth without overheads. Using CoRQ, these reconfigurations take 358,036 cycles[7] which corresponds to a reconfiguration bandwidth of 381.22 MiB/s. Thus, CoRQ is only 0.066% (or 236 cycles) slower than the theoretical maximum.

Figure 10 shows the reconfiguration bandwidth results as measured by the CPU. The results were obtained for reconfigurations using the CoRQ-internal memory (Int. Mem.), as well as main memory over the shared AHB system bus (Main. Mem.). “Stalling” leaves the CPU idle during reconfiguration, whereas “polling” means that the CPU repeatedly reads CoRQ’s status register to check whether reconfiguration has completed (producing traffic on the AHB). “Bus conflicts” uses a simple DMA unit that repeatedly initiates maximum length (256 words) AHB burst transactions to provoke system bus and main memory conflicts during

[6]More precisely: $(4 \cdot 1024^{-2})/10^{-8} = 381.4697265625$ MiB/s.

[7]Sum of latencies of the individual commands: $4 + 25 \cdot (9 + \lceil 57{,}248/4\rceil) + 4 + 3$ cycles.

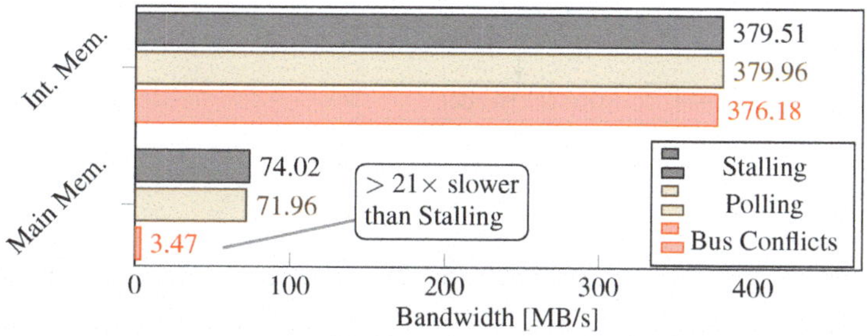

Fig. 10 A high variance in reconfiguration bandwidth is revealed when using main memory (measured by CPU, average of 50 measures, maximum error <1%)

reconfiguration. The small variance in measurements when using CoRQ-internal memory (<1%) stems from the overhead of measuring. CoRQ's commands itself always have exactly the same latency when using internal memory.

When using main memory for reconfiguration, accesses from the CPU, the DMA, and CoRQ are in conflict. This results in a strong variance in reconfiguration bandwidth between the measurements: The measurement under DMA bus conflicts reports only 4.69% of the stalling bandwidth. This shows that reconfiguration controller design is crucial in runtime-reconfigurable real-time systems. Simply utilizing a shared memory for reconfiguration can lead to a slowdown of more than 21× in reconfiguration bandwidth.

This section demonstrated how reconfiguration delay guarantees can be achieved to enable runtime reconfiguration in real-time systems using CoRQ. In another work, CoRQ formed the basis to design a reconfiguration controller that enables preemptable runtime reconfiguration in Xilinx Zynq-based multi-priority real-time systems [36]. Once reconfiguration delay guarantees are established, the following section shows how WCET estimates are obtained for tasks that leverage runtime-reconfigurable accelerators for predictable performance.

4.2 *Worst-Case Execution Time Analysis*

To obtain a safe worst-case execution time (WCET) estimate, timing analysis needs to be performed on the reconstructed control-flow graph (CFG) of the application binary [43]. The analysis of WCET estimates in the presence of CIs that are reconfigured using software emulation (as explained in Sect. 1.1.2, see Fig. 3) is achieved as follows. Reconfiguration of accelerators that speed up an upcoming kernel is initiated using CoRQ's reconfiguration commands (see Sect. 4.1) in a basic block immediately before entering the respective kernel. Analysis of this basic block yields the guaranteed reconfiguration delay per CI. Stalling the execution for the whole reconfiguration delay of all CIs and only then entering the kernel was mentioned in the previous section as a simple technique to perform analyzable

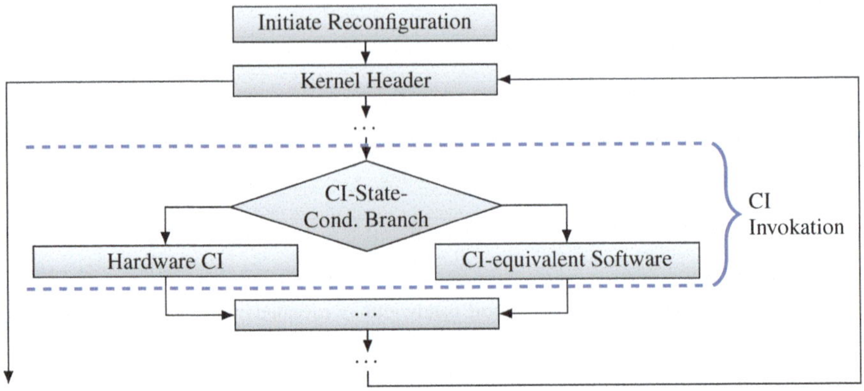

Fig. 11 *i*-Core kernel model for static timing analysis in the presence of runtime reconfiguration of custom instructions

reconfiguration. In this case, however, the reconfiguration delay (of several milliseconds) is often not amortized, especially in applications that switch between multiple kernels. Instead of stalling, this section presents a worst-case analysis that enables the RISC pipeline of the *i*-Core to be used to execute a software emulation of the CIs during the reconfiguration delay in hard real-time systems. This is achieved by (1) the reconfiguration controller presented in Sect. 4.1 and (2) by worst-case analysis of the *CI Invocation* construct shown in Fig. 11. The CI Invocation construct introduces a conditional branch that executes either the CI on the fabric or functionally equivalent cISA code on the processor pipeline.

CI Invocations realize software emulation in an analyzable way: First, execution of a kernel starts in software only (without accelerators) and at some point in time (during some iteration) accelerators finish their configuration and succeeding kernel iterations are sped up. An example of this process is shown in the timeline on Fig. 3 (bottom) for a kernel with n iterations utilizing two CIs. When entering the kernel, all CI Invocations of the first iterations are executed using cISA code. Reconfiguring the accelerators required by CI_1 finishes during iteration $i_1 - 1$. Beginning with iteration i_1, CI Invocations of CI_1 use accelerators on the fabric and benefit from a much lower runtime per iteration. In parallel, reconfiguration proceeds. During iteration $i_2 - 1$ all accelerators for CI_2 become available. Beginning with iteration i_2, the remaining iterations of the kernel are sped up even more as additionally to CI_1, all CI Invocations of CI_2 are now executed in hardware.

4.2.1 Timing Anomaly of Runtime-Reconfigurable Systems

The challenge in obtaining a precise WCET estimate is to statically determine the worst-case iteration i at which a CI can be guaranteed to be readily configured in hardware. A safe, but imprecise execution time bound can be obtained by assuming that no CI ever finishes configuration and all CI Invocations always branch to the unaccelerated cISA code. While this would result in speedups at runtime, the WCET bound would be as high as not using accelerators at all. To obtain a safe and precise bound yielding speedups, the worst-case iteration i_k, in which reconfiguration finishes and the CI Invocation utilizes the *i*-Core fabric to execute the CI, needs to be determined statically for each CI_k. We do so by determining execution time bounds for all basic blocks in a kernel with existing timing analysis tools extended by analysis for CI latencies (we use the commercial AbsInt aiT [2] and the open-source OTAWA [5] timing analyzers in our evaluations). Once the time bounds for all basic blocks are known, the time bounds for one iteration of the whole kernel can be determined. These time bounds depend on whether specific CI Invocations branch to the CI or equivalent software. Starting with a time bound of WCET_1 for an iteration with cISA code only and a reconfiguration delay of r_1 for CI_1, Fig. 3 (bottom) seems to suggest that i_1 can be determined as: $\lceil r_1/\mathrm{WCET}_1 \rceil + 1$, i.e., CI_1 is unavailable for $\lceil r_1/\mathrm{WCET}_1 \rceil$ iterations and in the following iteration we can assume it to be available. It turns out, however, that *executing every iteration of the kernel in worst-case time during reconfiguration does not necessarily result in the worst-case value for* i_1 as the following example demonstrates.

The timeline in Fig. 12 shows an example for a kernel executing 6 iterations and configuring a single CI. Figure 12a shows a possible runtime behavior of the

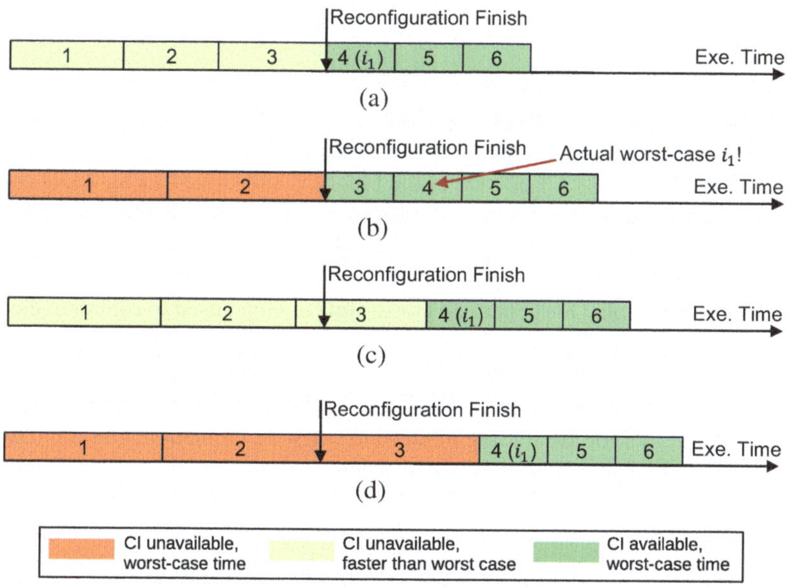

Fig. 12 Different cases for execution times of kernel iterations before the CI becomes available. (**a**) Faster than worst case, assuming no iterations overlap reconfiguration finish. (**b**) Worst case, assuming no iterations overlap reconfiguration finish. (**c**) Faster than WCET before reconfiguration finish. Iteration overlapping reconfiguration finish leads to extended execution time (*timing anomaly*). (**d**) Applied case for safe WCET bounds

kernel, where iterations 1–3 are each executed faster than worst-case time (CI not yet available) and iterations 4–6 execute in worst-case time (CI available, because reconfiguration finished). To obtain the worst-case execution for the whole kernel, it seems intuitive to assume all iterations before the reconfiguration finishes to execute in WCET_1 (software emulation of CI_1) and in WCET_2 (CI_1 available in hardware) after the reconfiguration finishes (Fig. 12b). However, when executing slightly faster iterations than WCET_1 before the reconfiguration finishes, we end up with the execution sequence of Fig. 12c: Iteration 3 cannot benefit from CI_1 and delays following iterations, because it "overlaps" the reconfiguration finish. This *timing anomaly* of runtime reconfiguration was first discovered and safely bounded in [17], it was shown that i_1 (the worst-case iteration in which CI_1 is guaranteed to be available) is determined as:

$$i_1 = \lceil r_1/\text{WCET}_1 \rceil + 2$$

That is, one additional iteration needs to be accounted for in which the reconfiguration might not yet have finished (Fig. 12d). Once i_k is determined for each CI_k, constraints are generated for the worst-case path analysis approach implicit path enumeration technique (IPET) [32]. [17] provides a detailed explanation of how constraints are generated for the IPET for multiple CIs with support for nested loops, conditional execution, and multi-context analysis (supporting, e.g., caches).

4.2.2 Results

We evaluated the timing analysis approach on *i*-Core by generating constraints for AbsInt aiT. aiT is closed-source software, thus, CI support could not be directly integrated. Instead, every CI opcode in the binary is substituted by an ADD opcode and a constraint in aiT's AIS2 constraint language to set the delay for the new ADD instruction to the delay of the specific CI. aiT outputs an XML report, which is parsed to determine every i_k for every kernel and generate the constraints described in the previous section. Our generated constraints are then used to calculate the final WCET bound using aiT.

We evaluated our analysis with an H.264 encoder application which uses 9 CIs covering the most compute-intensive kernels. In the following, we show results for the loop filter kernel for the stalling model and software emulation. Observed worst-case execution time results are obtained using our SystemC-based cycle-accurate simulator of *i*-Core.

For the analysis we use results obtained from performing timing analysis on a binary that executes the loop filter kernel of H.264 on 99 macroblocks (QCIF resolution). The loop filter is the kernel of lowest complexity in the H.264 encoder, it contains a single CI (in-loop deblocking edge filter on 4 pixels) and allows detailed analysis of worst-case CI availability. The guaranteed time bounds are compared to results obtained by executing the same binary in our simulator. f_{fabric} stays constant at 100 MHz and we choose multiples of it for f_{CPU} which resemble realistic setups

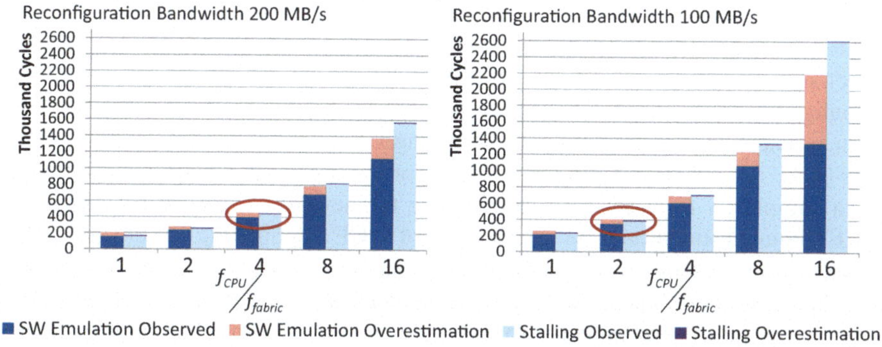

Fig. 13 Observed runtimes and guaranteed WCET bounds (*in cycles*) for loop filter. Fabric frequency $f_{\text{fabric}} = 100\text{MHz}$. Red circle marks the point from which software emulation is beneficial over stalling. Software only WCET bound is >4.5 million cycles

(rounded to the next power of two). For example, the LEON3 processor, which the *i*-Core implementation is based on, is advertised as running at 400 MHz when implemented as an ASIC, its successor the LEON4 is advertised running at 1500 MHz. The commercially available Xilinx Zynq-7000 SoC couples an ARM Cortex A9 at 866 MHz with a Xilinx 7-Series reconfigurable fabric.

All results shown in Fig. 13 are measured in *cycles of the CPU pipeline*. The left and right graph show the results for a reconfiguration bandwidth of 200MB/s and 100MB/s, respectively. When increasing the minimum evaluated CPU pipeline frequency of 100 MHz by a factor of c for a fixed f_{fabric} (x-axis), the runtime benefit of hardware CIs compared to software emulation decreases. Thus, the execution time in CPU cycles increases. In the observed runtime, software emulation is always beneficial over stalling. However, the WCET overestimation (execution time difference of WCET over observed runtime) is higher for software emulation than for stalling. As a result, stalling is beneficial over software emulation for $f_{\text{CPU}}/f_{\text{fabric}} \in \{1, 2\}$ at a reconfiguration bandwidth of 200MB/s, as well as $f_{\text{CPU}}/f_{\text{fabric}} = 1$ at a reconfiguration bandwidth of 100MB/s for obtaining a low WCET bound. In our experiments we observed that software emulation benefits from slow reconfiguration bandwidths or high CPU frequencies.

This section detailed how WCET estimates are obtained for tasks utilizing runtime-reconfigurable processors like *i*-Core for a given selection of CIs. In the following section we present how WCET-optimizing CIs are selected for a constrained reconfigurable area.

4.3 Worst-Case Execution Time Optimization

This section presents an approach of *selecting WCET-optimizing sets of CIs* for computational kernels that seamlessly integrates into state-of-the-art timing analysis. The approach does not target the reduction of overestimation of a task's WCET bound or resolving the problem of timing anomalies, but to statically select subsets from a bigger set of possible CI implementations for a constrained reconfigurable area, with the aim of optimizing the task's WCET bound. The main problem in selecting WCET-optimizing CIs is *the instability of the worst-case path*, i.e., when reducing the latency of the worst-case path by inserting a CI, a completely different path can become the new worst-case path. Therefore, WCET bound estimation is an integral part of WCET-optimizing CI selection. The problem bears resemblance to other static optimizations targeting the worst-case path like instruction cache locking or scratchpad memory allocation of program code, but requires new models [15]. CI selection, also referred to as *instruction set selection*, is the second of the two main steps in the so-called *instruction set extension problem* [22]. The first step is the *CI generation* that is performed when compiling the application source code. In this step, kernels are identified in the application and partitioned into segments of code to execute in software and segments to execute using hardware accelerators. For the segments to execute in hardware, several alternatives that differ in resource demands as well as latencies are generated and then synthesized into configurations for the reconfigurable fabric (see Fig. 2). Which CIs are implemented in hardware instead of the original software code and how many reconfigurable containers to allocate for accelerators of a respective CI are determined by the CI selection according to an optimization goal, e.g., average-case performance. Several approaches to CI generation exist that can provide CIs and implementation alternatives as input to CI selection [22]. Different from existing CI selection approaches targeting average-case performance, WCET-optimizing selection requires the application binary, as it is the only way to be able to obtain precise WCET bound estimates (see Sect. 4.2). To obtain a finished binary with generated CIs while keeping the flexibility to execute the original software, we introduced the *CI Invocation* construct in Sect. 4.2 (see Fig. 11). CI Invocations are further extended to *CI super blocks* that allow multiple choices of hardware implementations for a CI (instead of just the binary choice between hardware and software).

4.3.1 CI Super Blocks and Optimization Goal

CI super blocks are a concept used to enable static WCET optimization. As shown in Fig. 14, CI super blocks begin with a conditional before every CI which jumps to the functionally equivalent software code when the CI is not implemented in hardware. This way, their implementation in an application behaves just like the CI Invocation construct (see Sect. 4.2). During WCET optimization, however, CI super blocks capture all the information about implementation alternatives for the respective CI that should be invoked. For each implementation j of CI k that is

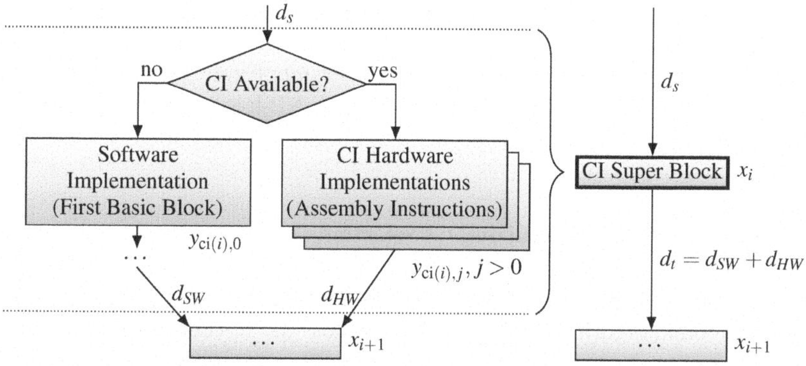

Fig. 14 CI super block as part of a CFG

invoked by a CI super block i, the following information are stored:

- reconfiguration delay $r_{k,j}$
- worst-case execution time $e_{i,j}$
- reconfigurable container demands $a_{k,j}$ (area).

By convention, $j = 0$ denotes the software implementation of a CI k, thus $r_{k,0} = 0$ and $a_{k,0} = 0$. Effectively, we obtain a CFG that is parameterized by the chosen implementation for each CI using CI super blocks. Thus, the result of the selection is a matrix $y \in \{0, 1\}^{|\mathcal{CI}| \times M}$ that maps each CI $k \in |\mathcal{CI}|$ to an implementation $j \in M$, where $|\mathcal{CI}|$ is the total number of CIs and M the maximum number of implementations per CI, respectively. The worst-case path analysis IPET [32] formulates an ILP objective function over basic blocks i with constant execution times c_i that finds the execution counts x_i of the respective basic block that lead to the maximum execution time ($\max_x \sum_i c_i x_i$). On top of that, the WCET-optimizing selection needs to find the selection y that minimizes the maximum execution time:

$$\min_y \left(\max_x \left(\underbrace{\sum_{i \in \text{BB}} c_i x_i}_{\text{Basic Blocks (BB)}} + \underbrace{\sum_{i \in \text{SB}} \sum_j e_{i,j} y_{\text{ci}(i),j} x_i}_{\text{CI Super Blocks (SB)}} \right) + \underbrace{\sum_k \sum_j y_{k,j} r_{k,j}}_{\text{Reconfiguration Delay}} \right) \quad (1)$$

where $\text{ci}(i)$ maps a CI super block i to the CI it invokes. Similar to flow networks, IPET formulates ILP constraints on the variables (x_i) that model the control flow and capture relative execution counts of basic blocks. One of the additional constraints for WCET-optimizing CI selection is that the total number of reconfigurable containers A must not be exceeded by the selection y ($\sum_k \sum_j a_{k,j} y_{k,j} \leq A$). Even if the reconfigurable area was infinitely large, it is not necessarily beneficial to select the highest-performance implementation for each CI: it might also incur the most reconfiguration delay.

The objective function (Eq. (1)) and its constraints formulate an NP-hard combinatorial optimization problem (already IPET is NP-hard). In the following, we discuss approaches that solve this problem optimally and heuristically.

4.3.2 Solving WCET-Optimizing CI Selection

An optimal solution to the WCET-optimizing CI selection can be obtained using a branch and bound algorithm that generates all possible selections of CI implementations and prunes branches that do not fit onto the reconfigurable area [15]. The problem with this approach is that the WCET of each possible selection needs to be evaluated. The number of possible selections grows exponentially in the number of reconfigurable containers A of the reconfigurable fabric and the number of CIs used in the task under optimization. Thus, we designed a greedy algorithm that proceeds as follows:

1. Start with an empty reconfigurable area (choose the software implementation $j = 0$ for all CIs)
2. Obtain a WCET estimate for the current selection y
3. For each CI k, calculate the profit on *current* worst-case path of "upgrading" from selected implementation j to the next implementation $j+1$ in the ascending order of reconfigurable container demand $a_{k,j}$ as:

$$\text{profit}(k, j+1, x) := \underbrace{\sum_{\substack{i \in \text{SB} \\ \text{ci}(i)=k}} (e_{i,j} - e_{i,j+1}) x_i}_{\text{latency reduction on current worst-case path x}} - \underbrace{(r_{k,j+1} - r_{k,j})}_{\text{additional reconfiguration cost}} \tag{2}$$

4. If exists, select upgrade $j + 1$ (instead of j) for CI k with highest positive profit (increasing allocated reconfigurable containers by at least one, possibly changing the worst-case path) and go to 2, terminate otherwise

Instead of requiring a number of WCET estimates that grows exponentially in the number of accelerators that fit onto the reconfigurable fabric A and the number of CIs used in the task under optimization, this greedy algorithm performs at most A WCET estimates: After each estimate (2) at least one accelerator is added to the selection (4) until all reconfigurable containers are occupied or no further candidate for upgrading exists. In the following we will show how this greedy algorithm compares to the optimal solution.

4.3.3 Results

While our timing analysis of tasks on reconfigurable processors as detailed in Sect. 4.2 could be evaluated using the commercial timing analyzer AbsInt aiT [2], we extended WCET analysis as an integral part of WCET optimization in this section. Therefore, the optimal branch and bound as well as heuristic selection algorithms were implemented "processors" within the open-source WCET estimation framework OTAWA [5]. We extended the existing analysis support for the LEON3 CPU in OTAWA to support CI opcodes, CI super blocks with configuration-dependent latency, and reconfiguration delay. In the following, results are shown for a reconfiguration bandwidth of 400 MB/s (maximum bandwidth, see Sect. 4.1), running the CPU at 400 MHz (which the LEON3 processor is advertised as running at when implemented as an ASIC) and the reconfigurable fabric at 100 MHz (which corresponds to the current *i*-Core design on Xilinx Virtex-7).

Figure 15 shows the runtime of our optimization approaches when optimizing the most complex kernel of the H.264 encoder "encode macroblock." The kernel uses six different CIs for which the algorithms need to select WCET-optimizing implementations that fit onto the reconfigurable area. Results are shown for different area sizes (number of accelerators that fit onto the reconfigurable fabric). As can be seen in the graph, even when no area is available for selecting CI implementations (and no candidates but a complete software evaluation are evaluated), the runtime is almost 5 s for both approaches. The reason is that reconstructing the CFG from the binary and analyzing all basic blocks already takes around 4.6 s. The runtime of the optimal algorithm first increases exponentially when the amount of area is increased, but then flattens out because branch and bound finds more opportunities to prune the search space. The runtime of the greedy algorithm seems to stay constant. It rises slightly, however, until an area of 10 accelerators. The amount of WCET estimates that need to be performed to find the final result grows linearly in the available area.

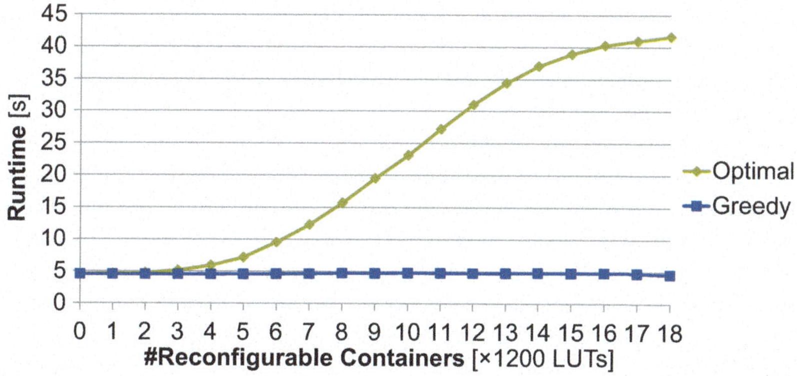

Fig. 15 Runtime of WCET-optimizing CI selection when optimizing H.264's encode macroblock kernel, which invokes six different CIs

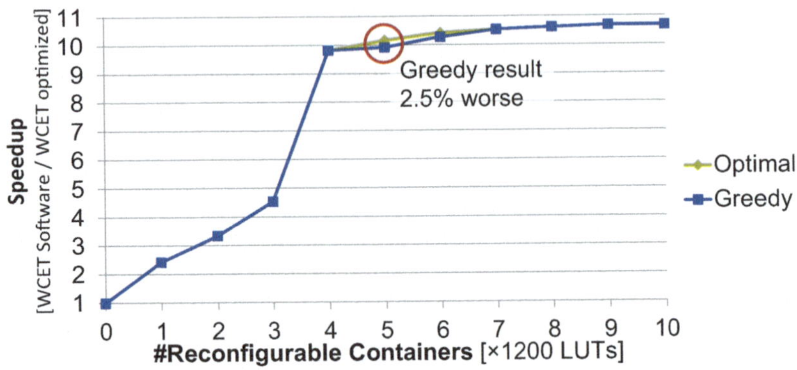

Fig. 16 Resulting speedup of WCET-optimizing CI selection when optimizing H.264's encode macroblock kernel compared to executing without acceleration

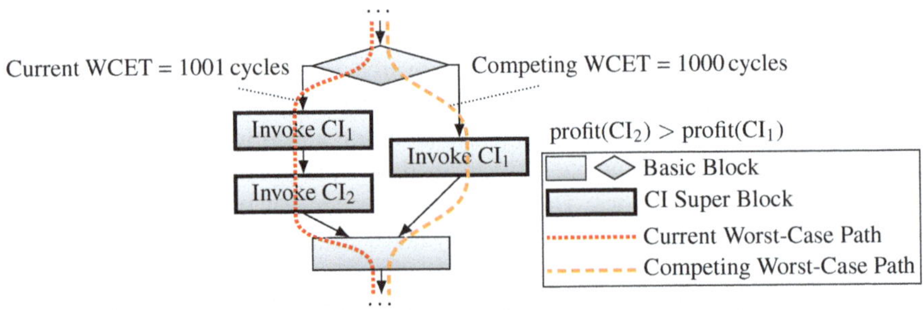

Fig. 17 Greedily selecting the CI with biggest profit on *current* worst-case path (CI_2) is not always optimal. The worst-case path changes during optimization

On average, an additional estimate adds only 40ms to the algorithm's runtime. Once the area reaches 10, there are still CIs that could be upgraded but the profit (Eq. (2)) becomes negative, because the additional reconfiguration delay increase is bigger than the latency reduction of the current worst-case path. Therefore, the runtime does not increase any further.

Figure 16 shows that the speedup results of the CI selections obtained by the greedy algorithm compared to execution without any accelerators are on par with the optimal solution for most cases. However, for an area of 5 and 6 accelerators the speedup obtained by greedy is up to 2.5% worse than the speedup of the optimal solution. Figure 17 shows a simplified example of the case in which greedy performs suboptimal. In this example, the current worst-case path (left path through the CFG) contains two CI super blocks that could be upgraded. CI_2 provides a bigger profit (see Eq. (2)) than CI_1, therefore, greedy will select CI_2. However, no matter how big the profit of CI_2 actually is on the current worst-case path the WCET will only be reduced by a single cycle, because the right path through the CFG will immediately become the new worst-case path and reduce the WCET from 1001

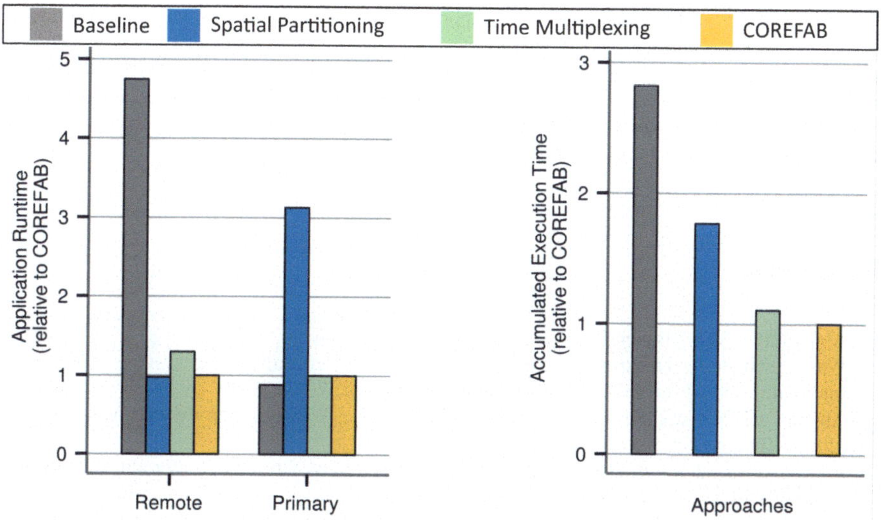

Fig. 19 Runtime results comparing state-of-the-art approaches to COREFAB. Baseline: a reconfigurable processor and three GPPs that cannot execute CIs

summary, COREFAB utilizes the reconfigurable resources more effectively than state-of-the-art approaches to optimize the performance of reconfigurable multi-core SoCs.

6 Conclusion

This chapter presented an overview of how non-functional requirements of cyber-physical systems are addressed by the runtime-reconfigurable processor platform *i*-Core. The benefits of a very tight integration of reconfigurable fabric and CPU (*i*-Core attaches an FPGA-based reconfigurable fabric directly to the CPU pipeline, see Sect. 1) were shown for several non-functional requirements. Not only were increases in performance shown by reconfiguration-aware task scheduling (Sect. 2) and providing access to *i*-Core's accelerators to general-purpose processors in the same SoC (Sect. 5), but also improvements in reliability (Sect. 3) and worst-case execution time guarantees (Sect. 4). In summary, our research in the context of *i*-Core leverages a processor with a runtime-reconfigurable instruction set to provide a holistic approach that targets the requirements of cyber-physical systems. Our future work focuses on security aspects of cyber-physical systems, where early results have shown that runtime reconfiguration is an effective countermeasure to side-channel attacks.

Acknowledgements This work was partly supported by the German Research Foundation (DFG) as part of the Transregional Collaborative Research Center "Invasive Computing" (SFB/TR 89) and as part of the priority program "Dependable Embedded Systems" (SPP 1500). Parts of the research summarized in this chapter were conducted together with Dr. Artjom Grudnitsky and Dr. Hongyan Zhang who earned their PhDs under the supervision of Prof. Dr. Jörg Henkel in 2015 and 2017, respectively. We want to thank them for fruitful discussions.

References

1. Abdelfattah, M.S., Bauer, L., Braun, C., Imhof, M.E., Kochte, M.A., Zhang, H., Henkel, J., Wunderlich, H.J.: Transparent structural online test for reconfigurable systems. In: IEEE International On-Line Testing Symposium (IOLTS), pp. 37–42 (2012)
2. AbsInt: aiT Worst-Case Execution Time Analyzers. Website: http://www.absint.com/ait/ (2018). [Online; accessed 10-Aug-2018]
3. Achronix: Speedchip FPGA chiplets. Website: https://www.achronix.com/product/speedchip/ (2018). [Online; accessed 10-Aug-2018]
4. Axer, P., Ernst, R., Falk, H., Girault, A., Grund, D., Guan, N., Jonsson, B., Marwedel, P., Reineke, J., Rochange, C., Sebastian, M., Hanxleden, R.V., Wilhelm, R., Yi, W.: Building timing predictable embedded systems. ACM Trans. on Embed. Comput. Syst. **13**(4), 82:1–82:37 (2014)
5. Ballabriga, C., Cassé, H., Rochange, C., Sainrat, P.: OTAWA: An open toolbox for adaptive WCET analysis. In: IFIP International Workshop on Software Technologies for Embedded and Ubiquitous Systems (SEUS), pp. 35–46. Springer (2010)
6. Bauer, L., Braun, C., Imhof, M.E., Kochte, M.A., Schneider, E., Zhang, H., Henkel, J., Wunderlich, H.J.: Test strategies for reliable runtime reconfigurable architectures. IEEE Transactions on Computers (TC), Special Section on Adaptive Hardware and Systems **62**(8), 1494–1507 (2013)
7. Bauer, L., Grudnitsky, A., Shafique, M., Henkel, J.: PATS: a performance aware task scheduler for runtime reconfigurable processors. In: Field-Programmable Custom Computing Machines (FCCM), pp. 208–215 (2012)
8. Bauer, L., Henkel, J.: Run-time Adaptation for Reconfigurable Embedded Processors. Springer Science+Business Media, LLC (2011)
9. Bauer, L., Shafique, M., Henkel, J.: A computation-and communication-infrastructure for modular special instructions in a dynamically reconfigurable processor. In: Int. Conf. on Field Programmable Logic and Applications, pp. 203–208 (2008)
10. Bauer, L., Shafique, M., Henkel, J.: Cross-architectural design space exploration tool for reconfigurable processors. In: Conference on Design, Automation and Test in Europe (DATE), pp. 958–963 (2009)
11. Bauer, L., Shafique, M., Henkel, J.: Concepts, architectures, and run-time systems for efficient and adaptive reconfigurable processors. In: NASA/ESA Conference on Adaptive Hardware and Systems (AHS), pp. 80–87 (2011)
12. Benkrid, K., Liu, Y., Benkrid, A.: A highly parameterized and efficient FPGA-based skeleton for pairwise biological sequence alignment. IEEE Transactions on Very Large Scale Integration (VLSI) Systems **17**(4), 561–570 (2009)
13. Biondi, A., Balsini, A., Pagani, M., Rossi, E., Marinoni, M., Buttazzo, G.: A framework for supporting real-time applications on dynamic reconfigurable FPGAs. In: Real-Time Syst. Symp., pp. 1–12 (2016)
14. Chen, L., Mitra, T.: Shared reconfigurable fabric for multi-core customization. In: Design Automation Conference (DAC), pp. 830–835 (2011)
15. Damschen, M., Bauer, L., Henkel, J.: Extending the WCET problem to optimize for runtime-reconfigurable processors. ACM Trans. on Archit. and Code Optim. **13**(4), 45:1–45:24 (2016)

16. Damschen, M., Bauer, L., Henkel, J.: CoRQ: Enabling runtime reconfiguration under WCET guarantees for real-time systems. IEEE Embedded Systems Letters **9**(3), 77–80 (2017)
17. Damschen, M., Bauer, L., Henkel, J.: Timing analysis of tasks on runtime reconfigurable processors. IEEE Trans. on Very Large Scale Integration Syst. **25**(1), 294–307 (2017)
18. Das, A., Nguyen, D., Zambreno, J., Memik, G., Choudhary, A.: An FPGA-based network intrusion detection architecture. IEEE Transactions on Information Forensics and Security **3**(1), 118–132 (2008)
19. De Schryver, C., Shcherbakov, I., Kienle, F., Wehn, N., Marxen, H., Kostiuk, A., Korn, R.: An energy efficient FPGA accelerator for Monte Carlo option pricing with the Heston model. In: International Conference on Reconfigurable Computing and FPGAs (ReConFig), pp. 468–474. IEEE (2011)
20. Dennl, C., Ziener, D., Teich, J.: On-the-fly composition of FPGA-based SQL query accelerators using a partially reconfigurable module library. In: IEEE International Symposium on Field-Programmable Custom Computing Machines (FCCM), pp. 45–52 (2012)
21. Edwards, S.A., Lee, E.A.: The case for the precision timed (PRET) machine. In: Proc. of Design Automat. Conf., pp. 264–265. ACM (2007)
22. Galuzzi, C., Bertels, K.: The instruction-set extension problem: A survey. ACM Trans. on Reconfig. Technol. and Syst. **4**(2), 18 (2011)
23. Grudnitsky, A., Bauer, L., Henkel, J.: COREFAB: concurrent reconfigurable fabric utilization in heterogeneous multi-core systems. In: Int. Conf. on Compilers, Architecture and Synthesis for Embed. Syst. (CASES), pp. 1–10 (2014)
24. Grudnitsky, A., Bauer, L., Henkel, J.: MORP: Makespan optimization for processors with an embedded reconfigurable fabric. In: ACM/SIGDA International Symposium on Field-Programmable Gate Arrays (FPGA), pp. 127–136 (2014)
25. Grudnitsky, A., Bauer, L., Henkel, J.: Efficient partial online synthesis of special instructions for reconfigurable processors. IEEE Transactions on Very Large Scale Integration (VLSI) Systems **25**(2), 594–607 (2017)
26. Henkel, J., Bauer, L., Becker, J., Bringmann, O., Brinkschulte, U., Chakraborty, S., Engel, M., Ernst, R., Härtig, H., Hedrich, L., Herkersdorf, A., Kapitza, R., Lohmann, D., Marwedel, P., Platzner, M., Rosenstiel, W., Schlichtmann, U., Spinczyk, O., Tahoori, M., Teich, J., Wehn, N., Wunderlich, H.J.: Design and architectures for dependable embedded systems. In: IEEE International Conference on Hardware/Software Codesign and System Synthesis (CODES+ISSS), pp. 365–374 (2011)
27. Henkel, J., Bauer, L., Dutt, N., Gupta, P., Nassif, S., Shafique, M., Tahoori, M., Wehn, N.: Reliable on-chip systems in the nano-era: Lessons learnt and future trends. In: IEEE/ACM Design Automation Conference (DAC) (2013)
28. Henkel, J., Herkersdorf, A., Bauer, L., Wild, T., Hübner, M., Pujari, R.K., Grudnitsky, A., Heisswolf, J., Zaib, A., Vogel, B., et al.: Invasive manycore architectures. In: Asia and South Pacific Design Automation Conference (ASP-DAC), pp. 193–200 (2012)
29. Hoang, T., Nguyen, V.L.: An efficient FPGA implementation of the advanced encryption standard algorithm. In: IEEE Int. Conf. on Computing Communication Technologies, Research, Innovation, and Vision for the Future (RIVF), pp. 1–4 (2012)
30. Honegger, D., Oleynikova, H., Pollefeys, M.: Real-time and low latency embedded computer vision hardware based on a combination of FPGA and mobile CPU. In: IEEE/RSJ International Conference on Intelligent Robots and Systems (IROS), pp. 4930–4935 (2014)
31. Lam, S.K., Srikanthan, T., Clarke, C.: Selecting profitable custom instructions for area-time-efficient realization on reconfigurable architectures. IEEE Transactions on Industrial Electronics **56**(10), 3998–4005 (2009)
32. Li, Y.T.S., Malik, S.: Performance analysis of embedded software using implicit path enumeration. In: Workshop on Languages, Compilers, & Tools for Real-time Syst., pp. 88–98 (1995)
33. Lockwood, J.W., McKeown, N., Watson, G., Gibb, G., Hartke, P., Naous, J., Raghuraman, R., Luo, J.: NetFPGA–an open platform for gigabit-rate network switching and routing. In: IEEE International Conference on Microelectronic Systems Education (MSE), pp. 160–161 (2007)

34. Puschner, P., Koza, C.: Calculating the maximum execution time of real-time programs. Real-Time Syst. **1**(2), 159–176 (1989)
35. Putnam, A., Caulfield, A.M., Chung, E.S., Chiou, D., Constantinides, K., Demme, J., Esmaeilzadeh, H., Fowers, J., Gopal, G.P., Gray, J., et al.: A reconfigurable fabric for accelerating large-scale datacenter services. ACM SIGARCH Computer Architecture News **42**(3), 13–24 (2014)
36. Rossi, E., Damschen, M., Bauer, L., Buttazzo, G., Henkel, J.: Preemption of the partial reconfiguration process to enable real-time computing with FPGAs. ACM Transactions on Reconfigurable Technology and Systems (TRETS) **11**(2), 10 (2018)
37. Sabeghi, M., Sima, V.M., Bertels, K.: Compiler assisted runtime task scheduling on a reconfigurable computer. In: Field Programmable Logic and Applications (FPL), pp. 44–50 (2009)
38. Schoeberl, M.: Time-predictable computer architecture. EURASIP Journal on Embed. Syst. pp. 2:1–2:17 (2009)
39. Tessier, R., Burleson, W.: Reconfigurable computing for digital signal processing: A survey. Journal of VLSI signal processing systems for signal, image and video technology **28**(1-2), 7–27 (2001)
40. Tessier, R., Pocek, K., DeHon, A.: Reconfigurable Computing Architectures. Proceedings of the IEEE **103**(3), 332–354 (2015)
41. Thiele, L., Wilhelm, R.: Design for timing predictability. Real-Time Syst. **28**(2-3), 157–177 (2004)
42. Watkins, M., Albonesi, D.: ReMAP: a reconfigurable heterogeneous multicore architecture. In: International Symposium on Microarchitecture (MICRO), pp. 497–508 (2010)
43. Wilhelm, R., Engblom, J., Ermedahl, A., Holsti, N., Thesing, S., Whalley, D., Bernat, G., Ferdinand, C., Heckmann, R., Mitra, T., Mueller, F., Puaut, I., Puschner, P., Staschulat, J., Stenström, P.: The Worst-case Execution-time Problem—Overview of Methods and Survey of Tools. ACM Trans. Embed. Comput. Syst. **7**(3), 36:1–36:53 (2008)
44. Wilhelm, R., Grund, D., Reineke, J., Schlickling, M., Pister, M., Ferdinand, C.: Memory hierarchies, pipelines, and buses for future architectures in time-critical embedded systems. Trans. on Comput.-Aided Design of Integrated Circuits and Syst. **28**(7), 966–978 (2009)
45. Zhang, H., Bauer, L., Henkel, J.: Resource budgeting for reliability in reconfigurable architectures. In: IEEE/ACM Design Automation Conference (DAC) (2016)
46. Zhang, H., Bauer, L., Kochte, M.A., Schneider, E., Braun, C., Imhof, M.E., Wunderlich, H.J., Henkel, J.: Module diversification: Fault tolerance and aging mitigation for runtime reconfigurable architectures. In: IEEE International Test Conference (ITC), pp. 1–10 (2013)
47. Zhang, H., Bauer, L., Kochte, M.A., Schneider, E., Wunderlich, H.J., Henkel, J.: Aging resilience and fault tolerance in runtime reconfigurable architectures. IEEE Transactions on Computers (TC), Special Section on Innovation in Reconfigurable Computing Fabrics from Devices to Architectures **66**(6), 957–970 (2017)
48. Zhang, H., Kochte, M.A., Imhof, M.E., Bauer, L., Wunderlich, H.J., Henkel, J.: GUARD: Guaranteed reliability in dynamically reconfigurable systems. In: IEEE/ACM Design Automation Conference (DAC) (2014)
49. Zhang, H., Kochte, M.A., Schneider, E., Bauer, L., Wunderlich, H.J., Henkel, J.: STRAP: Stress-aware placement for aging mitigation in runtime reconfigurable architectures. In: IEEE/ACM International Conference on Computer-Aided Design (ICCAD) (2015)

Color Primary Correction of Image and Video Between Different Source and Destination Color Spaces

Santanu Dutta

Abstract This article presents an introductory review of color correction—a color remapping of image and video between different source and destination color spaces. The review specifically focuses on two main aspects of color remapping—primary color space conversion and gamut mapping—and outlines the requirements, algorithms, methods, and possible implementation options.

1 Introduction

Trichromatic color vision [1] is the ability of humans to see different colors based on the interaction between three types of color-sensing *cone* cells in the human eye. At a basic level, the trichromatic theory states that the cone cells are sensitive to red, green, and blue light, and our brain mixes the colors together.[1] This observation has led to the *RGB* color model, where *R* (Red), *G* (Green), and *B* (Blue) light are added together in different amounts to reproduce different colors. The *RGB* color model is very useful for the electronic representation and display of images in both input devices (e.g., color TV, video cameras, and image scanners) and output devices (e.g., TV sets of various technologies, computer and mobile phone displays, and video projectors). There are a number of variants of the basic *RGB* color model as well, e.g., the well-known *sRGB* [3] color model that HP and Microsoft created for monitors, printers, and the Internet.

[1]Color perception is three-dimensional. In an *RGB* camera, or on a computer monitor, those attributes are the intensity of *Red (R)*, *Green (G)*, and *Blue (B)*; in the Munsell color system, the attributes are *Lightness*, *Saturation*, and *Hue*; in the CIELAB system, the three coordinates are L^*, a^*, and b^*. Thus, in any color space, there are three [2]. It is important to note, however, that *cones* actually do not detect any specific color, but respond to *Long (L)*, *Medium (M)*, and *Short (S)* wavelengths; LMS can be thought of as RGB.

S. Dutta (✉)
NVIDIA, Santa Clara, CA, USA

S. S. Bhattacharyya et al. (eds.), *Embedded, Cyber-Physical, and IoT Systems*,
https://doi.org/10.1007/978-3-030-16949-7_2

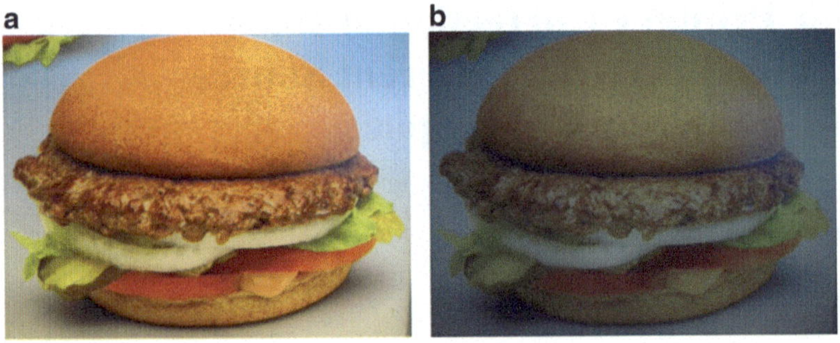

Fig. 1 Discoloration of food image on an uncorrected display. (**a**) Appetizing look. (**b**) Non-appetizing look

Primary color correction arises out of the need to convert pixels in the source *RGB* color space to the destination *RGB* color space, when the source video or image and the destination display have different color primaries. To understand why this is important, let us consider the traditional cathode ray tube (CRT) monitors first and look at the situation where both the source and destination conform to the *sRGB* color space. Even though *sRGB* is an output-referred[2] standard, geared towards CRTs, it is a common experience that when the same *RGB* data is displayed on two different monitors side by side, it is possible to get a noticeably different color balance on each. Each monitor produces a slightly different shade and intensity of red, green, and blue light. Not only do the two monitors look different, they often do not represent the source content accurately. It is, therefore, not much of a surprise that for flat panel displays (FPDs), which have very different color characteristics, the problem is more acute. Now throw in different types of FPDs (e.g., LCD—liquid crystal display, PDP—plasma display panel, DLP—digital light processing, OLED—organic light-emitting diode, etc.) and projectors, and different manufacturers[3] for each, and the situation quickly gets out of hand. The result can be quite disturbing: non-appetizing color of food on LCD screens at drive-through lanes of fast-food restaurants (an extreme example of which is shown in Fig. 1 to illustrate the point), lost differentiation of tissue type(s) on diagnostic medical displays, and miscolored movies on the standard monitor(s) in a typical household!

At the heart of the previously mentioned problem is the fact that *RGB* is a *device-dependent* color model, i.e., different devices detect or reproduce a given *RGB* value differently, since the color elements (such as phosphors) and their response to the individual *R*, *G*, and *B* levels vary not only from one manufacturer to another, but even in the same device over time [4]. When the source and the display (or, different displays) have different *RGB* color ranges that they span or can faithfully reproduce,

[2]sRGB is a normalized reference standard designed to match the color performance of an output device, e.g., a CRT monitor under typical viewing conditions.

[3]Note that some manufacturers achieve cost downs and/or a higher light output for mobile displays by producing non-standard displays (i.e., displays with non-standard specifications).

we say that the devices have different *color gamuts*; an *RGB* value that looks correct on the source may not look correct *as is* on the display; in fact, even if the same *RGB* values are sent to different displays with different color gamuts, the resulting colors will be different, unless the devices are identical and process the values identically. Before we can talk about solutions, however, it is important to understand what *color gamut* of a device means; since it has its roots in the CIE [5] *chromaticity diagram* that is where we start from.

2 Chromaticity Diagram

The *CIE XYZ* color space [6] encompasses all colors that are visible to a person with average eyesight. In the *XYZ* color space, the tristimulus[4] values are called *X*, *Y*, and *Z*, and these are roughly equivalent to the red, green, and blue, respectively, of an *RGB* model. The *CIE XYZ* color space was deliberately designed so that the *Y* parameter is a measure of the luminance of a color. Since the human eye has three types of cones that respond to different ranges of wavelengths, a full plot of all visible colors calls for a three-dimensional graph. This is inconvenient to represent in two dimensions. So, for convenience, the CIE transformed the three-dimensional *XYZ* color space into two artificial dimensions (x, y) of color (collectively called *chromaticity*) and one of intensity (Y), and then took a two-dimensional slice through this space at an arbitrary level of intensity. The resultant horseshoe curve, like the one shown in Fig. 2, is called the *chromaticity diagram*; it is essentially a

Fig. 2 Chromaticity diagram

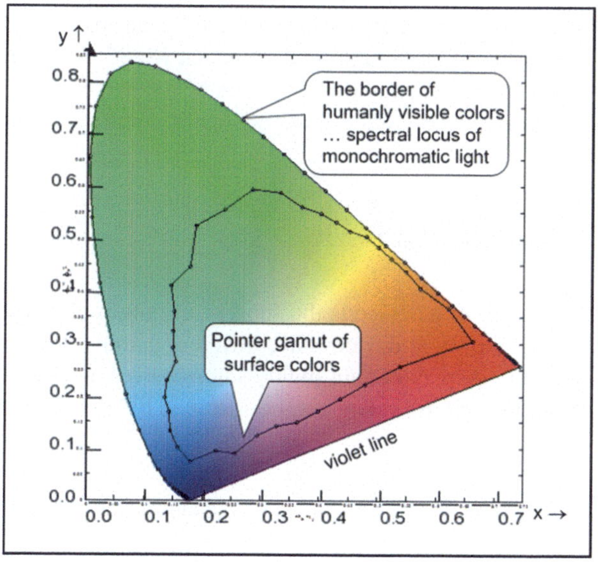

[4]The term *tristimulus* comes from the fact that color perception results from the retina of the eye responding to three types of stimuli.

two-dimensional (2D) projection of the three-dimensional (3D) *XYZ* color space onto the *x*–*y* chromaticity plane. In this diagram, the chromaticity of a color is specified by the two derived parameters *x* and *y*, which are functions of all three tristimulus values *X*, *Y*, and *Z*. All physical colors reside inside the horseshoe, i.e., the horseshoe is the range of all visible colors on the CIE plot. As Poynton [7] points out, the sensation of purple cannot be produced by a single wavelength—to produce purple requires a mixture of shortwave and longwave light; the line of purples or the violet line on a chromaticity diagram joins extreme blue to extreme red. All physical colors are thus contained in the area in (*x*, *y*) bounded by the violet line and the spectral locus (meaning that the horseshoe separates spectral colors[5] that lie on the horseshoe from non-spectral colors that lie inside the horseshoe and non-existent or hyper-saturated colors that reside outside). Note in particular the irregular shape drawn inside horseshoe—it is called the "pointer set" of surface colors, the largest set of surface colors that has been found in nature and art.

XYZ is a *device-independent* color model developed by the CIE for categorizing the world of colors. Theoretically, based on this system, every color we see can be described in terms of the (*x*, *y*) coordinates. The *XYZ* model (space) also offers the following interesting properties [8]:

- Two colors with the same *XYZ* values look the same,
- Two colors with different *XYZ* values look different,
- *Y* value represents the luminance information, and
- *XYZ* color of any object can be objectively measured by a colorimeter [9].

The *CIE XYZ* is the fundamental basis of all color management such as calibration, color space conversions, and color matching. Figure 3 shows the location of the *X*, *Y*, and *Z* primaries on the CIE chromaticity diagram. These *X*, *Y*, and *Z* primaries can be thought of as "special *RGB*" values that extend much beyond the visible range, and any real color can be defined by a linear combination of *X*, *Y*, and *Z* values.

3 Gamut of a Device

The *gamut* of a device is the subset (or, range) of visible colors that the device can display. Different devices have different gamuts. It is very possible that one device is able to display certain highly saturated colors that another device cannot. Since an additive *RGB* system can be specified by the chromaticities of its primaries and

[5]A spectral color is the color sensation created by a monochromatic (single wavelength) light in the visible spectrum.

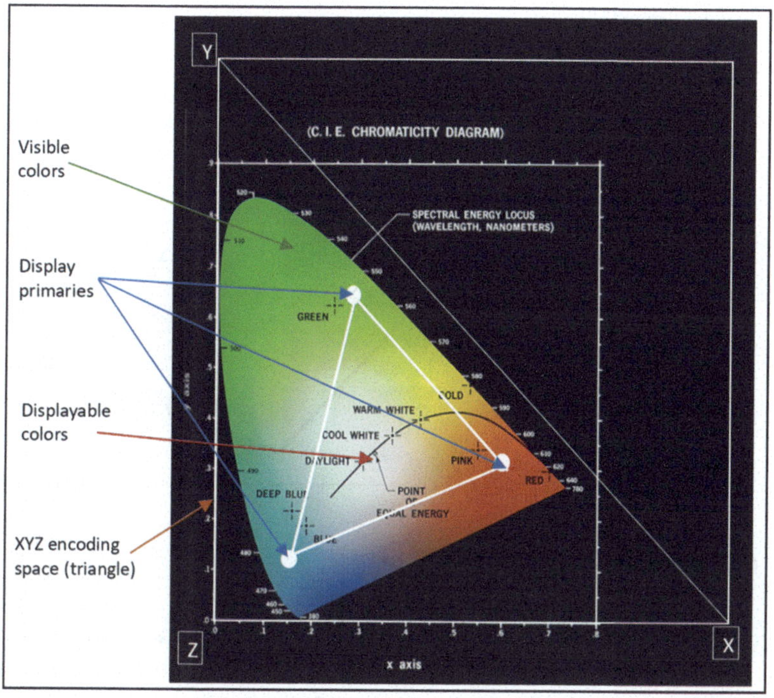

Fig. 3 *XYZ* and visible color spaces on a CIE diagram

its white point,[6] the gamut of a particular *RGB* device can be determined from its *R*, *G*, and *B* chromaticities and the white point. Gamuts are commonly represented as areas in the (x, y) chromaticity diagram. These areas typically have triangular shapes because most color reproduction is done with three primaries. The gamut of the colors that can be mixed from a given set of *RGB* primaries is shown in the chromaticity diagram by a triangle whose vertices are the chromaticities of the primaries, e.g., the white triangle inside the horseshoe in Fig. 3.

It is to be noted that the accessible color gamut also depends on the *brightness* (because all three display primaries must be turned on in order to achieve higher luminance); hence, a full and true-color gamut must be represented in a 3D space, with luminance or brightness on the third (vertical) axis, as illustrated in Fig. 4. This cone shape of the 3D gamut indicates that bright reflecting surfaces are necessarily not very colorful because they must reflect a large part of the spectrum!

[6] *White point* [10] is the tristimulus values or chromaticity coordinates that define a chosen color of "white," often D65 [11].

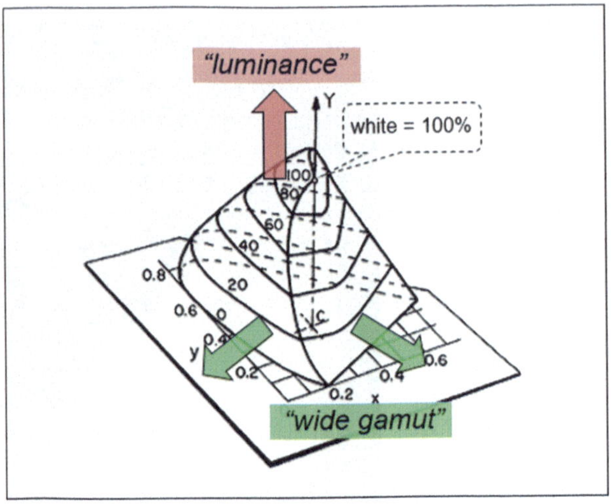

Fig. 4 3D color gamut [12]

3.1 Color Gamut Examples

Two common color gamuts for representative image sources—standard TV and digital cinema—are shown in Fig. 5. As discussed before, the horseshoe shape represents every color that the human eye can possibly see, i.e., it is the complete color gamut of the human eye reduced to the two dimensions of *saturation* and *hue*. The smaller triangle inside the horseshoe is the device color gamut of today's standard ITU-Rec.709 television; it does not give full coverage of the pointer set, but most TVs can reproduce this color gamut easily enough because that is how the gamut was originally defined (from a first European color CRT display). The Digital Cinema P3 standard has a color gamut represented by the bigger triangle; it is considerably larger than that of the standard television but still does not cover the entire pointer set.

Figure 6 shows four example color gamuts for common displays. The ITU-Rec.709 gamut was once the standard for every color TV in the world; originally defined by the prevalent phosphors in CRT displays, it is also the standard for computer monitors under the name of *sRGB*. Digital cinema has a wider color gamut than Rec.709, and it is interesting to note that this standard is based on the DLP projector engine from Texas Instruments illuminated with a Xenon arc lamp through dichroic filters.[7] As can be seen, a certain example LED-backlit LCD panel exhibits an even wider gamut—an even larger triangle—with very good coverage of red, green, and cyan. However, the hypothetical three-primary laser display represents the largest triangle and thus does even better—it is quite extreme in terms of how large a color gamut it spans.

[7] These filters selectively pass light of a narrow range of colors and reflect all other colors.

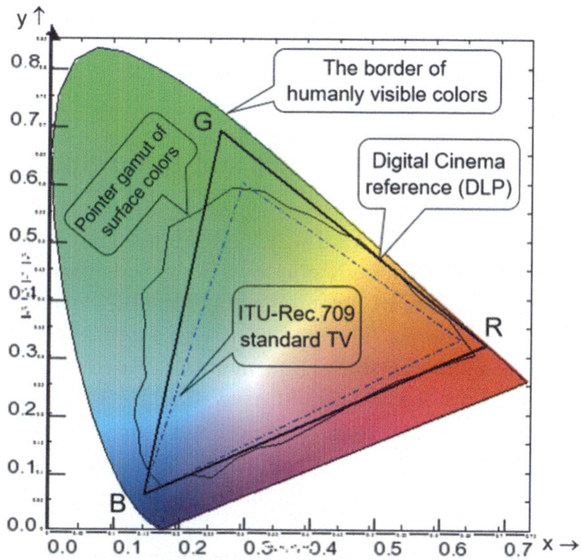

Fig. 5 Example color gamuts of image sources

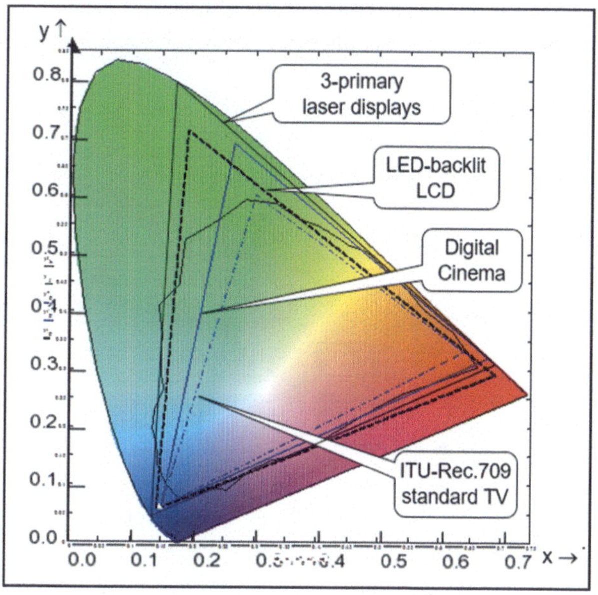

Fig. 6 Example color gamut of displays

Given that different *RGB* source images and *RGB* destination displays have different color gamuts, different values must be sent to the different display devices to faithfully reproduce the source image on the display device. What is needed is a way to take the values that represent the desired color on the source, and from them produce the corresponding values that reproduce the "same" color on the destination. This correction calls for a color transformation that is commonly referred to as *primary color space conversion.*

4 Device Independence and Color Space Conversion

The most straightforward way to go from a source color space to a different display color space is to directly employ a dedicated color transformation. However, due to the large number of potential sources and destinations, this method does not scale well. For example, if there are n source and m destination devices, we need mxn possible color transformation methods; but if we were to add a new destination device to the list, we would have to add n new transforms, one for each source. Conversely, a new source would dictate that we make m new transforms, one for each destination [13]. A common way to simplify the situation, as shown in Fig. 7, is to introduce an intermediate color space such that you have one transform per source to convert the source to an intermediate standard, and one transform per destination to convert the intermediate standard to each destination. Thus, it now becomes an additive problem, and the introduction of a new device only requires the introduction of one new transform either to or from the standard. When both the source video and

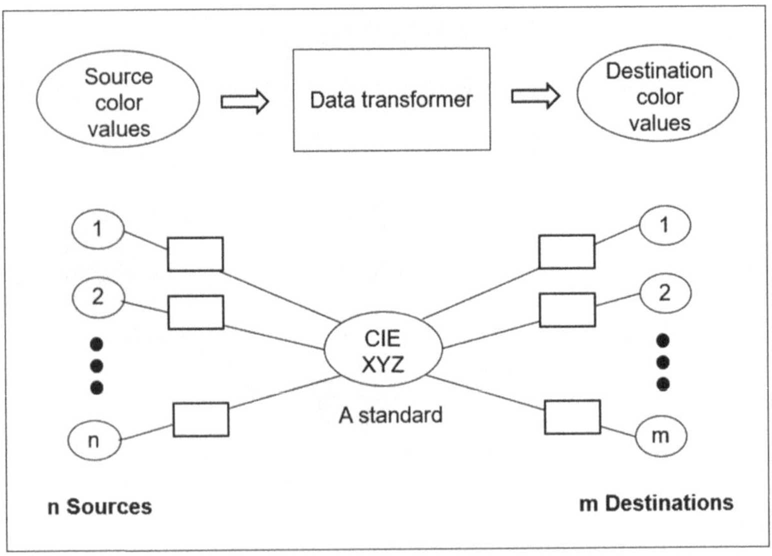

Fig. 7 Color transformation between source and destination [13]

the destination display constitute *RGB* spaces,[8] as is the case for standard video sources and displays, the aim is to find a fixed, standard, intermediate, device-independent space through which the source and the destination *RGB* values can be related. The CIE 1931 *XYZ* color space discussed earlier offers this desired independence; since its proposal in 1931, the primaries of color spaces are usually specified in (x, y) coordinates.

4.1 Color Transformation

Given the device independence of *CIE XYZ*, the current color transformation method used widely is to have an $RGB_{\text{source}} \rightarrow XYZ \rightarrow RGB_{\text{destination}}$ transform mechanism that is customized according to the *color space definitions* as shown in Fig. 8. The International Color Consortium (ICC) [14] has standardized the color space definitions and called them *profiles*. The transforms are implemented in what is commonly called a color matching module or color matching method (CMM). In an ICC-based color management system, the standard reference space, into or out of which the color data is transformed, is called a profile connection space (PCS). *CIE XYZ* is one of the most widely used PCSs in the ICC system.

5 *CIE XYZ* Chromaticity Coordinates

The *CIE XYZ* is a 3D linear color space,[9] where one of the channels, *Y*, equals luminance, and the *X* and the *Z* values add chromaticity information. Chromaticity coordinates (denoted usually with a lowercase letter) are tristimulus values (denoted usually with an uppercase letter) that are normalized by the total energy [15]:

$$\left.\begin{aligned} x &= \frac{X}{X+Y+Z} \\ y &= \frac{Y}{X+Y+Z} \\ z = \frac{Z}{X+Y+Z} &= 1 - x - y \end{aligned}\right\} \tag{1}$$

[8]Note that RGB is a *color model* and the range of colors that can be represented by the *color model* is its corresponding *color space*. Slightly different primaries (i.e., primaries with slightly different chromaticities) within the same *RGB* color model can give rise to different *RGB* color spaces. In this article, however, the terms *model* and *space* are used somewhat interchangeably.

[9]Linearity allows saturation to be defined in terms of additive color mixing in (*X*,*Y*,*Z*) or any (*R*,*G*,*B*). Note that both the photopic luminance sensitivity (i.e., how light *gray* is) and the luminous sensitivity of the human eye (i.e., how light a *color* is) are linear.

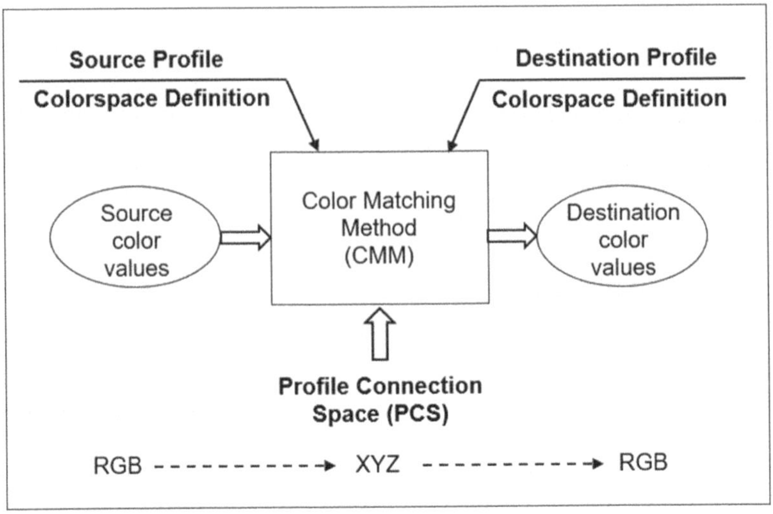

Fig. 8 Color matching flow [13]

The previous representation of the *CIE XYZ* color space is called **xyY**. Due to the normalization, we get $x + y + z = 1$. Since $z = 1 - x - y$, only two values are needed to describe the chromaticity. The tristimulus values can, however, be easily recovered from the chromaticity values when the absolute luminance Y is known:

$$\left.\begin{aligned} X &= \left(Y/_{y}\right) x \\ Z = \left(Y/_{y}\right) z &= \left(Y/_{y}\right) (1 - x - y) \end{aligned}\right\} \quad (2)$$

Any real color, hence any of the primaries, can be described by the *XYZ* tristimulus coordinates (X_R, Y_R, Z_R) for R, (X_G, Y_G, Z_G) for G, and (X_B, Y_B, Z_B) for B; so the tristimulus coordinates *XYZ* of any combination color of *RGB* can be calculated using the linear transformation shown next [15]:

$$\begin{bmatrix} X \\ Y \\ Z \end{bmatrix} = \begin{bmatrix} X_R & X_G & X_B \\ Y_R & Y_G & Y_B \\ Z_R & Z_G & Z_B \end{bmatrix} \times \begin{bmatrix} R \\ G \\ B \end{bmatrix} \quad (3)$$

The "white point" of a display is the chromaticity (x, y) of the display's nominal white, i.e., the color produced when $R = G = B = Y$. It is customary to specify the display colors by specifying the (x, y) chromaticities of the individual color components R, G, and B, plus the (x, y) of the white point of the display. The white point allows one to infer the relative brightness of the three color components; the brightness cannot be determined from the chromaticities alone. It is simple to compute the *XYZ* coordinates for the display's white, red, green, and blue points from their chromaticity values (as explained next). An intuition, articulated here but used later, that would help is that $4 \times (x, y) = 8$ values, but the matrix needs 9, so we set $Y = 1$ for white!

6 *XYZ* to *RGB* Color Conversion

The conversion between any two trichromatic color spaces is the same as solving a set of linear matrix equations. Note that for this conversion, the source data must be linear—i.e., inverse-gamma compensated if necessary—in order to obtain a natural and a linear intensity representation. For a fully described color space, the chromaticity coordinates (of the three primary colors *R*, *G*, and *B*) and the white point coordinates are used for the conversion. From Eq. (1), the chromaticity coordinates of *R*, *G*, and *B* can be written as

$$\left.\begin{aligned}
x_r &= \frac{X_R}{X_R+Y_R+Z_R} = \frac{X_R}{A_R} \Rightarrow X_R = x_r A_R \\
y_r &= \frac{Y_R}{X_R+Y_R+Z_R} = \frac{Y_R}{A_R} \Rightarrow Y_R = y_r A_R \\
z_r &= \frac{Z_R}{X_R+Y_R+Z_R} = \frac{Z_R}{A_R} \Rightarrow Z_R = z_r A_R \\
x_g &= \frac{X_G}{X_G+Y_G+Z_G} = \frac{X_G}{A_G} \Rightarrow X_G = x_g A_G \\
y_g &= \frac{Y_G}{X_G+Y_G+Z_G} = \frac{Y_G}{A_G} \Rightarrow Y_G = y_g A_G \\
z_g &= \frac{Z_G}{X_G+Y_G+Z_G} = \frac{Z_G}{A_G} \Rightarrow Z_G = z_g A_G \\
x_b &= \frac{X_B}{X_B+Y_B+Z_B} = \frac{X_B}{A_B} \Rightarrow X_B = x_b A_B \\
y_b &= \frac{Y_B}{X_B+Y_B+Z_B} = \frac{Y_B}{A_B} \Rightarrow Y_B = y_b A_B \\
z_b &= \frac{Z_B}{X_B+Y_B+Z_B} = \frac{Z_B}{A_B} \Rightarrow Z_B = z_b A_B
\end{aligned}\right\} \tag{4}$$

where $A_R = X_R + Y_R + Z_R$, $A_G = X_G + Y_G + Z_G$, and $A_B = X_B + Y_B + Z_B$.

Combining Eqs. (3) and (4), we get

$$\begin{bmatrix} X \\ Y \\ Z \end{bmatrix} = \begin{bmatrix} x_r A_R & x_g A_G & x_b A_B \\ y_r A_R & y_g A_G & y_b A_B \\ z_r A_R & z_g A_G & z_b A_B \end{bmatrix} \times \begin{bmatrix} R \\ G \\ B \end{bmatrix}$$

$$\text{or,} \begin{bmatrix} X \\ Y \\ Z \end{bmatrix} = \begin{bmatrix} x_r & x_g & x_b \\ y_r & y_g & y_b \\ z_r & z_g & z_b \end{bmatrix} \times \begin{bmatrix} A_R & 0 & 0 \\ 0 & A_G & 0 \\ 0 & 0 & A_B \end{bmatrix} \times \begin{bmatrix} R \\ G \\ B \end{bmatrix} \tag{5}$$

$$\text{or,} \begin{bmatrix} X \\ Y \\ Z \end{bmatrix} [\mathbf{K}] \times [\mathbf{A}] \times \begin{bmatrix} R \\ G \\ B \end{bmatrix}$$

This is the desired conversion equation from *RGB* to *XYZ*, where **K** is the chromaticity matrix obtained from the chromaticity coordinates of the source or the destination. But, how do we find the unknown diagonal matrix **A**?

Since Eq. (5) holds for any color value, **A** can be found by solving Eq. (5) for the white point, say of the source, for which three of the four matrices in Eq. (5) are known. Note that the white point is a *neutral hue* (also referred to as *gray* or *achromatic*). Since the white point is defined to be the hue that is shown when all three of the channels or electron guns of the monitor are set to be equal, we can set them to their maximum values without any loss of generality, i.e., $R = G = B = 1$.[10] This implies that the luminance of the source white point (Y_{sw})[11] may be arbitrarily set to 1 (because $R = G = B = Y$ defines the white point as mentioned earlier). As with the color primaries, the source white point chromaticity coordinates ($x_{\text{sw}}, y_{\text{sw}}$) are also given as the normalized (x, y) values. For conversion purposes, the absolute tristimulus values are needed. Since Y_{sw} has to be 1 for the source white point, we get from Eq. (2):

$$\left.\begin{aligned} X_{\text{sw}} &= \left({}^{Y_{\text{sw}}}\!/_{y_{\text{sw}}}\right) x_{\text{sw}} = {}^{x_{\text{sw}}}\!/_{y_{\text{sw}}} \\ Y_{\text{sw}} &= 1 \\ Z_{\text{sw}} &= \left({}^{Y_{\text{sw}}}\!/_{y_{\text{sw}}}\right) z_{\text{sw}} = {}^{z_{\text{sw}}}\!/_{y_{\text{sw}}} = \frac{(1 - x_{\text{sw}} - y_{\text{sw}})}{y_{\text{sw}}} \end{aligned}\right\} \tag{6}$$

Because *RGB* and *XYZ* are both physical descriptions of color, a color conversion only requires linear scaling of the chromaticities around the white point, and then solving the 3 × 3 linear equations in order to convert the axis system.

The chromaticity matrix **K** for the source is written as[12]

[10]The RGB values here are assumed to be in the range from 0 to 1, which is merely a convenience, and typical of a 100% filled gamut, but not essential.

[11]The subscripts *s* and *w* refer to *source* and *white*, respectively.

[12]x_{sr}, in the matrix, denotes the *x* chromaticity (coordinate) of the **r**ed (*r*) primary and corresponds to the **s**ource (*s*) color space.

$$[\mathbf{K}] = \begin{bmatrix} x_{sr} & x_{sg} & x_{sb} \\ y_{sr} & y_{sg} & y_{sb} \\ z_{sr} & z_{sg} & z_{sb} \end{bmatrix} \tag{7}$$

and the white point matrix **W**, by virtue of Eq. (6), is written as

$$[\mathbf{W}] = \begin{bmatrix} X_{sw} \\ Y_{sw} \\ Z_{sw} \end{bmatrix} = \begin{bmatrix} x_{sw} / y_{sw} \\ 1 \\ z_{sw} / y_{sw} \end{bmatrix} \tag{8}$$

If we define the chromaticities in the *RGB* axis system (under a given color temperature) such that $R = G = B = 1$ yields the white point, **A** must satisfy Eq. (5), i.e.,

$$\begin{bmatrix} X \\ Y \\ Z \end{bmatrix} = [\mathbf{K}] \times [\mathbf{A}] \times \begin{bmatrix} R \\ G \\ B \end{bmatrix}$$

$$\text{or, } \begin{bmatrix} X_{sw} \\ Y_{sw} \\ Z_{sw} \end{bmatrix} = \begin{bmatrix} x_{sr} & x_{sg} & x_{sb} \\ y_{sr} & y_{sg} & y_{sb} \\ z_{sr} & z_{sg} & z_{sb} \end{bmatrix} \times \begin{bmatrix} A_{sR} & 0 & 0 \\ 0 & A_{sG} & 0 \\ 0 & 0 & A_{sB} \end{bmatrix} \times \begin{bmatrix} 1 \\ 1 \\ 1 \end{bmatrix} \quad \text{(using Eqs.(7) and (8))}$$

$$\text{or, } \begin{bmatrix} X_{sw} \\ Y_{sw} \\ Z_{sw} \end{bmatrix} = \begin{bmatrix} x_{sr} & x_{sg} & x_{sb} \\ y_{sr} & y_{sg} & y_{sb} \\ z_{sr} & z_{sg} & z_{sb} \end{bmatrix} \times \begin{bmatrix} A_{sR} \\ A_{sG} \\ A_{sB} \end{bmatrix}$$

or, $\mathbf{W}=\mathbf{KA}$,

or, $\mathbf{A} = \mathbf{K}^{-1}\,\mathbf{W}$

(9)

Equation (9) can be solved easily and robustly for **A** on knowing the chromaticity coordinates of the three *RGB* primaries (that yield **K**) and the white point (**W**) of the source. The solution for **A** is thus a 3×1 column matrix:

$$[\mathbf{A}] = \begin{bmatrix} A_{sR} \\ A_{sG} \\ A_{sB} \end{bmatrix} \tag{10}$$

With **A** known, the chromaticity matrix **K**, scaled by **A** (often called the achromatic correction), now provides the conversion matrix **M**, under the desired white point. We thus have $\mathbf{M} = \mathbf{K} \times \mathbf{A}$, and using Eqs. (7) and (9) to yield **K** and **A**, we can express **M** as

$$[\mathbf{M}] = \begin{bmatrix} x_{sr}A_{sR} & x_{sg}A_{sG} & x_{sb}A_{sB} \\ y_{sr}A_{sR} & y_{sg}A_{sG} & y_{sb}A_{sB} \\ z_{sr}A_{sR} & z_{sg}A_{sG} & z_{sb}A_{sB} \end{bmatrix} \tag{11}$$

The *RGB* to *XYZ* conversion for the source can therefore be written as

$$\begin{bmatrix} X \\ Y \\ Z \end{bmatrix} = [\mathbf{M}] \times \begin{bmatrix} R \\ G \\ B \end{bmatrix} \tag{12}$$

7 Conversion from *RGB* Source to *RGB* Destination

Once the color profiles (i.e., the chromaticity of the primaries) are known for the *RGB* source video and the *RGB* destination display, a conversion from the source to the destination color spaces can be performed via an intermediate *XYZ* color space. To see this, let **M** be the matrix that converts the source *RGB* pixels to the intermediate *XYZ* space, and let **N** be the matrix that converts the destination *RGB* pixels to the intermediate *XYZ* space. We can then write

$$\begin{aligned} \begin{bmatrix} X \\ Y \\ Z \end{bmatrix} &= [\mathbf{N}] \times \begin{bmatrix} R_{\text{destination}} \\ G_{\text{destination}} \\ B_{\text{destination}} \end{bmatrix} \\ \text{or,}\quad [\mathbf{N}]^{-1} \times \begin{bmatrix} X \\ Y \\ Z \end{bmatrix} & \begin{bmatrix} R_{\text{destination}} \\ G_{\text{destination}} \\ B_{\text{destination}} \end{bmatrix} \end{aligned} \tag{13}$$

But since we already have from Eq. (12)

$$\begin{bmatrix} X \\ Y \\ Z \end{bmatrix} = [\mathbf{M}] \times \begin{bmatrix} R_{\text{source}} \\ G_{\text{source}} \\ B_{\text{source}} \end{bmatrix}$$

we can substitute in Eq. (13) to obtain

$$\begin{aligned} [\mathbf{N}]^{-1} \times [\mathbf{M}] \times \begin{bmatrix} R_{\text{source}} \\ G_{\text{source}} \\ B_{\text{source}} \end{bmatrix} &= \begin{bmatrix} R_{\text{destination}} \\ G_{\text{destination}} \\ B_{\text{destination}} \end{bmatrix} \\ \text{or,} \begin{bmatrix} R_{\text{destination}} \\ G_{\text{destination}} \\ B_{\text{destination}} \end{bmatrix} &= [\mathbf{C}] \times \begin{bmatrix} R_{\text{source}} \\ G_{\text{source}} \\ B_{\text{source}} \end{bmatrix} \end{aligned} \tag{14}$$

where $[\mathbf{C}] = [\mathbf{N}]^{-1} \times [\mathbf{M}]$ is the primary color space conversion matrix from a source *RGB* color space to a destination *RGB* color space. Multiplication by this matrix allows a linear transformation of the color primary coordinates (between source and destination) so that a source color can be represented perfectly in the destination space.[13]

7.1 Rationale for $R = G = B = Y = 1$

In the previous derivations of how we calculated the matrices **M**, **N**, and **C**, we made use of the fact that "white point" of a display is the chromaticity (x, y) of the display's nominal white, i.e., the color produced when $R = G = B = Y = 1$. The rationale is that we had a matrix with nine coefficients, but only eight equations for its calibration:

- Three for each of the x and y locations of the R, G, B primaries
- One for each of the x and y locations of white or the relative luminance of the R, G, B primaries (white balance)

The idea is that there is no ninth input for the absolute luminance, so we make one up, and we are happy if input and output luminances are equal. As the matrix is linear, one needs to use only one value; it does not matter whether Y is chosen to be 1 or not. And if it is chosen wrongly for **M** and **N**, it still cancels out in the calculation of **C**. So, to have the ninth degree of freedom, we just state/define that $R = G = B = 1 \Rightarrow Y = 1$. It is the best choice because it gives a more neutral **C**. We could have chosen a different value $R = G = B = Y < 1$, but then it is gray and not (peak) white.

In reality, the ninth degree of freedom is the display's brightness knob. Hence, one may choose to also use only six coefficients of a 3×3 matrix for the R, G, B color space conversion, and leave the white point adjustment (if any) to the existing three contrast gains on the R, G, and B signals[14] that any monitor or TV already has (and uses for the adjustment of the white point or the color temperature of the display as well). However, adjusting the three white point gains is not a perceptually correct method for simulating a different color of the illumination for the scene because chromatic adaptation [16] is ignored; so, ideally, one should modify all nine coefficients of the entire 3×3 matrix after all.

[13]In this sense, the matrix **C** functions as a primary color-space conversion matrix.

[14]The gain is actually applied to R', G', and B'—the non-linear gamma-corrected (following color space conversion and gamut mapping) R, G, and B signals that are sent to the display.

7.2 A Numerical Example of Color Conversion

Let us consider an example *RGB* conversion from the Rec.709 (standard) source to the NTSC TV (standard) destination. For the Rec.709 source, we consider the CIE standard D65 daylight illuminant as the white point (corresponding to the average mid-day light in Europe in the way it was defined). For the destination, we consider the CIE standard illuminant *C* as the white point.

Table 1 shows the (*x*, *y*) chromaticity coordinates of the Rec.709 *RGB* primaries and the (*x*, *y*) coordinate of the D65 white point (*W*) for the Rec.709 source. Based on the values provided and the equations shown earlier, one can calculate source-side value of **A** and derive the $RGB_{\text{source}} \rightarrow XYZ$ conversion matrix **M** to be

$$[A] = \begin{bmatrix} 0.6444 \\ 1.1919 \\ 1.2032 \end{bmatrix} \quad [M] = \begin{bmatrix} 0.4124 & 0.3576 & 0.1805 \\ 0.2126 & 0.7152 & 0.0722 \\ 0.0193 & 0.1192 & 0.9505 \end{bmatrix}$$

Table 2 shows the (*x*, *y*) chromaticity coordinates of the NTSC *RGB* primaries and the (*x*, *y*) coordinate of the illuminant *C* white point (*W*).

Based on the values provided and the equations shown earlier, one can calculate destination-side value of **A** and derive the $RGB_{\text{destination}} \rightarrow XYZ$ conversion matrix **N** to be

$$[\mathbf{A}] = \begin{bmatrix} 0.9060 \\ 0.8259 \\ 1.4327 \end{bmatrix} \quad [\mathbf{N}] = \begin{bmatrix} 0.6070 & 0.1734 & 0.2006 \\ 0.2990 & 0.5864 & 0.1146 \\ 0.0000 & 0.0661 & 1.1175 \end{bmatrix}$$

Using **M** and **N**, we can now do the composition and calculate the *RGB*-source-to-*RGB*-destination color space conversion matrix **C** as

Table 1 R709 and D65 white point chromaticities of source

	x	*y*
R	0.6400	0.3300
G	0.3000	0.6000
B	0.1500	0.0600
W	0.3127	0.3290

Table 2 NTSC primaries and white point for an *RGB* display

	x	*y*
R	0.6700	0.3300
G	0.2100	0.7100
B	0.1400	0.0800
W	0.3100	0.3160

$$[\mathbf{C}] = [\mathbf{N}]^{-1} \times [\mathbf{M}] = \begin{bmatrix} 0.6688 & 0.2678 & 0.0323 \\ 0.0185 & 1.0746 & -0.0603 \\ 0.0162 & 0.0431 & 0.8542 \end{bmatrix}$$

8 Implementation of Color Space Conversion

The primary color space conversion of a source image targeted to a destination display (implemented on any computation or processing device according to the scheme outlined in the previous section) is depicted in Fig. 9. It has the following characteristics:

- It is a color space conversion from one *RGB* space to another, essentially a 3D coordinate transformation.
- It comprises a linear 3 × 3 matrix operation on linear-light *R*, *G*, and *B* signals, and needs inverse-gamma and back-to-gamma conversion steps based on the opto-electrical transfer function (OETF) of the source and the electro-optical transfer function (EOTF) of the display.
- The matrix is calculated from the two sets (source and destination) of color primaries and is a concatenation of *RGB* → *XYZ* and *XYZ* → *RGB* conversions.
- The matrix coefficients are related to the white point coordinates of the source and the destination color spaces; hence, if the color temperature (or the white balance) of an image (or, video) is altered by changing one of the reference white points, the matrix coefficients should be recomputed using the new white point(s). This changes the color temperature back to the reference white; if one wants to deliberately change the colors of a scene, then using a different desired white point (x, y) is the way to go.
- The resultant display colors are mathematically correct, as long as linearity holds, i.e., the matrix operation is done in linear-light domain, and none of the (R, G, B) signals need to be *clipped* (more on this later).

8.1 A Pictorial Example of Color Conversion

Figure 10 tries to provide an intuitive insight into color space conversion from one gamut to another. The arrows in Fig. 10a show the problem that would be incurred if color correction is not applied when a source with a larger gamut, e.g., digital cinema (reference), is to be displayed on a destination with a smaller color gamut, e.g., a standard TV. If we do not do anything, we will get (and see) the wrong color gamut of the standard TV display, where red and green tints will be shifted to yellow-white. Figure 10b shows that by applying color space correction, we shift all colors back to where they came from, and they are now displayed correctly.

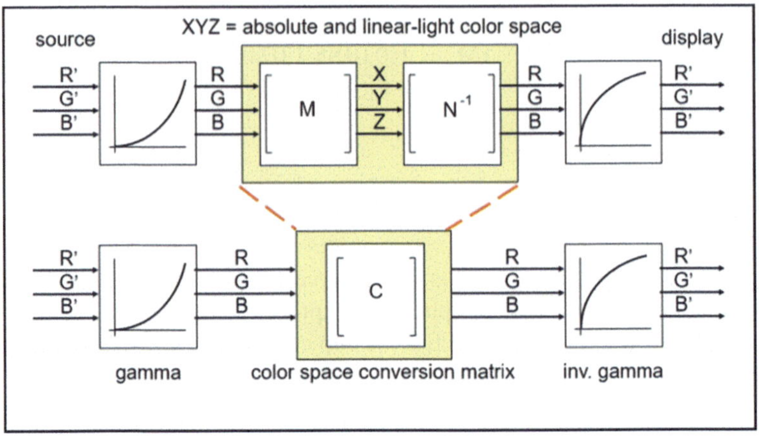

Fig. 9 Color conversion in linear light

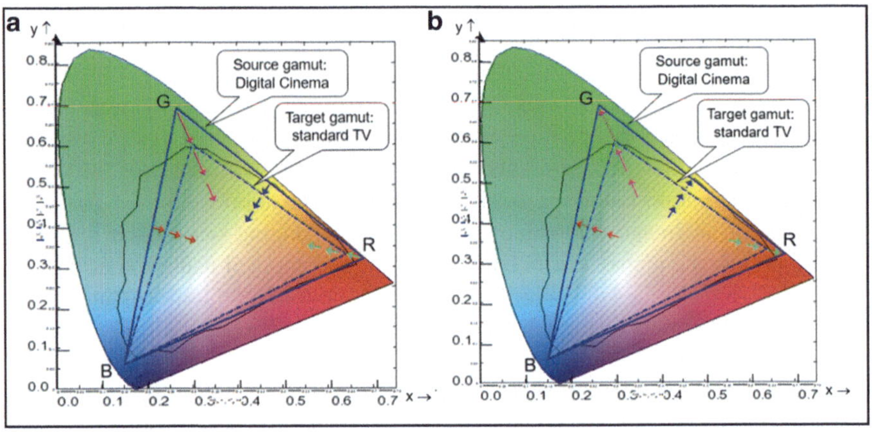

Fig. 10 Color correction via color space conversion [12]. (**a**) Without correction. (**b**) With correction

8.2 *Out-of-Range Colors*

As we have discussed earlier, *color space conversion* is calculated in two steps: from *RGB* input space to an intermediate *CIE XYZ* space, and then from *XYZ* to an *RGB* output space. In this process, the derived color conversion matrix **C** can end up with both negative and positive coefficients. In this context, two situations need special attention when converting from a larger to a smaller gamut, i.e., the display has a smaller color gamut than the source:

Fig. 11 Color conversion matrix for digital cinema to standard TV [12]

$$\begin{bmatrix} R_{out} \\ G_{out} \\ B_{out} \end{bmatrix} = \begin{bmatrix} 1.225 & -0.225 & 0.000 \\ -0.042 & 1.042 & 0.000 \\ -0.020 & -0.079 & 1.098 \end{bmatrix} \times \begin{bmatrix} R_{in} \\ G_{in} \\ B_{in} \end{bmatrix}$$

1. There are off-diagonal negative coefficients in **C**: this has the potential to convert legal input *RGB* values (*RGB* > 0) into illegal output *RGB* values (*RGB*[15] < 0). These negative coefficients can thus cause negative output values, which is an *out-of-gamut* condition that needs to be handled.
2. There are coefficient values greater than 1 along the diagonal of **C**: this can cause overflow (*RGB* > 1) that needs to be handled.

Conversely, when a display has a wider color gamut than a source, meaning that we are converting from a smaller to a wider gamut, we will have positive coefficients (between 0 and 1) in **C** that can convert illegal *RGB* values (*RGB* < 0) into legal values (*RGB* > 0). All row sums in **C** in this situation add up to 1, so unless we change the white point or boost the contrast, underflow and overflow are not too likely.

Figure 11 shows the color conversion matrix for the example in Fig. 10. This matrix corresponds to a conversion from a wider Digital Cinema P3 gamut to a narrower Rec.709 standard TV gamut, and can produce both *illegal RGB signals* (underflow: *RGB* < 0) and *problematic RGB signals* (overflow: *RGB* > 1). For example, the conversion matrix wants to create a −23% red light output on a standard display for a 100% green source, e.g., $R_{out} = -0.225 \times G_{in} = -0.225$ (when $R_{in} = 0$, $G_{in} = 1$, and $B_{in} = 0$).[16] After the destination *RGB* values are calculated, we must get rid of any negative output values, because displaying negative light is physically impossible, and no display can handle it. One solution is clipping the final negative values to zero, and accepting the color error and some loss of information too. The general solution is *gamut mapping* described in the next section.

Note that with the above conversion matrix, we can also get output values higher than 100% for some input colors, which means that the display will be overdriven. For example, the matrix wants to produce 123% red light output on a standard display for a 100% red source, e.g., $R_{out} = 1.225 \times R_{in} = 1.225$ (when $R_{in} = 1$, and $G_{in} = B_{in} = 0$).[17] This is not a physical problem (since a brighter display can accommodate higher signal values); hence, it can be solved by a simple attenuation

[15]Implies a calculated output *RGB* value where $R < 0$ or $G < 0$ or $B < 0$.

[16]Given the matrix coefficients, B_{in} can be any value. So, the observation holds for green ($B_{in} = 0$) and cyan ($B_{in} \neq 0$) inputs.

[17]Once again, B_{in} can be any value. So, the observation holds for red ($B_{in} = 0$) and magenta ($B_{in} \neq 0$) inputs.

of the signal when converted back to the gamma domain, and preferably feeding a brighter display.

Based on the previous discussions, it is important to understand, distinguish between, and handle the two special cases that may occur after color space conversion depending on the color conversion matrix coefficients:

1. *MIN (R, G, B) < 0:* one or more of the calculated *R*, *G*, and *B* values is negative; it is a *physical problem* and requires *gamut mapping*, i.e., a reduction of saturation to make MIN (*R*, *G*, *B*) ≥ 0.
2. *MAX (R, G, B) > 1:* one or more of the calculated *R*, *G*, and *B* values is greater than 1; this is a *scale problem* and requires *tone mapping*, i.e., a reduction of contrast and/or saturation to make MAX (*R*, *G*, *B*) ≤ 1.

9 Gamut Mapping

Color space conversion converts one color gamut to another and reproduces the correct colors. But, how do we make sure that the colors of an input image when transformed to the colors of a target device exploit the full potential of the rendering device in terms of color rendition? That is the problem that *gamut mapping* wants to solve. To understand how it works, let us take the same example as before of a wider digital cinema source gamut to be shown on a standard TV display with its smaller color gamut, as illustrated in Fig. 12.

The corresponding color conversion matrix, shown earlier in Fig. 11, already hints at the problem we are about to encounter in color space conversion—negative coefficients in all of the three rows of the matrix! Consequently, for each side of the

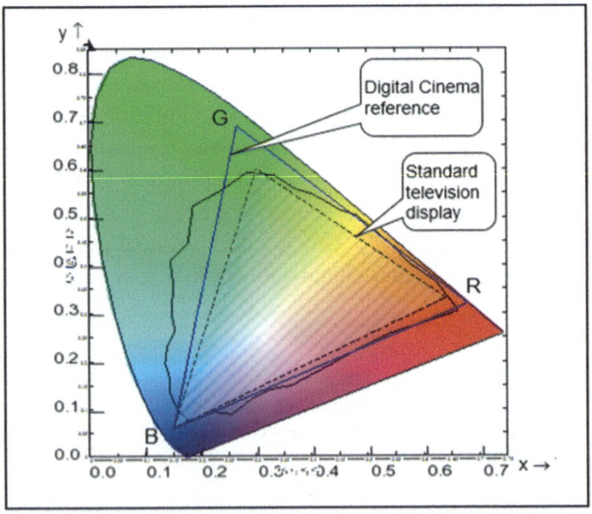

Fig. 12 Gamut mapping example

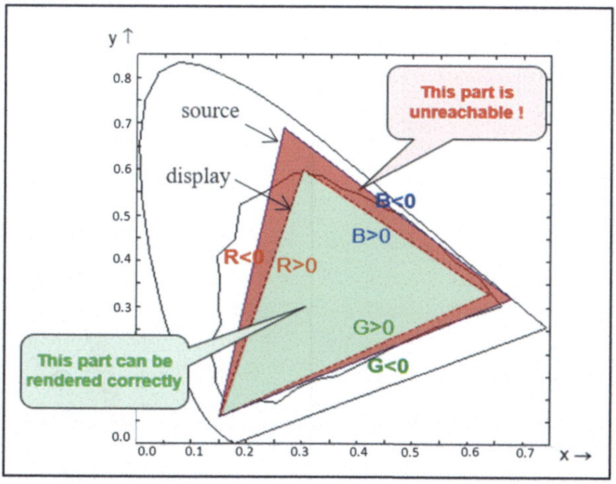

Fig. 13 Colors that need gamut mapping [12]

triangle of the display gamut, one of the three *RGB* values will go negative (two for each corner)! Since the display cannot suck in light, these negative values cannot be sent to the display but need to be substituted. The concept is illustrated next with pictures. After we apply the color space correction, for going from a wider to a smaller gamut, we get into a situation as shown in Fig. 13. All colors in the green area are perfectly shown, but the colors in the red part (where the *RGB* values become negative) can never be shown by the display; hence, we must substitute some other colors from the display gamut for these unreachable source colors. This substitution is "gamut mapping."

In the current example, gamut mapping can be implemented by shifting the red region colors towards the display gamut, and substituting in-gamut colors that are perceptually similar. This is illustrated in Fig. 14. Gamut mapping is thus more of an art than science because the choice of the substitute colors is subjective. So, whereas color space conversion is objective math, gamut mapping is largely a subjective choice! There is no single perfect solution. Generally we are happy when the output color gamut stays closely inside the display's color space/gamut, which for a 3-primary display is a simple *RGB* cube. Other devices, like multi-primary displays, have different gamut shapes (not cubes), and then the gamut mapping is much more challenging.

9.1 A Second Example of Gamut Mapping

Let us consider the illustration in Fig. 15 where the source gamut (digital cinema reference) is the same as before, but the display—a rare LCD with *RGB* LED backlight—now has a wider gamut than the standard TV display of the last example. The color-space-correction matrix in this case takes the form shown in Fig. 16.

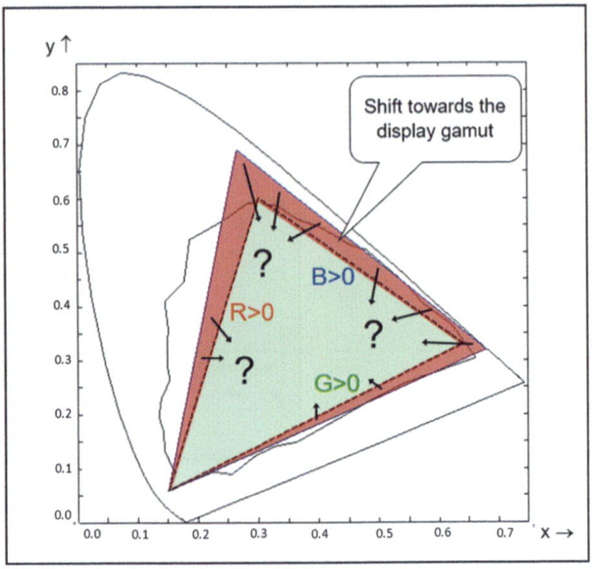

Fig. 14 Shifting unreachable source colors towards display gamut [12]

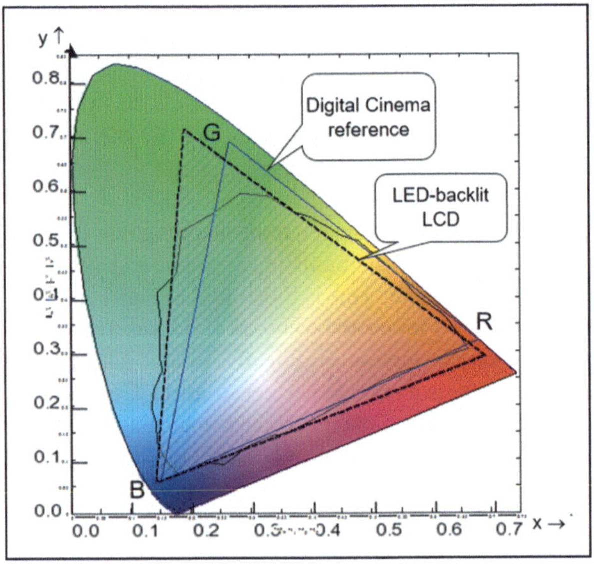

Fig. 15 A second example of gamut mapping

An immediate problem is obvious from the matrix—the negative coefficients in the third row of the matrix can cause a negative blue output value for certain inputs, i.e., for the top side of the triangle of the display gamut, the output B signal could go negative, which is an out-of-gamut condition and needs to be handled; the problem is more manageable though than with the narrow-gamut display of the earlier example where we had coefficients negative for all three colors.

Fig. 16 Color conversion matrix for digital cinema to LED LCD [12]

$$\begin{bmatrix} R_{out} \\ G_{out} \\ B_{out} \end{bmatrix} = \begin{bmatrix} 0.810 & 0.168 & 0.022 \\ 0.041 & 0.958 & 0.001 \\ -0.017 & -0.049 & 1.066 \end{bmatrix} \times \begin{bmatrix} R_{in} \\ G_{in} \\ B_{in} \end{bmatrix}$$

Once we apply the color space correction, we get a situation as shown in Fig. 17:[18]

1. All colors in the green overlap area are perfectly guided by the mathematics of *color space conversion*. This covers almost all used surface colors, except for some yellow and orange.
2. The colors in the red part on the edges of the source gamut (where the negative coefficients make B_{out} go negative for some of the input values) cannot be showed with this display, and if the source actually transmits such colors then there is a problem; we must perform *gamut mapping*[19] and substitute some other colors from the display gamut. Fortunately, for the current example of the digital cinema source gamut shown in Fig. 15, we only need to approximate the more extreme yellow and orange colors—which rarely occur—from the source.
3. The colors in the yellow part on the edges of the display gamut are reachable, but they do not exist in the source gamut and are therefore never transmitted. These colors may have been gamut-mapped away at the transmitter side but it still seems like a waste to not be able to use them. It is a great temptation to try and reach these displayable colors anyway, by means of *gamut extension* because that would make the image look *extra colorful*. Gamut extension is also a subjective process—it is like trying to invert the gamut mapping that may have been applied to the input image when it was limited to the source gamut!

9.2 Gamut Reduction

Gamut reduction of out-of-gamut colors when going from a wide-gamut source to a display with a narrower gamut—where source will transmit colors that the display cannot show—can be implemented in multiple ways. Clipping of out-of-gamut values to the in-gamut border, for example, is one of the simplest ideas that we discussed before. Shifting towards in-gamut colors is another technique we also touched upon briefly. The literature abounds with different methods to go about it.

In one paper, Bronner et al. [17] have proposed two common projection techniques for gamut reduction: towards white point (TWP) and closest. The *TWP*

[18]Recall that the *RGB* values are positive inside the destination gamut triangle, and negative outside.

[19]*Gamut mapping* also goes by the name of *gamut compression* or *gamut reduction*.

Fig. 17 Three color-mapping cases [12]

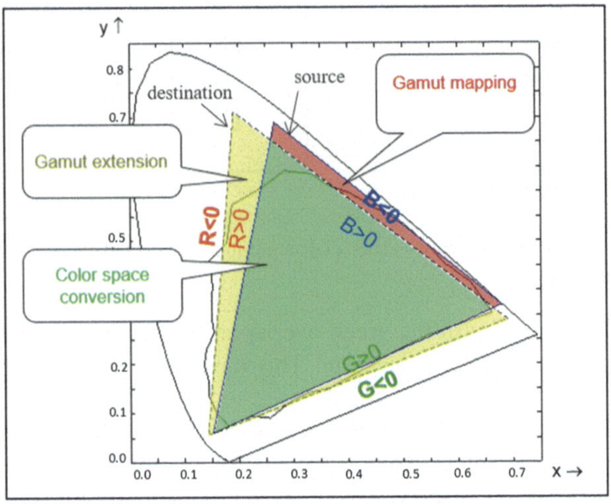

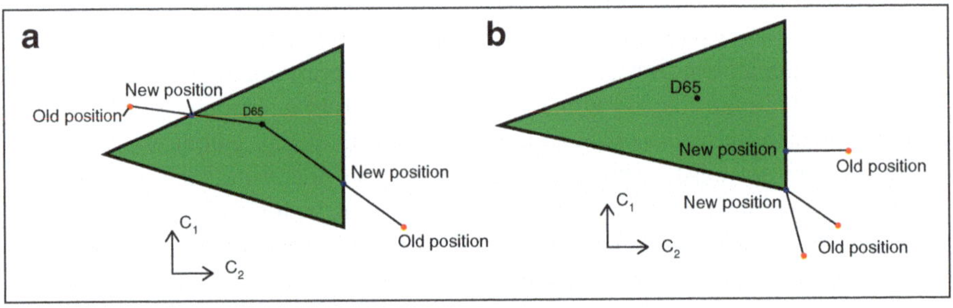

Fig. 18 Gamut mapping via projections [17]. (**a**) Towards white point. (**b**) Closest

projection technique, as illustrated in Fig. 18a, aims at keeping the theoretical hue (not necessarily the perceived hue) unchanged while altering only the saturation value. To do so, it maps chroma values (in coordinate space C1, C2) lying outside of the gamut (green triangle) to the intersection (new position) between the gamut's boundary and the line segment joining the white point (D65) and the original source color value (old position). This is a hard clipping, so many colors along the mapping vector end up on the same point on the edge of the target gamut.[20] The *closest projection technique*, on the other hand, tries to minimize, as illustrated in Fig. 18b, the Euclidean distance between the original color and the mapped color by mapping chroma values (in coordinate space C1, C2) lying outside of the gamut (green triangle) to the coordinates that correspond to the smallest Euclidean distance on the chromaticity plane. This gives a concentration of colors mapped to the *R*, *G*, *B* corners of the target gamut.

[20] Also, this 2D illustration gives no hint on what to do with the third dimension—lightness—of colors.

In another paper, Zamir et al. [18] have provided a detailed categorization of different gamut mapping and gamut extension algorithms and frameworks; the paper also provides an overview of the main concepts of retinex [19], which has received considerable attention in the context of gamut mapping. Some other recent examples of published gamut mapping algorithms are SGDA [20] and CARISMA [21], where lightness and chroma conversions are performed simultaneously by mapping colors towards a single point. Interestingly, as described in Addari's thesis [22], there are also examples [23, 24] where the source gamut is divided into multiple regions that are handled differently: the *core* where colors are kept the same, the *area around the cusp* where each color is mapped towards the point in the middle, the *region with lowest lightness* where the points are mapped towards the point with maximum lightness of the destination gamut, and the *brightest area* where each color is compressed towards the darkest point in the destination gamut. If one now throws high dynamic range (HDR) imaging in the mix, the obvious problem that comes up is for tone reproduction and mapping to explicitly consider the target display gamut; Sikudova et al. [25] have addressed this in their paper where they discuss a gamut mapping framework for accurate color reproduction of HDR images.[21]

9.3 Gamut Extension

Gamut extension has received less attention in the past compared to gamut reduction, but is catching up fast with respect to research work and publications owing to the recent introduction of various types of wide-gamut display standards and displays, e.g., ITU has released Rec.2087 that defines the standard to convert Rec.709 content to the Rec.2020 gamut [26]. There are five common gamut extension algorithms mentioned in the literature [27, 28]:

1. *True-color*: This algorithm maps the color information of the input image into the destination gamut without applying any sort of extension, meaning that the output of *true-color* is basically color space conversion applied without any gamut processing after that. It is nothing other than the application of the color conversion formula we derived before, i.e., it is just a representation of the input image in a wide-gamut color space.

$$\begin{bmatrix} R \\ G \\ B \end{bmatrix}_{\text{true-color}} = [N]^{-1} \times [M] \times \begin{bmatrix} R \\ G \\ B \end{bmatrix}_{\text{source}} = [C] \times \begin{bmatrix} R \\ G \\ B \end{bmatrix}_{\text{source}}$$

[21] Tone mapping from HDR to SDR, and gamut mapping for MAX $(R, G, B) > 1$ are essentially the same problem.

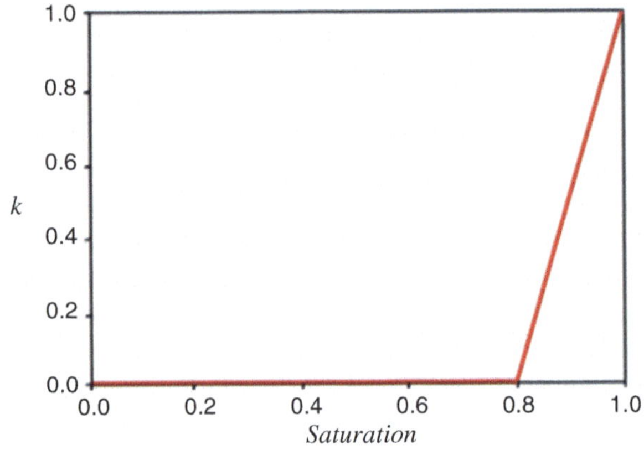

Fig. 19 Mixing function for HCM

2. *Same Drive Signal (SDS)*: The *RGB* primaries of the input are linearly mapped to the *RGB* primaries of the display device, enabling the algorithm to make full use of the wide-gamut of the display without regard to correct values of *lightness*, *saturation*, or *hue*.
3. *Hybrid Color Mapping (HCM)*: This algorithm linearly combines the output of the true-color and SDS algorithms based on the saturation of the input image as illustrated by the next equation, where "*k*" is a mixing factor and a function of saturation as shown in Fig. 19.

$$\begin{bmatrix} R \\ G \\ B \end{bmatrix}_{\text{true-color}} = (1-k)\cdot\begin{bmatrix} R \\ G \\ B \end{bmatrix}_{\text{true-color}} + k\cdot\begin{bmatrix} R \\ G \\ B \end{bmatrix}_{\text{SDS}}$$

4. *Chroma Extension (CE)*: The algorithm maps colors of the source gamut to the reproduction gamut along lines of the chroma axis in the CIE LCh color space [29] while keeping lightness and hue constant, thereby trying to avoid potential hue shifts produced by the SDS algorithm. However, because of the lightness and hue restrictions, the target color gamut will not be used entirely. [Note: Any affordable formula to calculate a hue value does not necessarily represent the *perceived hue* exactly, so it will be a mistake to assume that extending chroma, while keeping some hue value constant, will not alter the perceived hue.]
5. *Lightness Chroma Adaptive (LCA)*: The algorithm alters both lightness and chroma while keeping the hue constant; both CE and LCA algorithms make use of a high chroma boost (HCB) function which smoothly maps colors of an input image in a manner that the high-chroma objects get more boost in saturation than the low chroma ones. The target gamut will be filled more, but

still not entirely. Also, it is hard to achieve a constant perceived hue (as opposed to some mathematical hue metric), and non-linear processing may introduce amplification of noise and/or false contours (from quantization errors).

One of the major drawbacks of typical gamut extension algorithms is that they tend to overdo their task and produce images that are highly saturated and artificial in appearance. To this end, Gang et al. [30] have produced an algorithm that first transforms the standard *RGB* value to the device *RGB* value, and then extends the gamut with the saturation ratio of gamut boundaries and a non-linear function that reduces the oversaturation problem. In his thesis [28], Zamir also cites some of the usual problems with common gamut extension algorithms, and then proposes special extension methods that comply with the global and local perceptual properties of human vision. Note that some of the gamut reduction algorithms can also be applied to gamut expansion via an inverse transformation, e.g., Hirokawa [31] has suggested an extension algorithm where colors are mapped along the lines of constant lightness with a non-linear function.

10 An Age-Old Art

We talked about *gamut reduction* (aka *gamut mapping*) and *gamut extension*. Both are subjective art in the sense that there are subjective options and opinions in what to do when it comes to generating the "new" colors. However, it is possibly obvious from our discussions so far that *gamut mapping is a necessity* (mapping out-of-range colors present in and transmitted by the wide-gamut source to the colors of the narrow-gamut display), whereas *gamut extension is a luxury* (utilizing colors in the wide-gamut display that are not present in or not sent by the narrow-gamut source or gamut-mapped away[22] at the transmitter). Further, it may be intuitively obvious that gamut mapping is an ill-defined problem and is subject to artistic interpretation. The usage of the word "artistic" is a conscious choice. Gamut mapping is an age-old art—artists and painters have been doing gamut mapping for thousands of years! Recall that gamut mapping essentially deals with different color ranges being available to different instances of an image. We have practiced this art for long; as soon as humans created the first images, gamut mapping took place! This is well articulated in Morovic's book [32] on color gamut mapping. As Morovic points out, the Paleolithic cave paintings (e.g., Altamira in Spain), that recorded various hunting experiences, made use of only three ochres—charcoal, red, and yellow—and tried their best to represent the *colorful* hunting experience(s) with a limited range of colors. From these Paleolithic beginnings, there has been a gradual but

[22]Trying to invert a previous gamut mapping operation, without having metadata available that characterizes the operation, is guesswork. Luckily, there is now a trend towards invertible tone mapping and gamut mapping, both described by metadata that is transmitted with the down-mapped images.

continuous expansion of the color palette, thereby enabling the creation of images with greater color gamut. For example, earlier cave paintings with three to five color palettes gave way to more extensive color palettes for the Greek frescoes in fifteenth century BC and dry-wall paintings in the tombs of the Egyptian rulers in twenty-third century BC. The frescoes and the tomb paintings, the painted and enameled tiles in Babylon in sixth century BC, and the painted pottery of ancient Greece, all made use of gamut mapping in reproducing different real-world color imagery with a limited palette of colors. In his posts [33–35], James Gurney has provided a great guide to mapping out the variety of colors that is to be used for a painting task, and how the used colors can ultimately be made to appear warm or cool.

11 Rendering Intent and ICC Profiles

As we have discussed before, colors that are available in the source color space but cannot be converted into the display color space are called *out-of-gamut colors* and somehow need to be remapped to other colors that are reproducible on the display. To aid the process, the ICC has defined *rendering intents* in its profiles to make it easier for various software (that uses ICC profiles) to use the correct mapping algorithm. An ICC profile is a set of data that describes the properties of a color space, i.e., the range of colors that a monitor can display or a printer can output. There are four rendering intents specified by ICC for mapping out-of-gamut colors from the source color space into the gamut of the destination color space. The intent and the mapping depend on the source, the destination, the out-of-gamut colors, and the nature of the imagery [36–38]:

1. *Match or Absolute Colorimetric Intent*: *Absolute colorimetric intent* preserves the white point of the original color space; any color that falls outside the range that the output device can render is adjusted to the closest color that can be rendered (i.e., out-of-gamut colors are clipped), while all other colors are left unchanged (which is perfect if those colors do not occur anyway, or we do not care about rare colors).
2. *Proof or Relative Colorimetric Intent*: *Relative colorimetric intent* does not preserve the white point of the original color space (often the Photoshop "working space") but matches it to that of the output (typically a printer profile); other colors are scaled accordingly. Any color that falls outside the range that the output device can render is adjusted to the closest color that can be rendered, while all other colors are left unchanged. Although not as accurate as absolute rendering, relative rendering can be used in hard-proofing workflows [39].
3. *Picture or Perceptual Intent*: *Picture intent* comes into play when the destination gamut is smaller; it causes the full gamut of the image to be compressed or expanded to fill the gamut of the destination device, so that gray balance is preserved but colorimetric accuracy may not be preserved. The idea here is to try to render colors in a way that is natural to the human vision. The gamut of

the source image color space (typically a "working space" or camera profile) is scaled to the output color space (usually a printer profile) in a way that pulls out-of-gamut colors into the gamut. Colors at or near the edge of the gamut are also pulled in. In other words, if certain colors in an image fall outside of the range of colors that the output device can render, the *picture intent* will cause all the colors in the image to be adjusted so that every color in the image falls within the range that can be rendered so that the relationship between colors is preserved as much as possible. This intent is not recommended for hard proofing but is most suitable for display of photographs and images, and is generally the default intent.

4. *Graphic or Saturation Intent*: *Saturation intent* pulls saturated colors out to the edge of the gamut, thereby increasing saturation, or color strength. It is intended to be used for vector graphics, i.e., logos, line-art, etc., to preserve the chroma of colors in the image at the possible expense of hue and lightness. This intent is most suitable for business graphics such as charts, where it is more important that the colors be vivid and contrast well with each other.

The effects and differences of various rendering intents are colorfully illustrated on the *Cambridge in Color* website [40].

12 Final Word on Color Spaces

A color space is defined as *perceptually uniform* if a difference in value anywhere in the color space corresponds to the same difference in perception. This property is illustrated by the MacAdams ellipses [17] which predict when two different color values will be differentiable by a human observer. Figure 20 represents these ellipses in the CIE chromaticity diagram. Dots in the center of ellipses represent the reference color, while the outlines of the ellipses mark the boundaries of the smallest difference of values in every direction before a human observer would rate those two colors as different. The MacAdams ellipses indicate the eye's sensitivity to color—smaller means more sensitive, e.g., small variations near green are less distinguishable than small variations near blue. In an ideal perceptually uniform color space, all ellipses are circles and have the same radius as shown in Fig. 21.

Recall that the *XYZ* color space is formed via the use of a set of *XYZ* non-real[23] primaries that can be combined to produce any color, including spectrum colors that are not possible with any real set of *RGB* primaries. As is clear from Fig. 20, equal distances in this space do not represent equally perceptible differences, i.e., the relative distribution of colors is not perceptually uniform for the eye. In particular, this color space actually stretches and overemphasizes the eye's resolution of greens and compresses the reds and blues. It is for this reason that uniform color spaces (UCS)—defined by the well-known $L^*a^*b^*$ and $L^*u^*v^*$ coordinates—were

[23]Non-realizable in reality, but mathematically very real, e.g., $(x, y) = (1, 0)$, $(0, 1)$, and $(0, 0)$

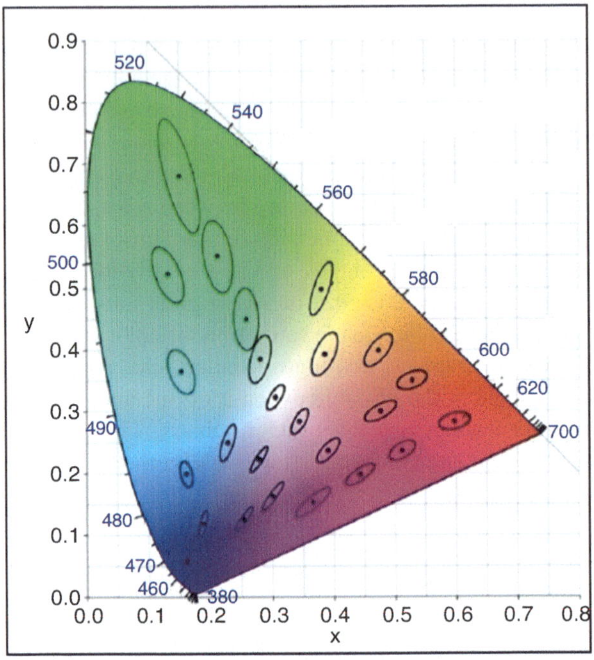

Fig. 20 MacAdams ellipses in CIE chromaticity diagram [17]

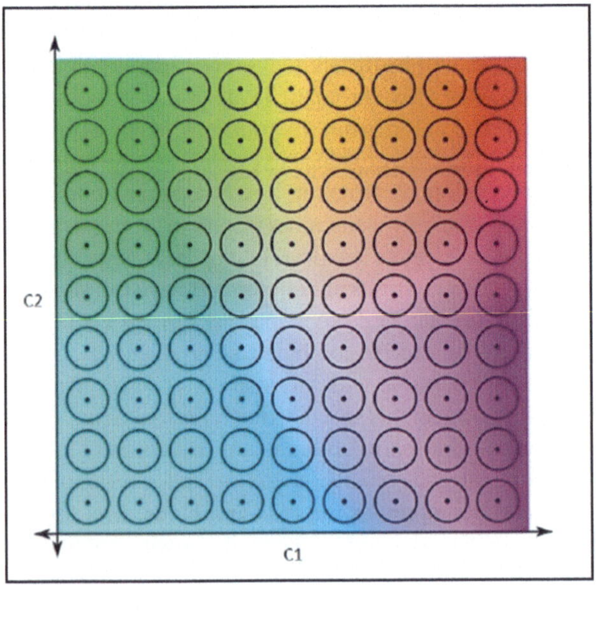

Fig. 21 MacAdams ellipses in a hypothetical perceptually uniform chromaticity plane [17]

developed to provide perceptually more-uniform color spaces and allow the basis for a color difference metric calculated as the Euclidean distance between coordinates within the space.[24] Though these color spaces do not offer directly displayable formats, their gamuts encompass the entire visible spectrum and can accurately represent colors of any display, print, or input device.

The selection of the mapping color space is a crucial aspect of both the conversion and the gamut mapping process. Some of the color spaces commonly used are [42]:

- CIE *xyY*
- CIE *u'v'Y*
- CIE *L*a*b**
- Simplified Lab
- Uniform color space based on CIECAM02

In the CIE *xyY* and *u'v'Y* color spaces, linear color mixing allows the implementation of simpler gamut mapping algorithms that utilize linear transforms; however, perceived lightness, chroma, and hue of colors may change during the mapping process. The CIE *L*a*b** color space was, therefore, designed and utilized to have a better visual uniformity than the *xyY* or *u'v'Y* color spaces.[25] The Simplified Lab color space is a simplification of the CIE *L*a*b** color space in which color gamuts have simpler shapes with approximately planar surfaces. One advantage of this *L*a*b** color space is its perceptual uniformity that takes lightness into account; Euclidean distances in the three-dimensional color space are nominally proportional to the perceived color differences, and linear mapping paths, when possible, preserve lightness, chroma, and hue.

13 Color Lookup Tables

An alternative to real-time color-mapping calculations is to devise a 3D lookup table (3D LUT) which can implement the pre-computed mapping functions (for any mapping algorithm) for the red, green, and blue input and output. In fact, a lookup table also allows combining multiple color space conversion rules and gamut mapping rules into one mapping rule. According to this scheme, the desired gamut-transformation function is loaded into a three-dimensional lookup table in the setup mode. Thereafter, in the normal operation mode, the table generates transformed Rout, Gout, and Bout data for each Rin, Gin, and Bin pixel input. A lookup table allows the gamut-transformation rules to be loaded into the table and updated as

---

[24]One step further than perceptually uniform color spaces is to introduce (near-)straight lines of constant perceived hue, leading to color appearance models [41].

[25]Note that these color spaces are good for quantitative analysis and color difference metrics, but not for image processing.

and when needed. B&H has an interesting article [43] describing three main types of LUTs:

- *Calibration LUT*: *Calibration LUTs* are used to correct inaccuracies in displays, and ensure that a reference monitor or projector adheres to a standard color space, e.g., Rec.709; by creating an appropriate Calibration LUT for each monitor, it is ensured that images are displayed in a consistent and accurate way.
- *Display LUT*: *Display LUTs* are used to convert from one color space to another, e.g., for on-set monitoring of log footage; instead of flat, desaturated log images, mapping of the log footage to the color space of the monitor enables viewing images with contrast and saturation resembling that of the final product.
- *Creative LUT*: *Creative LUTs* are used to normalize a footage to a specific color space by also making creative adjustments, e.g., mimicking the look of a particular film stock, popular effect, or other stylized look. A lot more details can be found in Jonny Elwyn's article [44] on matching film look using LUTs.

While LUTs offer great control and variation in mapping input color values to output color values, a general disadvantage of using LUTs, however, is the high memory requirements for implementing the complete gamut mapping function in the lookup table. For example, just for 12-bit data (even though a common number in the movie world is 17), the size of the required table for full mapping is $4096 \times 4096 \times 4096 \times 12 \times 3$ bits = 309,237,645,312 Bytes! Such a large memory is often not conducive to implementation. Therefore, it is important to look for ways to speed up the computation or reduce the size of the lookup tables by storing fewer points and resorting to some form of interpolation [45, 46]. In one of the SMPTE papers, Frohlich et al. [47] have evaluated various gamut mapping algorithms for digital cinema, and also analyzed the losses introduced by using 3D lookup tables for gamut mapping.

Even though 3D LUTs have received a lot of attention in the literature, it is the author's opinion that they offer flexibility in mapping effects but do not provide a good or practical solution for gamut mapping, especially if the mapping needs to be changed frequently. A quality 3D LUT is expensive in terms of size, and not much smaller than the image itself. 3D LUTs happen to shift the load (of filling the LUT) to the CPU, problem to the applicant, and decision to the future! A well designed gamut mapping algorithm, based on a few simple rules, and running on dedicated hardware or a fast GPU could be superior and sufficient.

14 Conclusion

In this article, we have presented an introductory review of color correction and gamut mapping. Given the physical limitations of the ranges of color that can be represented by the source image and the destination display, a linear-light transformation is called for to correctly reproduce the source on the display. The transformation has two[26] steps: (1) *Color space conversion*—an *objective process* where the source colors that can be shown on the display are correctly represented based on the white point and *RGB* chromaticities of the source and the display, and (2) *Gamut mapping*—a *subjective process* where out-of-range colors are clipped or substituted by other in-range colors when the source and the display have different color gamuts. *Gamut compression* happens when the display has a smaller gamut, and *gamut extension* happens when the display has a wider gamut. The color gamut of a scene is limited by the source,[27] and due to the variety of displays available today, it is inevitable that some gamut mapping will occur on the display side. With the advent of LED LCD TVs and digital cinema, it is expected that the application of gamut compression will diminish as the color gamuts of consumer displays continue to grow larger; the new tendency here will be to expand the color gamut by the receiver to make use of the wider range of the display.

Acknowledgment The author is grateful to Jeroen Stessen for valuable and insightful discussions over the years that have led to many of the explanations presented in this article. Jeroen has been extremely gracious in proof reading of the material presented, and in permitting usage of diagrams from his 2007 presentation on extended color gamut at the HPA Tech Retreat [12]. At the time of the collaboration, Jeroen was a Senior Scientist at Philips Group Innovation, Intellectual Property & Standards, Eindhoven, Netherlands; he is currently a Principal Engineer at V-Silicon.

References

1. "Trichromacy," [Online]. Available: https://en.wikipedia.org/w/index.php?title=Trichromacy&oldid=878916396.
2. "John the Math Guy - Applied math and color science with a liberal sprinkling of goofy humor.," [Online]. Available: http://johnthemathguy.blogspot.com/2017/01/the-long-medium-and-short-of-cones.html.
3. "List of color spaces and their uses," Wikipedia, [Online]. Available: https://en.wikipedia.org/w/index.php?title=List_of_color_spaces_and_their_uses&oldid=847070964.
4. "RGB color model," Wikipedia, [Online]. Available: https://en.wikipedia.org/w/index.php?title=RGB_color_model&oldid=879201669.
5. "CIE," [Online]. Available: http://www.cie.co.at/.
6. "CIE 1931 color space," [Online]. Available: https://en.wikipedia.org/w/index.php?title=CIE_1931_color_space&oldid=873140583.
7. C. Poynton, "Color FAQ - Frequently Asked Questions Color," [Online]. Available: https://poynton.ca/notes/colour_and_gamma/ColorFAQ.html.
8. "PNG (Portable Network Graphics) Specification, Version 1.2," [Online]. Available: http://www.libpng.org/pub/png/spec/1.2/PNG-ColorAppendix.html.

[26]There are *four* steps if the conversions to and from the linear-light domain are counted.

[27]The camera (that captured the scene) does not necessarily have a fundamental gamut limitation, but the grading display does. Therefore, the limitation actually stems from the gamut-mapping processing at the source because content producers like to see the colors that they sign off on; so they make the gamut of the content smaller or equal to the gamut of the grading display so that the *display is faithful*.

9. "HOW IS OBJECTIVE COLOUR MEASUREMENT ACHIEVED?," Admesy - ADVANCED MEASUREEMNT SYSTEMS, [Online]. Available: https://www.admesy.com/objective-colour-measurement-achieved/.
10. "White point," [Online]. Available: https://en.wikipedia.org/w/index.php?title=White_point&oldid=875229464.
11. "Illuminant D65," [Online]. Available: https://en.wikipedia.org/w/index.php?title=Illuminant_D65&oldid=874275113.
12. J. Stessen, "Extended Color Gamut for Consumers - the Display side," in *HPA Technology Retreat*, 2007.
13. King, James C;, "Why Color Management?," Adobe Systems Incorporated, [Online]. Available: http://www.color.org/whycolormanagement.pdf.
14. "International Color Consortium," [Online]. Available: http://www.color.org/.
15. A. Ford and A. Roberts, "Colour Space Conversions," [Online]. Available: https://poynton.ca/PDFs/coloureq.pdf.
16. "Chromatic adaptation," [Online]. Available: https://en.wikipedia.org/w/index.php?title=Chromatic_adaptation&oldid=867388112.
17. T. Bronner, R. Boitard, M. Pourazad, P. Nasiopoulos and I. Ebrahimi, "Evaluation of Color Mapping Algorithms in Different Color Spaces," in *Proc. SPIE 9971, Applications of Digital Image Processing*, 2016.
18. S. Zamir, J. Vazquez-Corral and M. Bertalmio, "Gamut Mapping in Cinematography through Perceptually-based Contrast Modification," *IEEE Journal of Selected Topics in Signal Processing*, vol. 8, no. 3, 2014.
19. E. H. Land and J. J. McCann, "Lightness and retinex theory," *Journal of the Optical Society of America*, 1971.
20. P. Zolliker and K. Simon, "Continuity of Gamut Mapping Algorithm," *Journal of Electronic Imaging*, vol. 15, no. 1, 2006.
21. A. Johnson, "Colour Appearance Research for Interactive System Management and Application (CARISMA), Report WP2-19 Color Gamut Compression," 1992.
22. G. Addari, *Colour Gamut Mapping for Ultra-HD TV*, M.S. Thesis, Dept. of Electronic and Electrical Engineering, University of Surrey, 2016.
23. L. Neumann and A. Neumann, "Gamut Clipping and Mapping based on Coloroid System," in *Proceedings of Second European Conference on Colour in Graphics, Imaging and Vision*, 2004.
24. M. Ito and N. Katohs "Gamut Compression for Computer Generated Images," in *Extended Abstracts of SPSTJ 70th Anniversary Symposium on Fine Imaging*, 1995.
25. E. Sikudova, T. Pouli, A. Artusi, A. Aky, A. Akyuz, F. Banterle, Z. Mazlumoglu and E. Reinhard, "Gamut Mapping Framework for Color-Accurate Reproduction of HDR Images," *IEEE Computer Graphics and Applications*, vol. 36, no. 4, 2016.
26. "Colour conversion from Recommendation ITU-R BT.709 to Recommendation ITU-R BT.2020," International Telecommunication Union, 2015.
27. J. Laird, R. Muijs and J. Kuang, "Development and evaluation of gamut extension algorithms," Wiley Online Library, 2009. [Online]. Available: https://onlinelibrary.wiley.com/doi/pdf/10.1002/col.20537.
28. S. W. Zamir, "Perceptually-Inspired Gamut Mapping for Display and Projection Technologies," *PhD Thesis, Universitat Pompeu Fabra Barcelona,* 2017.
29. "Introduction to Colour Spaces," Graphic Quality Consultancy, [Online]. Available: https://www.colourphil.co.uk/lab_lch_colour_space.shtml.
30. S. Gang, L. Yihui and L. Hua, "A gamut extension algorithm based on RGB space for wide-gamut displays," in *IEEE 13th International Conference on Communication Technology*, 2011.
31. T. Hirokawa, M. Inui, T. Morioka and Y. Azuma, "A Psychophysical Evaluation of a Gamut Expansion Algorithm Based on Chroma Mapping II: Expansion within Object Color Data Bases," in *NIP & Digital Fabrication Conference, International Conference on Digital Printing Technologies (Society for Imaging Science and Technology)*, 2007.

32. J. Morovic, Color Gamut, Wiley-IS&T Series in Imaging Science and Technology ed., vol. Book 10, John Wiley & Sons, 2008.
33. J. Gurney, "Gurney Journey Part 1: Gamut Masking Method," [Online]. Available: http://gurneyjourney.blogspot.com/2011/09/part-1-gamut-masking-method.html.
34. J. Gurney, "Gurney Journey Part 2: Gamut Masking Method," [Online]. Available: http://gurneyjourney.blogspot.com/2011/09/part-2-gamut-masking-method.html.
35. J. Gurney, "Gurney Journey Part 3: Gamut Masking Method," [Online]. Available: http://gurneyjourney.blogspot.com/2011/09/part-3-gamut-masking-method.html.
36. "Colour Management - How It Works," Graphic Quality Consultancy, [Online]. Available: https://www.colourphil.co.uk/rendering_intents.shtml.
37. A. Rodney, "What is color management? Part 5: Rendering Intents and ICC profiles," [Online]. Available: http://digitaldog.net/files/05Rendering%20Intents%20and%20ICC%20profiles.pdf.
38. "Rendering Intents," Microsoft - Windows Dev Center, [Online]. Available: https://msdn.microsoft.com/en-us/library/windows/desktop/dd372183(v=vs.85).aspx.
39. D. Chinnery, "BLOG: HARD PROOFING TECHNIQUES – A PRACTICAL EXAMPLE," [Online]. Available: http://www.dougchinnery.com/hard-proofing/.
40. "CAMBRIDGE in COLOUR - A LEARNING COMMUNITY FOR PHOTOGRAPHERS," [Online]. Available: https://www.cambridgeincolour.com/tutorials/color-space-conversion.htm.
41. "Color appearance model," [Online]. Available: https://en.wikipedia.org/w/index.php?title=Color_appearance_model&oldid=851919289.
42. "Colour gamut conversion from Recommendation ITU-R BT.2020 to Recommendation ITU-R BT.709," International Telecommunication Union, 2017.
43. J. Dise, "An Introduction to LUTs," B&H, [Online]. Available: https://www.bhphotovideo.com/explora/video/tips-and-solutions/introduction-luts.
44. J. Elwyn, "https://jonnyelwyn.co.uk/film-and-video-editing/using-luts-to-match-a-real-film-look/," [Online].
45. Y. Kim, Y. Cho, C. Lee and Y. Ha, "Color Look-Up Table Design for Gamut Mapping and Color Space Conversion," in *IS&Ts International Conference on Digital Production Printing and Industrial Applications*, 2003.
46. J. Selan, "GPU Gems - Chapter 24. Using Lookup Tables to Accelerate Color Transformations," [Online]. Available: https://developer.nvidia.com/gpugems/GPUGems2/gpugems2_chapter24.html.
47. J. Frohlich, A. Schilling and B. Eberhardt, "Gamut Mapping for Digital Cinema," in *SMPTE 2013 Annual Technical Conference & Exhibition*, 2013.

Hardware/Software Interface Codesign for Cyber Physical Systems

Ahmed Amine Jerraya

Abstract Cyber physical systems (CPS) are a new generation of systems combining intensive connectivity and embedded computing. CPS allow to link physical and digital worlds. Despite all the previous research in the area of HW/SW interfaces design this is still a bottleneck in the design process. The key issue is that, in addition to the classical HW/SW interfaces brought by embedded systems, CPS bring another HW/SW interfaces to link the application SW and the environment. This paper explores HW/WS interfaces for cyber physical systems.

1 The Paradigm Shift

Up to 1990, system design used to be based on separate HW and SW design approaches where HW–SW interfaces are defined twice using 2 different models one representing the HW view and one representing the SW view. This double definition of HW/SW interfaces has created a great separation between HW and SW communities and left HW/SW interfaces as an unexplored no man's land. As soon as semiconductor advances allowed to integrate instruction set processor on chip to build SoC, in the 1990s, lots of research have been made and pioneers [1] advocated the importance of what was called HW–SW. Codesign and even several synthesis tools were created [2] based on logic view of the interfaces. Starting from 2000s a new generation of work started taking into account architecture and operating system in holistic approaches of HW–SW interfaces. These gave birth to all the developments on HW/SW co-simulation [3]. This research community is now

A. A. Jerraya (✉)
Commissariat à l'énergie atomique et aux énergies alternatives, Grenoble, France
e-mail: ahmed.jerraya@cea.fr

S. S. Bhattacharyya et al. (eds.), *Embedded, Cyber-Physical, and IoT Systems*,
https://doi.org/10.1007/978-3-030-16949-7_3

20 years old [4] and is dealing with multicore and multiprocessor SoCs (MPSoCs). More recently the advance of technology is allowing to integrate sensors and actuators in addition to wireless communication to build what is commonly called cyber physical systems [5]. Cyber physical systems (CPS) combine embedded computing and environment sensing, to create a link between physical and digital worlds and enable cooperation among systems. The design of CPS is creating a paradigm shift by introducing a new discontinuity between computing (HW and SW) and the environment. The focus becomes the overall function including both computing and communication.

Traditional approaches are based on serial methodologies. First the computing platform is designed including hardware and software, and then this is combined with additional HW to interface with the physical world and finally an application software is ported to the overall operating system and/or middleware. This means that the interface with the physical world can be finished only after finishing the computing platform. This often leads to poor global performances, since problems caught during application software development cannot be fixed in the HW platform. It also means that the design process takes too much time and cannot ensure the real time behavior required by CPS.

Efficient CPS design will need abstract models of the computing platform and the interaction with the physical world (sensing and wireless communication). Ideally one would like to have a set of SW tasks communicating with a set of HW subsystems interacting with the environment. This design process introduces a kind of hierarchical hardware software interface:

- Traditional HW/SW interfaces addressed in embedded computing. This interface hides the computing platform including HW and software.
- HW/SW interfaces specific to the communication between the application SW and the environment, this interface hides the complexity of the sensors/actuators architecture and also includes HW and software.

2 Traditional HW/SW Interfaces

The traditional HW/SW interfaces addressed in embedded computing are still a nonsolved problem even if it is quite old [6]. Because software components run on processors, the abstraction needed to describe the interconnection between software and hardware components is totally different from the existing abstraction of wires between hardware components as well as the function call abstraction used to describe software. The HW/SW interface is generally decomposed into two different layers: one on the software side using APIs and one on the hardware side using wires. This heterogeneity makes HW/SW interface design very difficult and time-consuming because the design requires the knowledge of both software and hardware and their interaction.

This HW/SW interface abstraction complexity is resulted from the fact that the ultimate hardware/software interface is the CPU and thus any abstraction of HW/SW interfaces requires the hiding of the CPU. A CPU is a hardware module that executes a software program. From a software side, the abstraction hides the CPU under a low level software layer; examples range from basic drivers and I/O functions to sophisticated operating systems and middleware. From the hardware point of view, HW/SW interface abstraction hides the details of the CPU bus through a hardware adaptation layer generally called the CPU interface. This may range from simple registers to sophisticated I/O peripherals including DMA queues and sophisticated data conversion and buffering systems. This double definition of HW/SW interfaces is still creating a separation between HW and SW communities.

The most massive recent work to solve this problem is done around the AUTOSAR standard [7]. A layered software architecture is defined as an integration platform for hardware independent software applications. Here again hardware and software development are still separated.

3 HW/SW Interfaces in Cyber Physical Systems

A similar discontinuity exists in CPS design where designers must also consider the interaction between application software and the environment. This is generally more coupled than the one in embedded computing because of the real time aspects of the environment. Here again traditional approaches suggest to model HW/SW interfaces twice. Embedded computing designers will use a functional model of the environment to test the performances of their platform. The application software designers use a HW/SW interface model (generally a protocol stack) to validate the functionality of their software. Using two separate models induces a discontinuity between the application software and the environment. The result is not only a waste of design time but also less efficient lower-quality design. This overhead in cost and loss in efficiency are not acceptable for CPS design [8].

4 The Limits of Modular Design

All the above-mentioned approaches are based on the concept of modular design that have been pushed in some safety based applications like automotive or aerospace. For example, today's cars and airplanes contain dozens or hundreds of computers [9]. Most of these have been designed for a specific function such as a radar. Figure 1 shows a typical automotive function. It is made of an MCU (micro controller unit) and a sensor with the associated HW interfaces and then several software layers. A specific automotive standardization layer (Autosar [7]) is used to abstract the environment in addition the SW layers used to abstract the MCU.

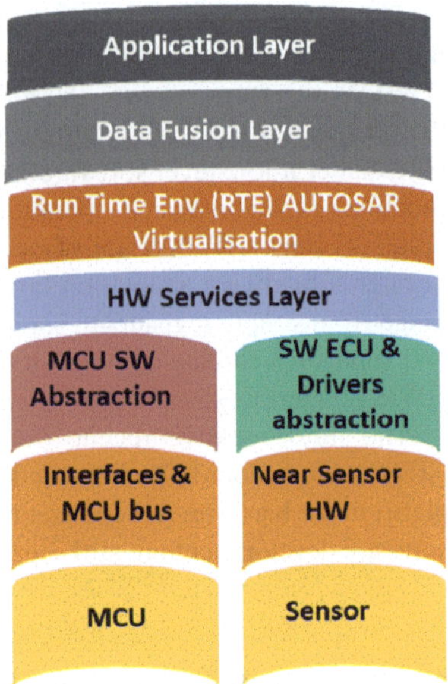

Fig. 1 Typical automotive function

Up to now the low complexity of the basic functions allowed to build safe and secure functions. But the clean separation between all these HW and SW components has led to an unprecedented complexity of SW in modern cars to reach 100 million lines of code [10]. It is obvious that this model does not scale for the design and optimization of complex functions such an ADAS (advanced driver assistance systems) that would require the integration of dozens of these functions [11].

It is clear that a global optimization of the whole system, including both kind of HW–SW interfaces, is required to ensure real time interaction with the physical world [12].

The solution is to abstract both embedded computing and the interface with the environment in a unified model. Such a model will be made of heterogeneous components interacting through abstract hardware/software interfaces. This concept opens new vistas that will bring fundamental improvements to the design process [3, 6, 12]:

- Codesign of both embedded computing and physical interfaces,
- Design of heterogeneous distributed systems including synchronous and asynchronous (analog) components,
- Easier global validation of embedded systems including embedded computing, interfaces with environment, and application software.

In summary, a single HW/SW reference model needs to be used to design embedded computing, interfaces with the environment, and application software in order to create the cyber and physical continuum required for efficient CPS design.

References

1. Wayne Wolf, "Hardware/Software Co-Design: Principles and Practice" Kluwer academic, 1997
2. R.Ernst al., "The COSYMA environment for hardware/software cosynthesis of small embedded systems", Microprocessors and Microsystems, Volume 20, Issue 3, May 1996, Pages 159-166
3. G. Nicolescu & Ahmed A. Jerraya "Global Specification and Validation of Embedded Systems: Integrating Heterogeneous Components", Springer 2007.
4. http://www.mpsoc-forum.org
5. Marylin Wolf, "High-Performance Embedded Computing: Applications in Cyber-Physical Systems and Mobile Computing", MK 2014
6. Jerraya, W. Wolf "Hardware-Software Interface Codesign for Embedded Systems", *Computer*, Volume 38, No.2, February 2005, pp.63-69
7. https://www.autosar.org/
8. ACATECH Report, « Integrated research agenda Cyber-Physical Systems (agenda CPS) Eva Geisberger/Manfred Broy (Eds.), March 2015.
9. Marylin Wolf, "Cyber-Physical Systems", IEEE Computer, March 2009.
10. https://informationisbeautiful.net/visualizations/million-lines-of-code/
11. Marilyn Wolf "Computers as Components: Principles of Embedded Computing System Design", Morgan Kaufmann 2016
12. Matthias Traub, Alexander Maier, Kai L. Barbehön: Future Automotive Architecture and the Impact of IT Trends, IEEE Software 2017 Vol. 34 Issue No. 03 - May-Jun.

Run-Time Security Assurance of Cyber Physical System Applications

Muhammad Taimoor Khan, Dimitrios Serpanos, and Howard Shrobe

Abstract We introduce a design methodology to assure run-time security of cyber physical system (CPS) applications. The methodology has two independent, but complementary, components that employ novel approaches to design run-time monitors that detect both computational and false data cyber-attacks to assure security of CPS at run-time. Based on the executable specification of a CPS application, the first component protects CPS computations through comparison of the application execution and the application-specification execution in real-time. The second component assures safety and integrity of CPS data through vulnerability analysis of the application specification for false data injection attacks based on non-linear verification techniques. We demonstrate our approach through its application to a typical CPS example application; we demonstrate that run-time monitors employing verification techniques are effective, efficient, and readily applicable to demanding real-time critical systems.

1 Introduction

A fundamental unit in the emerging industrial automation, Industry 4.0, is cyber physical systems (CPS), which are "engineered systems that are built from, and depend upon, the seamless integration of computational algorithms and physical components" [1]. These systems aim at improving quality of life by automating

M. T. Khan
Department of Computing and Information Systems, University of Greenwich, London, UK
e-mail: m.khan@gre.ac.uk

D. Serpanos (✉)
ISI/ATHENA RC and ECE, University of Patras, Patras, Greece
e-mail: serpanos@ece.upatras.gr

H. Shrobe
MIT CSAIL, Cambridge, MA, USA
e-mail: hes@csail.mit.edu

S. S. Bhattacharyya et al. (eds.), *Embedded, Cyber-Physical, and IoT Systems*,
https://doi.org/10.1007/978-3-030-16949-7_4

critical infrastructure-based industries, e.g., manufacturing, energy, health care, water management, and financial sectors [2]. Successful integration of CPS in these critical environments is bound by the control automation processes—the embedded software applications—that meet robust functional and real-time non-functional requirements of operations in these domains. Specifically, these applications have to assure (1) secure, reliable, and efficient execution of the operations, despite their uncertain physical components, and (2) fail-safe execution defending against cyber-attacks.

The security and reliability of CPS are critically important, considering the effects of CPS compromise or failure. A failed CPS can cause serious damage beyond the system itself, because it can violate safety requirements, e.g., releasing destructive chemicals or tampering with economic parameters that lead to a financial loss. An attack on the power-grid may disable transportation and medical systems. Such an effect was first demonstrated in the Aurora experiment where a pure cyber-attack destroyed a diesel generator [3]. Later, Stuxnet [4], the Ukrainian smart-grid attack [5] and other several less known incidents have demonstrated the potential of cyber-attacks to cause serious societal and economic damage. Recently, the Mirai attack compromised CPS converting them to a botnet [6].

Building defense systems to protect CPS against cyber-attacks is challenging because of different and uncertain behavioral characteristics of its computational and physical components; these components support discrete and continuous operations with linear and non-linear constraints respecting strict time bounds. There are two dimensions of CPS security: design-time security and run-time security. To assure design-time security of CPS applications, various attempts have been made, for instance, recently, a Coq library *VeriDrone* targets to ensure security of CPS models at different design levels, i.e., from high-level models to C implementations. In VeriDrone, the proof of security makes the assumption that the operating system and the run-time environment where the CPS application executes are secure and reliable, which does not hold typically. For instance, the run-time libraries supporting C may contain vulnerabilities that can be exploited to modify the code executing in a CPS controller. Furthermore, false data injection (FDI) attacks, corrupting the sensor output-data, can cause the control algorithm to issue commands that will have devastating effects, without modifying the controller application-code. Thus, while it is important to develop secure and reliable CPS controller application design, one must also monitor the run-time behavior of the CPS to ensure that the controller continues to behave correctly. Existing approaches to assure run-time security of CPS applications can be classified in four major classes as shown in Fig. 1. Class 1 monitoring systems detect attacks by matching system behavior with known bad behaviors, typically employing statistical techniques and supervised machine learning. For instance, such monitors build profile(s) of known bad system behavior/attacks [7] and then detect attacks by matching run-time system behavior with the build profiles. Since supervised machine learning techniques tend to generalize from the data presented, such monitors are more robust and informative. While Class 3 monitors detect deviation from expected behavior by building a statistical profile of normal (good) behavior [8] by employing unsupervised machine

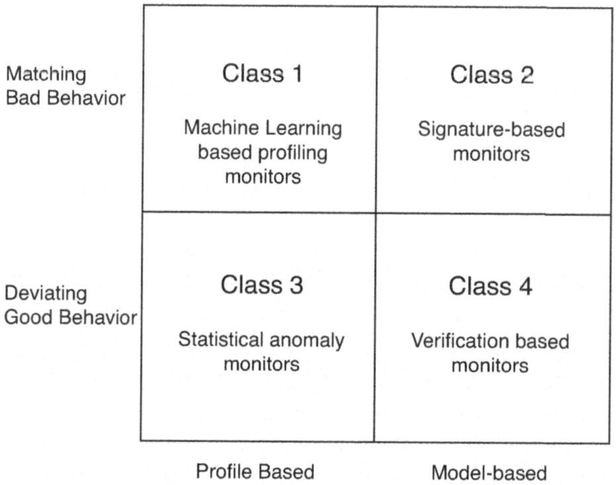

Fig. 1 Classification of run-time security monitors [12]

learning techniques. Such monitors are typically more stable because they are not retrospective and thus do not require knowledge of known attacks. Independent of the underlying machine learning technique, profile-based monitors suffer from high rate of false alarms mainly because whenever such monitors alarm an attack, it is only known that something unusual has happened and provide inadequate information to identify whether this is an error, actual malicious behavior, or just a variation of the normal behavior. The run-time monitors of the class 2 and 4 are widely used in highly secure environments, where successful attacks cause significantly high costs. Class 2 monitors (e.g., [9]) are more interpretative than class 3 monitors because whenever an attack is detected, these monitors provide adequate information to "understand" what was the failure but are retrospective being unable to detect unseen attacks. Class 4 security monitors (e.g., [10]) are even more interpretative (i.e., higher diagnostic resolution) because whenever such monitors detect deviates of system execution from the modeled good behavior, there is adequate information to infer the exact problem that caused the deviation (e.g., the affronted instruction or routine). However, building behavioral model of the legacy system is challenging and also observing irrelevant system behavior (to detect deviation) causes overhead that hinders run-time system performance, with high impact on real-time systems. Independent of the approaches, aforementioned monitors attempt to detect attacks with no information about physical components of CPS. Thus, these methods either produce false alarms or do not provide adequate information for diagnosis and recovery, when CPS fail or are attacked. Since these methods are typically retrospective expecting that something similar to a past incident will recur, attackers vary their attacks to avoid detection. A recent method [11] detects attacks in CPS by deriving process invariant for each CPS design stage from its initial design, and then monitoring them at run-time. Such approaches fail to detect advanced threats, e.g., computational and FDI.

We present a design methodology to assure CPS security at run-time based on CPS process behavior, where we characterize *behavior* in terms of software engineering, i.e., as the set of *functional* and *non-functional* (e.g., security, performance) characteristics of both the computational/cyber and physical components of the process. The cyber and physical components exhibit discrete and continuous behavior that operate at different granularity levels making behavioral description complex. To handle the complexity, we describe the *behavior* of a CPS process with an *abstract* executable specification. Importantly, we treat *computational* and *data* behaviors of CPS separately, since they are susceptible to attacks with different characteristics and thus require different defenses. From the specified behavior, we generate a monitor that assures the run-time security of CPS applications by checking consistency between the specification-execution and the implementation-execution of the applications. Our threat model includes both *computational attacks*—attacks that change code or data of the CPS application-execution and *false data injection (FDI) attacks*—attacks that fake sensor input values in a way that the control system (and monitor) fail that the input is wrong and sends in appropriate signals/commands as a result. The run-time monitor guarantees detection of any functional deviation (*computational attack*) but may not detect data-integrity attacks. Therefore, for data-integrity, we perform vulnerability analysis on the specification to identify potential FDI vulnerabilities. As a result, we obtain the values of the identified FDI attacks. Considering these values as attack-vectors, we either monitor input-data to detect the attacks or improve the design, adding constraints, to eliminate such attacks. Protection against computational and FDI attacks ensures highly secure CPS operations. For instance, our monitoring of computational attacks ensures that CPS operations are reliable and are secure against accidental or intentional insider threats. While our defense for handling of FDI attacks ensures that CPS operations have approximately exact estimate of the current state of the physical components and thus such operations are secure against intentional (e.g., man in the middle attacks) or system malfunctioning (e.g., interference and signal deterioration of the sensor).

We illustrate our approach [13] to handle computational and FDI attacks in the following sections, respectively.

2 Computational Attacks

Our run-time security monitor (RSM) takes as input both the specification (AppSpec) and implementation (AppImpl) of a CPS application and checks the consistency between the run-time behavior ("Wrapper" generated *observations*) and the expected behavior ("AppSpec" generated *predictions*) of the application [12], as Fig. 2a depicts. When an inconsistency is detected, the RSM raises an alarm. Since the specification and the implementation operate at different abstraction levels, we wrap the implementation such that the RSM gets run-time data that is directly comparable with the specification one.

a)

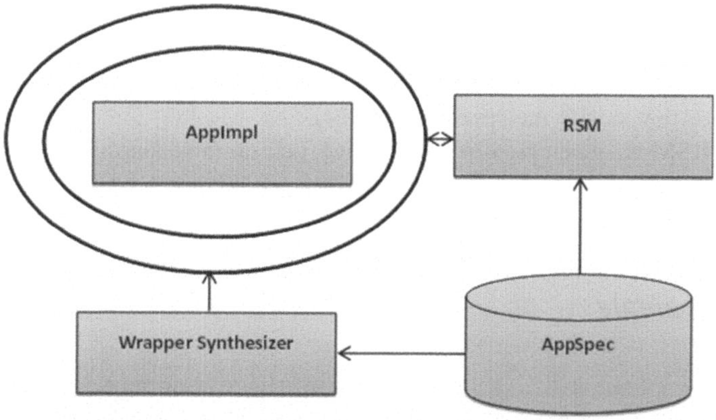

b)

Per 1 x 10^{-1} Cycle	CPU	Real-time
Full RSM	9.06 x 10^{-4}	9.07 x 10^{-4}
No Data Flow Checks	2.24 x 10^{-5}	3.19 x 10^{-5}
End to End (Only)	2.98 x 10^{-4}	3.01 x 10^{-4}
End to End (No Data Flow Checks)	1.04 10^{-5}	1.06 x 10^{-5}

Fig. 2 Run-time security monitor

The "AppSpec" is an active model that specifies behavior of cyber and physical resources of CPS as first class models through description of

- the *normal behavior* of the (cyber and physical) resources by decomposing their behavior into sub-modules, by encoding pre- and post-conditions and invariant for each sub-module,
- the *flow and control* of values, as data-flow and control-flow links connecting the sub-modules, and
- the *exceptional behavior* of the resources, *known attacks*, and suspected *attack-plans* to rigorously characterize the *misbehavior* of a module. The *attack-plans* are hypothetical attacks that describe ways of compromising a component. For robust defense, the monitor exploits them to detect any such run-time misbehavior.

The novelty of the RSM arises from the executable specification language (its elements, e.g., attack-plans, formalism, behavioral description of cyber and physical resources as first class models and discrete and continuous constraints) of the application it monitors. Monadic second order logic and event-calculus based rich formalism of the language enable us to describe system behavior at various levels of modular-abstraction. Semantically, such formalism-based specification language directly compiles into machine code, ensuring efficient behavioral comparison. Specifically, the formalism translates into a finite-automaton that recognizes only the words that satisfy the specification [14].

The RSM has been proven to be sound (i.e., no false alarms) and complete (i.e., no undetected *computational attacks*) [12] provided that the AppSpec runs securely.

2.1 Example

We have implemented an RSM prototype in Allegro Common Lisp on a MacBook-Pro with a 2.8 Ghz Intel Core-7 processor and have applied it to detect *computational* attacks to a PID controller that controls the water-level of a gravity-draining water-tank and a pump with an actuator. Our AppSpec includes a model of the controller's computations and a model of the sub-system (pump, tank, and actuator). A typical PID controller algorithm has four parameters: the setpoint, and three weighting factors K_p, K_i, and K_d; based on the integration and derivation of these factors, the algorithm computes the correct term. Listing 1 sketches a model of the algorithm's controller step, computation of an error-term, and an attack-plan for the controller-code.

```
(define-component-type controller-step
  :entry-events (controller-step)
  :exit-events (controller-step)
  :allowable-events (update-state accumulate-error)
  :inputs (controller observation dt)
  :outputs (command error)
  :behavior-modes (normal)
  :components (
    (estimate-error
        :type estimate-error
        :models (normal))
    (compute-derivative
        :type compute-derivative
        :models (normal))
    (compute-integral
        :type compute-integral
        :models (normal)))
  :dataflows (
    (observation
        controller-step observation estimate-error)
    (the-error
        estimate-error the-error compute-derivative)
    (derivative
        compute-derivative derivative
            compute-derivative-term)
    (weighted-proportional
            compute-proportional-term
                proportional compute-correction)
    (the-error
        estimate-error error controller-step) ) )
    ...
```

Listing 1 Example specification

Developing the model (AppSpec) and the application-implementation (AppImpl) in the same language enables immediate detection of any modification of the PID algorithm. For example, if parameter K_i is modified, then AppImpl computes an incorrect K_i-integral that is immediately detected as being different from the value predicted by the AppSpec. Such fine-grained modeling helps to characterize the attack immediately; in this case, we can deduce either that K_i or the code of K_i-integral computation was modified.

The RSM prototype is evaluated assuming that the PLC runs a control cycle on the water tank sub-system every 0.1 s. The application and the RSM were simulated for 100 s. When running fine-grained monitoring, the RSM consumes 8.93×10^{-4} s of CPU time and 9.07×10^{-4} s of real time for each cycle of the PID algorithm which is smaller than 0.1 s/cycle. Figure 2b shows the execution overhead of the RSM in four monitoring scenarios when: (1) the RSM monitors the full algorithm, (2) it disables dataflow integrity checks, (3) checking only pre- and post-conditions, and (4) disabling dataflow integrity and internal-conditions checks.

The RSM detects *computational attacks*. Recently, it has been shown that stealthy attacks (FDI-attacks) with tampered sensor readings may evade detection by the monitor. So, we present a method to analyze the monitor design to identify FDI vulnerabilities.

3 False Data Injection Attacks

FDI attacks are a special class of attacks where the adversary compromises external-input values to CPS control-systems without attacking the system itself. In such an attack, CPS systems operate correctly but on wrong input-data, leading the control-systems to take wrong decisions.

Current approaches to handle FDI attacks are mainly focused on power systems [15], where attacks are detected exploiting network-topology characteristics and statistical analysis of data.

We have developed an alternative method that identifies FDI vulnerabilities in a system-design (i.e., specification). Based on the identified vulnerabilities, we refine the design, introducing constraints that eliminate the identified attacks. The refinement is repeated until either all FDI vulnerabilities are eliminated or the vulnerabilities have been recorded for run-time detection.

We specify the system as a state function, which is amenable for verification to detect combinations of inputs that constitute FDI attacks. A CPS system implements a control-loop for a process P. The function $P(x_t)$ specifies the system-state at time t as input variables x. In the implementation, these variables are measured (as z_t) with sensors to estimate the system-state. A CPS monitor $mon(x, z)$ inputs the measurements z_t at every time instant t and evaluates them for acceptance.

Typically, the monitor accepts all measurements z and estimates the state $P(x, z)$. However, in a successful FDI attack with measurements z', the monitor accepts z', which are compromised values of z different from z. To detect such attacks, we specify system-state as a real function $f()$ and using a non-linear SMT solver, i.e., dReal [16], we ask whether there exist z' s.t. for the system $P(x, z)$, $mon(x, z')$ accepts z'. If the question is satisfiable, then z' constitutes an FDI attack. To defend against such an attack, we can consider the vulnerabilities as attack-vectors of values to be monitored for detection at run-time or can add constraints to the design, which reduces and eventually eliminates such attack.

3.1 Example

To demonstrate our approach, we consider a water-tank with two pumps, one for pumping-in and the other for pumping-out the water. Each pump has a sensor measuring the flow-rate (r_{in} and r_{out} as real numbers in range $\{0, 1\}$) and an actuator, for opening it to a specified rate. Another sensor measures the water-level in the tank, whose cross-sectional area is one. With these values, in the tank, the volume and the height of the water have the same absolute value.

Considering a discrete time operation of the tank, the state of the pumps may change at every time unit. The height H of the water at time $t + 1$ determines the system-state by

$$H(t + 1) = H(t) + r_{in}(t + 1) - r_{out}(t + 1) \tag{1}$$

Assuming that $H(t) = 7$, the flow $r_{in}(t + 1) = 0.5$ and $r_{out}(t + 1) = 0$, gives $H(t + 1) = 7.5$. If these values are measured correctly by sensors, the monitor will accept them. However, an attacker could compromise these measurements with fake values and if they are consistent with $H(t + 1)$, the monitor will accept them otherwise will reject them identifying an FDI attack. For example, the tampered values $r_{in}(t + 1) = 1$, $r_{out}(t + 1) = 0$, and $H(t + 1) = 4$ will be rejected being inconsistent with Eq. (1). However, the compromised values $r_{in}(t + 1) = 1$, $r_{out}(t + 1) = 0$, and $H(t + 1) = 8$ will be accepted as they satisfy Eq. (1), indicating an FDI vulnerability in the system. To detect such an attack, we provide Eq. (1) to the solver and ask if there are values (r_{in}, r_{out}, and H in defined-range), which are different from the real action. Listing 2 shows the input to dReal for our

```
(set-logic QF_NRA)
(declare-fun ht () Int)
(declare-fun ht1 () Int)
(declare-fun a () Int)
(declare-fun rin () Real)
(declare-fun rout () Real)
(assert (and (and (= ht 5) (= ht1 6)) (= a 1)))
(assert (and (<= 0 rin) (<= rin 1)))
(assert (and (<= 0 rout) (<= rout 1)))
(assert (= ht1 (+ ht (- (/ rin (^ a 2)) (/ rout (^ a 2))))))
(check-sat)
(exit)
```

Listing 2 Example specification and verification

described scenario and demonstrates the successful detection of FDI vulnerability. To eliminate such attacks, we may add constraints for parameters, e.g., considering r_{in} and r_{out} to be integers.

4 Conclusion

We have introduced a methodology to develop CPS application software that is secure against computational and false data injection attacks. For computational attacks, we have developed run-time security monitor that detects attacks by identifying the deviation of application execution from the specification of the application, and by avoiding false alarms. For false data injection attacks, we have used vulnerability analysis technique that identifies false (data injection attack) values in the given specification, which can later be monitored. We have demonstrated the effectiveness of our methodology by developing a simple CPS example. As a next step, our goal is to unify the two approaches in a tool that allows to (1) synthesize the security monitor from a given specification automatically and (2) detect and eliminate FDI vulnerabilities in the specification interactively.

References

1. "Nsf industry 4.0," https://www.nsf.gov/pubs/2014/nsf14542/nsf14542.htm.
2. "Nist cps," https://www.nist.gov/el/cyber-physical-systems.
3. M. Zeller, "Myth or reality – does the aurora vulnerability pose a risk to my generator?" in *2011 64th Annual Conference for Protective Relay Engineers*, April 2011, pp. 130–136.
4. R. Langner, "Stuxnet: Dissecting a cyberwarfare weapon," *IEEE Security and Privacy*, vol. 9, no. 3, pp. 49–51, May 2011.
5. B. Kang, K. McLaughlin, and S. Sezer, "Towards a stateful analysis framework for smart grid network intrusion detection," in *Proceedings of the 4th International Symposium for ICS & SCADA Cyber Security Research 2016*, ser. ICS-CSR '16, 2016, pp. 1–8.
6. U. Lindqvist and P. G. Neumann, "The future of the internet of things," *Communications of the ACM*, vol. 60, no. 2, pp. 26–30, January 2017.
7. V. Hodge and J. Austin, "A survey of outlier detection methodologies," *Artif. Intell. Rev.*, vol. 22, no. 2, pp. 85–126, 2004.
8. A. Lakhina, M. Crovella, and C. Diot, "Mining anomalies using traffic feature distributions," *SIGCOMM Comput. Commun. Rev.*, vol. 35, no. 4, pp. 217–228, Aug. 2005.
9. V. Paxson, "Bro: A system for detecting network intruders in real-time," in *Proceedings of the 7th Conference on USENIX Security Symposium - Volume 7*, ser. SSYM'98, Berkeley, CA, USA, 1998, pp. 2435–2463.
10. C. Watterson and D. Heffernan, "Runtime verification and monitoring of embedded systems," *Software, IET*, vol. 1, no. 5, pp. 172–179, 2007.
11. S. Adepu and A. Mathur, *Using Process Invariants to Detect Cyber Attacks on a Water Treatment System*. Springer, 2016, pp. 91–104.
12. M. Khan, D. Serpanos, and H. Shrobe, "A rigorous and efficient run-time security monitor for real-time critical embedded system applications," in *IEEE 3rd WF-IoT*, December 2016, pp. 100–105.

13. M. T. Khan, D. Serpanos, and H. Shrobe, “Armet: Behavior-based secure and resilient industrial control systems,” *Proceedings of the IEEE*, vol. 106, no. 1, pp. 129–143, Jan 2018.
14. B. Courcelle and J. Engelfriet, *Graph Structure and Monadic Second-Order Logic: A Language-Theoretic Approach*. Cambridge University Press, 2012.
15. G. Hug and J. A. Giampapa, “Vulnerability assessment of ac state estimation with respect to false data injection cyber-attacks,” *IEEE Transactions on Smart Grid*, vol. 3, no. 3, pp. 1362–1370, 2012.
16. S. Gao, S. Kong, and E. M. Clarke, “dReal: An SMT Solver for Nonlinear Theories over the Reals,” in *Proceedings of the CADE'13*. Springer, 2013, pp. 208–214.

Moving Camera Analytics: Computer Vision Applications

Chung-Ching Lin, Karthikeyan Natesan Ramamurthy, and Sharathchandra U. Pankanti

Abstract To date, billions of cameras have been actively used on moving platform. Video analytics applications for camera are emerging in diverse areas. Among various video analytics applications for moving cameras, we will discuss the application of unmanned aerial vehicles (UAVs). First, we present a system for summarizing videos by automatically creating a panorama for videos, detecting and tracking moving objects in the videos. Our video summarization experiments on the UAV dataset demonstrate that we can achieve efficient data reduction without losing significant activities of interest. Second, a distributed 3D reconstruction algorithm will be presented. Most methods for 3D reconstruction methods are either centralized or operate incrementally. The poor scalability affects the quality of solution for large-scale structure from motion (SfM). Our algorithm uses alternating direction method of multipliers (ADMM) to formulate a distributed bundle adjustment (BA) algorithm. The results are comparable to an alternate state-of-the-art centralized bundle adjustment algorithm on synthetic and real 3D reconstruction problems. The runtime of our implementation scales linearly with the number of observed points.

1 Introduction

In the instrumented, integrated internet of things, moving cameras are playing an increasingly major role as versatile sensors. Currently, billions of cameras are actively used on moving platforms such as cell phones, unmanned aerial vehicles (UAVs), cars, and people (e.g., wearable cameras), and the number is rapidly increasing. Various video analytics applications for moving cameras are emerging in many areas. Just a few of these include: retail, store-shelf monitoring and management with body cams, map and road updates with instrumented cars, car cameras that sense road conditions, event reconstruction with unstructured cell

C.-C. Lin (✉) · K. N. Ramamurthy · S. U. Pankanti
IBM Research AI, Yorktown Heights, NY, USA

S. S. Bhattacharyya et al. (eds.), *Embedded, Cyber-Physical, and IoT Systems*,
https://doi.org/10.1007/978-3-030-16949-7_5

phone cameras for public safety, and UAV video summarization. Each of these applications generates a huge dataset to be analyzed.

A powerful mobile video analytics system can be applied to various types of unmanned vehicles, including those that operate on the ground, in the air, and under water. For homeland security tasks such as border protection and control and coastline monitoring, UAVs that can monitor a very large area during an extended period of time are required. The UAVs can be equipped with optical and infrared cameras, lowlight cameras, radar, and biochemical sensors. Law enforcement departments in some countries plan to replace helicopters with medium-size UAVs with automatic video trackers. For post-disaster assessments such as those performed after hurricanes, earthquakes, floods, and forest fires, the UAVs need to be equipped with high-resolution cameras, telescopic lenses, and infrared sensors for accessing areas that are difficult or dangerous for humans to reach.

Each application of this nature collects large amounts of diverse types of data. In cases such as homeland security, large amounts of videos are generated daily. Because the human resources needed to screen and compare the videos are limited, a method to reduce video data effectively without losing important information is vital. A system that can summarize videos and reconstruct 3D scenes can help operators make decisions promptly and accurately.

There are two primary challenges to building this type of system. Providing useful operational 2D views and 3D views from unstructured, uncalibrated videos or images taken or streamed from mobile vantage points of view is difficult because of low spatial and temporal resolution, jitter, frequent undefined motion, non-ideal vantage points of view, frequently moving or changing background scenery, and the potential inclusion of misleading or tampered with sources of information.

Our approach addresses both of these challenges. Our system provides analytics for summarizing videos. In this chapter, we describe the system and its use with mobile videos acquired by UAVs.

Typically, most of the content of a particular UAV video does not contain any event of interest, such as a moving pedestrian or vehicle. Our solution's analytics can detect and track moving objects, and generate panoramas. At the end of analysis, the graphical user interface (GUI) presents a list of tracks from the videos, ranked based on their saliency [20]. Also, the panorama is shown in another window with every track registered. The panorama provides a global view of the area covered by videos from moving cameras, while the list of tracks can allow the operators to understand what events occurred in the videos.

For 3D reconstruction, we proposed a distributed bundle adjust algorithm. Most methods for 3D reconstruction methods are either centralized or operate incrementally. The poor scalability affects the quality of solution for large-scale structure from motion (SfM). Our algorithm uses alternating direction method of multipliers (ADMM) to formulate a distributed bundle adjustment (BA) algorithm [31]. Our approach is ideally suited for applications where image acquisition and processing must be distributed, such as in a network of unmanned aerial vehicles (UAVs). We assume that each UAV in the network has a camera and a processor; each camera acquires an image of the 3D scene, and the processors in the different

UAVs cooperatively estimate the 3D point cloud from the images. Therefore, we use the terms camera, processor, and UAV in an equivalent sense throughout the chapter. We also assume that corresponding points from the images are available (possibly estimated using a different distributed algorithm), and are only concerned about estimating the 3D scene points given the correspondences.

The system can enable new important applications (e.g., post-disaster assessments) and can also help existing applications, such as ones for homeland security, perform more effectively, efficiently, and cost-effectively. Every mobile platform is a resource-limited system and has a limited bandwidth available to communicate with other systems. We forecast that these types of applications will soon function primarily on server or cloud processors. This trend will be driven by the availability of more powerful processors and partly by the refinement of algorithms to make them less processor-intensive.

2 Related Work

As a result of the recent dramatic increase in the prevalence of mobile cameras and sensors, the amount of generated video data is growing rapidly. Accordingly, developing efficient and effective algorithms for video summarization has become increasingly beneficial. Currently, two main approaches exist for video summarization: keyframe-based summarization and object-based summarization. In the keyframe-based approach (e.g., [12]), the video is analyzed using relatively low-level visual properties, and a series of still images is generated as a summary of the entire video. In contrast, object-based summarization techniques (e.g., [9, 29]) typically first perform object segmentation and tracking and then compress the extracted trajectories (or tubes) to reduce spatio-temporal redundancy.

When a video sequence is captured from a moving camera, an important component in almost any analysis system is the video stabilizer. The stabilizer compensates for the motion of pixels on the image plane due to the motion of the camera (i.e., register the video frames). This process usually involves the selection of a motion model, followed by an estimation of the model parameters. Techniques for estimating the model parameters can be broadly divided into two classes: direct methods [16] and feature-based methods. In direct methods, the camera motion and other useful analytics are directly estimated from image pixel properties. In contrast, feature-based methods generally use a pipeline of keypoint (corner) detection, feature matching, and geometric registration. With the advances of the last decade in keypoint detection and matching (e.g., [24, 37]), feature-based methods are now remarkably robust, typically outperforming direct schemes. In our system, video data is registered using our robust feature-based algorithm, which automatically creates a mosaic [17] by stitching image frames together; this virtually increases the field of view of the camera.

For effective video summarization, reliably tracking moving objects is crucial. While object tracking [42] is a well-studied computer vision problem, and works very effectively for videos captured by a static camera (e.g., [43]), it is still difficult to effectively track objects in videos captured by a moving platform. To obtain reliable tracking results, we employ several efficient trackers [7, 34] and use a combination of their results.

For vision application to perform well in our three-dimensional world, it is usually essential to understand the 3D structure of a scene [14]. Estimating accurate poses of cameras and locations of 3D scene points from a collection of images obtained by the cameras is a classic problem in computer vision, referred to as *structure from motion* (SfM). Optimizing for the camera parameters and scene points using the corresponding points in images, known as *bundle adjustment* (BA), is an important component of SfM [13, 14, 35].

Many recent approaches for BA can be divided into three categories: (a) those that pose BA as non-linear least squares [18, 22, 35], (b) those that decouple the problem in each camera using a triangulation-resection procedure for estimation [25, 30], and (c) those that pose and solve BA in a linear algebraic formulation [11]. Some important considerations of these methods are reducing the computational complexity by exploiting the structure of the problem [1, 6, 22], incorporating robustness to outlier observations or correspondence mismatches [2, 44], distributing the computations or making the algorithm incremental [8, 15, 19, 39, 40], and making the algorithm insensitive to initial conditions [11]. In this chapter, we develop robust distributed BA over camera and scene points.

Robust approaches, such as [2, 44], are typically used to protect world point and camera parameter estimates from effects of *outliers*, which for BA are incorrect point correspondences that have gone undetected. In contrast, we use robust formulations to *accelerate consensus* in the distributed formulation. Depending on how distribution is achieved, every processor performing computation may see *only a small portion* of the total data, and attempt to use it to infer its local parameters. Small sample means can be extreme, even when the original sample is well-behaved (i.e., even when re-projection errors are truly Gaussian). In the limiting case, each processor may only base its computation on *one data point*, and therefore outliers are guaranteed to occur (from the point of view of individual processors) as an artifact of distributing the computation. Hence we hypothesize that using robust losses for penalizing re-projection errors and quadratic losses for enforcing consensus improves performance.

Our proposed robust BA approach supports a natural distributed parallel implementation. We distribute the world points and camera parameters as illustrated for a simple case of 2 cameras and 5 scene points in Fig. 1. The algorithm is developed using distributed alternating direction method of multipliers (D-ADMM) [5]. Each processor updates its copy of a set of parameters, while the updated estimates and dual variables ensure consensus. Distributing *both* the world points and the camera parameters yields iterations with $O(l)$ required operations in a serial setting, where l is the total number of 2D observations. In a fully parallel setting, it is possible to bring the time complexities down to $O(1)$ per iteration, a vast improvement compared to traditional and sparse versions of BA, whose complexities are $O((m+n)^3)$ and $O(m^3+mn)$, respectively [22] (with m and n the number of cameras and

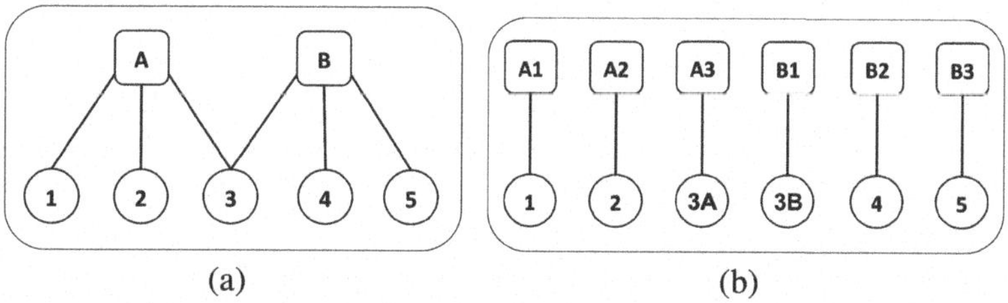

Fig. 1 (**a**) Original configuration of cameras A, B and scene points 1, 2, 3, 4, 5, (**b**) distributing both the camera parameter and scene point estimation with the constraints A1 = A2 = A3, B1 = B2 = B3, and 3A = 3B

3D scene points). We also exploit the sparsity of the camera network, since not all cameras observe all scene points.

Another optimization-based distributed approach for BA was recently proposed [8]. The authors of [8] distributed camera parameters, and performed synthetic experiments using an existing 3D point cloud reconstruction, perturbing it using moderate noise, and generating image points using known camera models. We go further, distributing both world points and camera parameters in a flexible manner, and we implement the entire BA pipeline for 3D reconstruction: performing feature detection, matching corresponding points, and applying the robust distributed D-ADMM BA technique in real data settings.

3 Video Summarization

In this section, we describe our method of video summarization for moving cameras. We implemented a robust approach for detecting and tracking moving objects. Our approach tackles challenges introduced by videos from moving cameras, such as rapidly moving and shaking platforms, irregular rotations, small object resolution, and low contrast of images. Our system also generates a panorama upon which the registered tracks are superimposed.

3.1 Object Detection and Tracking

Our approach first compensates for the camera motion by estimating the homography transformation using the detected and matched feature points between adjacent frames. After motion compensation, a foreground mask is obtained by comparing the aligned frames, and is used to identify potential moving objects. Due to low image quality, camera noise, intensity change, or parallax, this identification of potential moving objects may have false positive detections. We use the feature motion vectors of the potential moving objects to verify and reduce false detections. Potential moving objects with small feature motion vectors are considered to be steady objects and discarded. Detected objects are tracked by analyzing the variation of Bhattacharyya coefficient [3] within the mean shift framework [7]. Under a challenging dataset, this tracking method may not be robust enough. Therefore, we use the foreground mask and feature motion vectors to aid the tracking algorithm. Foreground masks provide possible locations of moving objects and the feature motion vectors are used to verify the movement of the objects, resulting in more accurate tracking results.

3.2 Panorama

To obtain a global view of the area covered by videos of moving cameras, our method generates panoramas for the videos. Due to the nature of the moving cameras, each video can contain several segments with dissimilar viewing angles and settings. Each segment is presented by a sub-panorama.

For more precise frame alignment, we compute the homography transformation between two consecutive frames using RANSAC (random sample consensus) [10]. We utilize SIFT (scale-invariant feature transform) [23] to obtain more precise feature point detection and matching. However, not every pair of adjacent frames has enough corresponding points to estimate homography transformations. In this case, we estimate an affine transformation. The estimated transformation is then further verified. If it is abnormal or cannot be found, the transformation between two consecutive frames is set to empty. After estimating all the transformation parameters, we then determine the first and the last frames of each segment by checking the consecutive normal transformations. To create sub-panoramas for each segment, we align every frame to the first frame within the segment. Next, we transform the frames to have the same coordinate system as the first frame. This can be done using the homographies H between each pair of frames, which have already been computed.

Using the transformation between adjacent frames, a frame can be aligned with the previous one using the inverse transformation

$$I_t \rightarrow I_{t-1} : x_{t-1,i} = H_{t-1\rightarrow t} x_{t,i} = H^{-1}_{t\rightarrow t-1} x_{t,i}, \tag{1}$$

where I_t is the frame at time t, $H_{t-1\rightarrow t}$ is the transformation from frame I_{t-1} to frame I_t, and $x_{t,i}$ is the position of i-th pixel in frame I_t. The transformation from the k-th frame to the first frame of the segment is the cascade of the inverse adjacent transformations between the first and k-th frames

$$I_t \rightarrow I_0 : x_{0,i} = H^{-1}_{1\rightarrow 0} H^{-1}_{2\rightarrow 1} \ldots H^{-1}_{t-2\rightarrow t-1} H^{-1}_{t\rightarrow t-1} x_{t,i}. \tag{2}$$

To avoid a blurred panorama caused by parallax, when each frame is transformed to the panorama coordinate system, we only update the uncovered area. We also developed a method to register each image to a single panorama. Every sub-panorama is generated from the frames with different viewing angles of video. When registering the sub-panoramas with large differences in viewing angle, 3D distortion causes dissimilarity in the extracted affine invariant descriptors of corresponding points in different views. Here, existing methods fail to find corresponding points. To overcome this, we developed a method to normalize 3D distortion with hypothesis transformation and find the accurate corresponding points. From our experience, affine transformation is sufficient to normalize 3D distortion.

3.3 Visualization

After detecting and tracking moving objects, we compute the saliency score of every track by evaluating the roughness of the object's trajectory [36]. To summarize the videos more efficiently, we generate keyframes that represent the video segments, along with their associated activities. Tracks that are overlapping in time are grouped together, and then registered and superimposed on a keyframe, as shown in Fig. 2 (left). In this way, the entire video is summarized by several keyframes. We also register and superimpose all tracks on the panorama to summarize the video in a single image [4].

3.4 Results

To demonstrate our video summarization, we used a challenging dataset, the aerial videos in version 2.0 of VIRAT (video and image retrieval and analysis tool) dataset [28], which captures real natural scenes.

Figure 2 shows an experimental video summarization. For each video, our system extracted a list of summarizing video segments, shown in Fig. 2 (left), and provided a video panorama, also shown in Fig. 2 (right). Each video segment has a starting and ending time and is also represented by one keyframe. Each keyframe summarizes a video segment, and the list of selected video segments is ranked by saliency scores. All the tracks in the video segment are superimposed on the keyframe of the video segment. These results show a full VIRAT aerial video of 18,575 frames summarized by 5 keyframes. All the tracks in the video are superimposed on the panorama of the whole video. This is equivalent to summarizing the entire video of 18,575 frames using only one image, demonstrating that our summarization method can achieve 10,000-fold data reduction. Our approach reaches an overall precision of 0.8 and recall of 0.92 in our test set of 8 VIRAT videos. The image quality of the VIRAT dataset is low, thus presenting difficulties in detecting and tracking moving

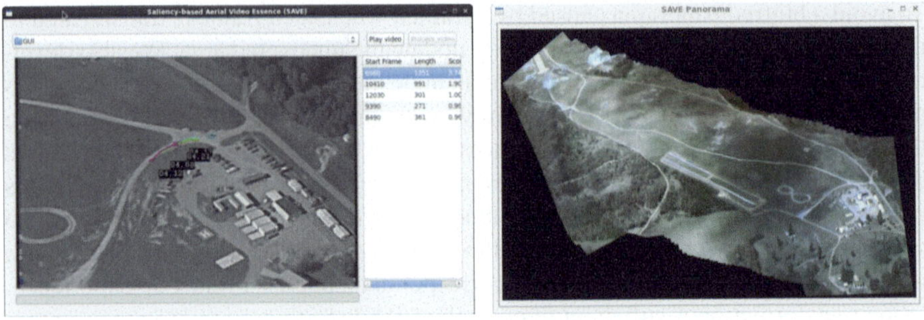

(a) (b)

Fig. 2 Main application windows: (**a**) a keyframe summarizing a video segment and the list of selected video segments ranked by saliency scores, showing that a full VIRAT aerial video of 18,575 frames was summarized by only 5 keyframes; (**b**) a panorama of the video

objects. If we can improve the image quality using super resolution, the detection and tracking performance will be significantly improved.

4 Distributed Bundle Adjustment

In this section, we introduce our proposed distributed bundle adjustment, that is, a robust and flexible approach to reconstruct 3D scene from 2D observations. We will discuss background, notations, and the proposed approach. Then, experiments and comparisons on synthetic and real data will be demonstrated.

4.1 Background

4.1.1 The Camera Imaging Process

We denote the m camera parameter vectors by $\{y_j\}_{j=1}^{m}$, the n 3D scene points as $\{x_i\}_{i=1}^{n}$, and the 2D image points as $\{z_{ij}\}$. Each 2D image point $z_{ij} \in \mathbb{R}^2$ is obtained by the transformation and projection of a 3D scene point $x_i \in \mathbb{R}^q$ by the camera $y_j \in \mathbb{R}^p$. BA is an inverse problem, where camera parameters and 3D world points are estimated from the observations $\{z_{ij}\}$. The forward model is a non-linear camera transformation function $f(x_i, y_j)$.

The number of image points is typically much smaller than mn, since not all cameras image all scene points. The camera parameter vector (y_j) usually includes position, Euler angles, and focal length. In this discussion, we assume focal length is known for simplicity, and $y_j \in \mathbb{R}^6$ comprises Euler angles α, β, γ and the translation vector $t \in \mathbb{R}^3$.

Denote the diagonal focal length matrix as $K \in \mathbb{R}^{3\times 3}$, with the first two diagonal elements set to the focal length and the last element set to 1. The rotation matrix is represented as $R = R_3(\gamma)R_2(\beta)R_1(\alpha)$, where R_1, R_2, R_3 are rotations along the three axes of $\mathbb{R}^3$. The camera transformation is now given as $\tilde{z} = Rx + t$. The final 2D image point z is obtained by a perspective projection, with coordinates given by

$$z_1 = \frac{\tilde{z}_1}{\tilde{z}_3}, z_2 = \frac{\tilde{z}_2}{\tilde{z}_3}. \tag{3}$$

4.1.2 Bundle Adjustment

Given the 2D points in multiple images that represent the same scene point, BA is typically formulated as a non-linear least squares problem:

$$\min_{\{x_i\},\{y_j\}} \sum_{j=1}^{m} \sum_{i \in S(j)} \|z_{i,j} - f(x_i, y_j)\|_2^2. \tag{4}$$

The set $S(j)$ contains i if the scene point i is imaged by the camera j. The number of unknowns in this objective is $3n + 6m$, and hence it is necessary to have at least this many observations to obtain a good solution; in practice, the number of observations is much larger. Problem (4) is solved iteratively, with descent direction (δ_x, δ_y) found by replacing f in (4) by its linearization

$$f(x + \delta_x, y + \delta_y) \approx f(x, y) + J(x)\delta_x + J(y)\delta_y,$$

where $J(x) = \partial_x f$, $J(y) = \partial_y f$. The Levenberg–Marquardt (LM) algorithm [26] is often used for BA.

The naive LM algorithm requires $O((m + n)^3)$ operations for each iteration, and memory on the order of $O(mn(m + n))$, since we must invert of an $O(m + n) \times O(m + n)$ matrix at each iteration. However, exploiting matrix structure and using the Schur complement approach proposed in [22], the number of arithmetic operations can be reduced to $O(m^3 + mn)$ and memory use to $O(mn)$. Further reduction can be achieved by exploiting secondary sparse structure [18]. The conjugate gradient approaches in [1, 6] can reduce the time complexity to $O(m)$ per iteration, making it essentially linear in the number of cameras.

Another popular approach to reduce the computational complexity involves decoupling of the optimization by explicitly estimating the scene point using back-projection in the *intersection* step and estimating the camera parameters in the *resection* step [30]. The resection step decouples into m independent problems, and hence the overall procedure has a cost of $O(m)$ per iteration. A similar approach, but with the minimization of ℓ_∞ norm of the re-projection error was proposed in [25]. It was shown to be more reliable and degraded gracefully with noise compared to ℓ_2 based BA algorithms. Recently Wu proposed an incremental approach for bundle adjustment [39], where a partial BA or a full BA is performed after adding each camera and associated scene points to the set of unknown parameters, again with a complexity of $O(m)$. We use the ADMM framework to develop our approach.

4.1.3 Alternating Direction Method of Multipliers

ADMM is a simple and powerful procedure well-suited for distributed optimization [21], see also [5]. In order to understand D-ADMM, consider the objective $h(x) := \sum_{i=1}^{n} h_i(x)$. We introduce local variables with a *consensus* equality constraint:

$$\begin{aligned} &\min_{\{x_i\},u} \sum_{i=1}^{n} h_i(x_i) \\ &\text{subject to } x_i - u = 0, i \in \{1, \ldots, n\}. \end{aligned} \tag{5}$$

To solve this problem, we first write down an *augmented Lagrangian* [32]:

$$\bar{l}_\phi(x, u, r, \rho) := \sum_{i=1}^{n} h_i(x_i) + r_i^T (x_i - u) + \frac{\rho}{2}\phi(x_i, u), \tag{6}$$

where $\rho > 0$ is the penalty parameter, r_i is the Lagrangian multiplier for the constraint, and $\phi(x_i, u)$ is the augmentation term that measures the distance individual variables x_i and the consensus variable u. We then find a saddle point using three steps to update $\{x_i\}$, u, and $\{r_i\}$. Typically $\phi(x_i, u)$ is chosen to the squared Euclidean distance in which case (6) becomes the proximal Lagrangian [32], but other distance or divergence measures can also be used.

4.2 *Algorithmic Formulation*

4.2.1 Distributed Estimation of Scene Points and Camera Parameters

We distribute the estimation among both the scene points and the camera parameters as illustrated in Fig. 1. We estimate the camera parameter and the scene point corresponding to each image point independently, and then impose appropriate equality constraints. Equation (4) can be written as

$$\min_{\{x_i^j\},\{y_j^i\},\{x_i\},\{y_j\}} \sum_{j=1}^{m} \sum_{i \in S(j)} \phi_m(z_{i,j} - f(x_i^j, y_j^i)), \tag{7}$$

$$\text{such that } x_i^j = x_i, \forall i, \text{and } \{j : i \in S(j)\}, \tag{8}$$

$$y_j^i = y_j, \forall j, \text{and } \{i \in S(j)\}. \tag{9}$$

The augmented Lagrangian, with dual variables r_i^j and s_j^i, is given by

$$\sum_{j=1}^{m} \sum_{i \in S(j)} \phi_m(z_{i,j} - f(x_i^j, y_j^i)) + r_i^{jT}(x_i^j - x_i) + s_j^{iT}(y_j^i - y_j)$$
$$+ (\rho_x/2)\phi_a(x_i^j - x_i) + (\rho_y/2)\phi_a(y_j^i - y_j). \tag{10}$$

Here ϕ_a measures the distance between the distributed world points and their consensus estimates, and distributed camera parameters and their consensus estimates. For ϕ_m we compare squared Euclidean and Huber losses, and ϕ_a is always the squared Euclidean loss.

The ADMM iteration is given by

$$(x_i^{j(k+1)}, y_j^{i(k+1)}) := \underset{\{x_i^j\},\{y_j^i\}}{\text{argmin}}\, \phi_m(z_{i,j} - f(x_i^j, y_j^i))$$
$$+ r_i^{j(k)T}(x_i^j - x_i^{(k)}) + s_j^{i(k)T}(y_j^i - y_j^{(k)})$$
$$+ (\rho_x/2)\phi_a(x_i^j - x_i^{(k)}) + (\rho_y/2)\phi_a(y_j^i - y_j^{(k)}), \tag{11}$$

$$x_i^{(k+1)} := \frac{1}{|j : i \in S(j)|} \sum_{j:i \in S(j)} \left(x_i^{j(k+1)} + (1/\rho_x) r_i^{j(k)}\right), \tag{12}$$

$$y_j^{(k+1)} := \frac{1}{|i \in S(j)|} \sum_{i \in S(j)} \left(y_j^{i(k+1)} + (1/\rho_y) s_j^{i(k)}\right), \tag{13}$$

$$r_i^{j(k+1)} := r_i^{j(k)} + \rho_x \left(x_i^{j(k+1)} - x_i^{(k+1)}\right), \tag{14}$$

$$s_j^{i(k+1)} := s_j^{i(k)} + \rho_y \left(y_j^{i(k+1)} - y_j^{(k+1)}\right). \tag{15}$$

Equation (11) has to be solved for all $j \in S(i), i \in \{1, \ldots, m\}$, and it can be trivially distributed across multiple processes. When ϕ_m is squared ℓ_2 distance, (11) can be solved using the Gauss–Newton method [27], where we repeatedly linearize f around the current solution and update (x, y). When ϕ_m is the Huber loss, we use limited memory BFGS (L-BFGS) [27] to update the distributed scene points. Upon convergence, we will obtain the consensus estimates x_i and y_j for all scene points and cameras.

4.2.2 Convergence Analysis

We show that under certain assumptions the proposed D-ADMM algorithm in Sect. 4.2.1 converges, using the non-convex and non-smooth framework developed by Wang et al. [38].

Theorem 1 *The D-ADMM algorithm proposed in Sect. 4.2.1 to the stationary point of the augmented Lagrangian in* (10) *when:*

1. $f(.,.)$ *is the perspective camera projection model,*
2. ϕ_m *is any convex, smooth loss function, and* ϕ_a *is the squared Euclidean loss.*
3. ρ_x *and* ρ_y *are sufficiently large.*

Proof Let d_{ij} be the stack of $\{x_i^j, y_j^i\}$, and $\hat{d} = [\hat{d}_{ij}]_{\forall i, \forall j}$. Similarly each pair of consensus variables are stacked as the vector $\hat{c}_{ij} = [x_i^T \, y_j^T]^T$ and $\hat{c} = [\hat{c}_{ij}]_{\forall i, \forall j}$. $\hat{d}$ and $\hat{c}$ are, respectively, equivalent to x and y in [38]. We show that the five assumptions (A1–A5) of [38, Thm. 1] are satisfied.

1. Given our assumptions, the objective function in (8) is coercive, i.e., it tends to ∞ as $\hat{d} \to \infty$ (A1).
2. The feasibility and sub-minimization path conditions are also satisfied since the constraint matrices are easily seen to be full rank (A2–A3).
3. Each additive part of the objective $\phi_m(z_{i,j} - f(x_i^j, y_j^i))$ is restricted prox-regular if ϕ_m is a smooth convex function and f is the perspective camera model. The gradient will be steep when $\tilde{z}_3$ in (3) is less than some $\epsilon > 0$ and $\phi_m(z_{i,j} - f(x_i^j, y_j^i))$ is prox-regular for $\epsilon > 0$; hence A4 in [38, Thm. 1] holds.
4. Our objective with respect to the consensus variable is identically 0, which is trivially regular (A5).

Since all the assumptions hold, the iterative algorithm in Eqs. (11)–(15) converges to a stationary point of the augmented Lagrangian for sufficiently large ρ_x and ρ_y.

4.2.3 Time Complexity

Optimizing (11) takes $O(l)$ time for each round of updates, since (11) must be solved l times, with each solve requiring constant time. The time complexity of the consensus steps for camera parameters and world points given by (12) and (13) are $O(m)$ and $O(n)$, respectively. For the Lagrangian parameter updates given by (14) and (15), the time complexity is $O(l)$. Hence the dominant time complexity of the proposed algorithm is $O(l)$ for each round. Since the algorithm can be trivially parallelized, the complexity can be brought down to $O(1)$ for each round, if we distribute all the observations to individual processors.

4.2.4 Communication Overhead

Considering a sparse UAV network, assume that each world point is imaged by d cameras. Each camera needs to maintain a copy of the consensus world points x_i. Therefore to update x_i using (12), each camera needs to obtain $d - 1$ individual estimates of x_i^j and send its version of x_i^j to $d - 1$ other cameras. Values r_i^j can be updated locally in each camera, given x_i^j, x_i and previous versions of r_i^j using (14). Hence, for each world point we have a communication overhead of $3(d - 1)d$ floating points per iteration (each world point is a 3D vector). Hence for n world points, the communication overhead is $3(d - 1)dn$ floating points per iteration, where d depends on the distance of the camera from the scene.

4.2.5 Generalized Distributed Estimation

The problem (11) requires each processor to estimate $p + q > 2$ parameters from a single 2D observation. To control the variability of individual estimates as the algorithm proceeds, we generalize the approach to use more than one observation and hence more than one scene point and camera vector during each update step. This generalized step provides flexibility to adjust the number of 3D scene points and cameras based on computational capability of each thread in a CPU or a GPU. We solve

$$\begin{aligned}
(X_i^{j(k+1)}, Y_j^{i(k+1)}) &:= \underset{\{X_i^j\},\{Y_j^i\}}{\operatorname{argmin}} \phi_m(Z_{i,j} - f(X_i^j, Y_j^i)) \\
&+ r_i^{j(k)T}(X_i^j - X_i^{(k)}) + s_j^{i(k)T}(Y_j^i - Y_j^{(k)}) \\
&+ (\rho_x/2)\phi_a(X_i^j - X_i^{(k)}) + (\rho_y/2)\phi_a(Y_j^i - Y_j^{(k)}),
\end{aligned} \tag{16}$$

where

$$\begin{aligned}
X_i^{j(k+1)} &:= \left[x_{i_1}^{j(k)}\ x_{i_2}^{j(k)}\ \ldots\ x_{i_\pi}^{j(k)}\right]^T, \\
Y_j^{i(k+1)} &:= \left[y_{j_1}^{i(k)}\ y_{j_2}^{i(k)}\ \ldots\ y_{j_\kappa}^{i(k)}\right]^T.
\end{aligned} \tag{17}$$

4.3 Experiments

We perform several experiments with synthetic data and real data to show the convergence of the re-projection error and the parameter estimates. We also compare the performance of the proposed approach to the centralized BA algorithm that we implemented using LM. The LM stops when the re-projection error drops

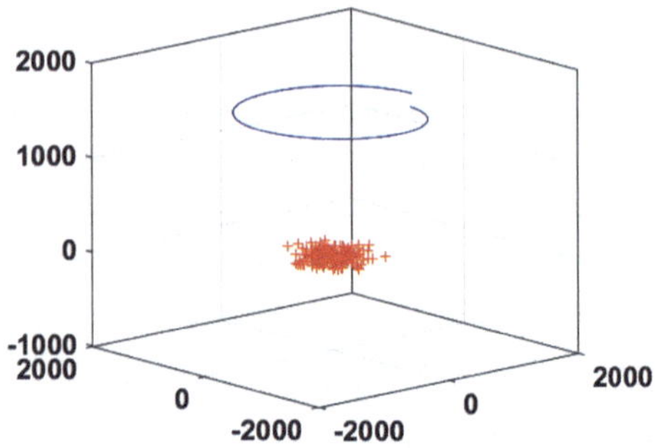

Fig. 3 Camera flight path (blue) and 3D scene points (red) for an example synthetic dataset

below 10^{-14}, or when the regularization parameter becomes greater than 10^{16}. We implement our distributed approach in a single multi-core computer and not in a sparse UAV network, but our architecture is well-suited for a networked UAV application.

4.3.1 Synthetic Data

We simulate a realistic scenario, with smooth camera pose transition, and noise parameters consistent with real-world sensor errors. Using the simulation, we evaluate the error in the estimated 3D scene point cloud and the camera parameters, and investigate how estimation error of camera pose affects the final tie points triangulation.

The camera positions are sampled around an orbit, with an average radius 1000 m and altitude 1500 m, with the camera directed towards a specific area. To each camera pose, a random translation and rotation is added as any real observer cannot move in a perfect circle while steadily aiming always in the same exact direction. The camera path and the 3D scene points for an example scenario are shown in Fig. 3. In practice, tie points are usually visible only within a small subset of the available views, and it is generally not practical to try to match all keypoints within each possible pair of frames. Instead, points are matched within adjacent frames. In our synthetic data, we create artificial occlusions or mis-detection so that each point is only visible on a few consecutive frames.

4.3.2 Convergence and Runtime

We investigate convergence of the re-projection error and parameters for D-ADMM BA, comparing the convergence when ϕ_m is squared ℓ_2 vs. Huber in (7), and ϕ_a always the squared ℓ_2. The number of cameras is 5, the number of scene points is 10, and the number of 2D image points (observations) is 50. We fix the standard deviation for the additive Gaussian noise during the initialization of the camera angles and positions to be 0.1. We vary the standard deviation of noise for the

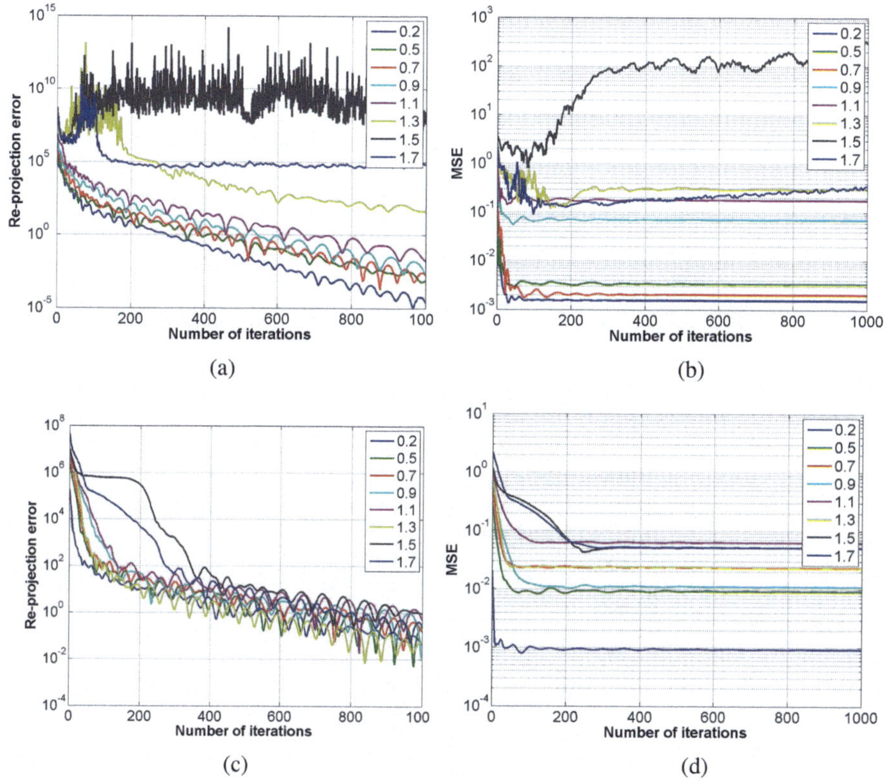

Fig. 4 Choosing ϕ_m loss to be Huber penalty leads to better performance in distributed BA, even when there are no outliers in the original data. Panels (**a**) and (**c**) compare re-projection errors, while (**b**) and (**d**) compare MSE of scene points. In all figures, curves correspond to values σ of scene variance, as shown in the legend. Consensus penalty ϕ_a is always ℓ_2. (**a**) Rep. errors: ϕ_m in (7) is ℓ_2. (**b**) MSE: ϕ_m in (7) is ℓ_2. (**c**) Rep. errors: ϕ_m in (7) is Huber. (**d**) MSE: ϕ_m in (7) is Huber

scene points from 0.2 to 1.7. Introducing robust losses for misfit penalty helps the convergence of the re-projection error significantly, see Fig. 4a vs. c. This behavior is observed with the convergence of the scene points, see Fig. 4b vs. d, and camera parameters. The Huber penalty is used to guard against outliers; here, outliers come from processors working with limited information. The performance degrades gracefully with noise, see Fig. 4c, d.

We also compare D-ADMM BA with the centralized LM BA and present the results in Fig. 5a, b. The number of camera parameters and 3D scene points are (10, 40), (15, 100), (25, 100), (30, 200), (100, 200), and (100, 250), with the number of observations increasing as shown in the x-axis of Fig. 5. In most settings, D-ADMM BA has a better parameter MSE than centralized LM BA. The runtime of the proposed approach with respect to the number of observations and parallel workers is shown in Fig. 5c. The parallel workers are configured in MATLAB, and

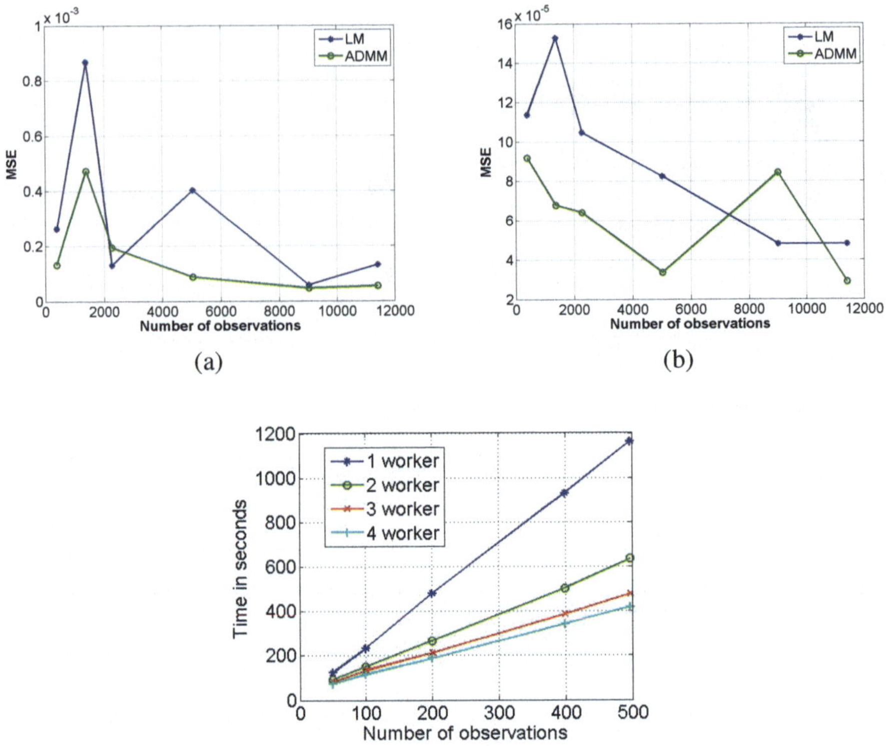

(c)

Fig. 5 (**a**) MSE between the actual and estimated camera parameters, (**b**) MSE between the actual and estimated scene points, (**c**) runtime of the proposed D-ADMM algorithm with increasing number of processor cores

the runtime is linear with respect to the observations and reduces with increasing workers. Our implementation is a simple demonstration of the capability of the algorithm—a fully parallel implementation in a fast language such as C can realize its full potential.

4.3.3 Real Data

To demonstrate the performance of D-ADMM BA, we conducted experiments on real datasets with different settings. All experiments are done with MATLAB on a PC with a 2.7 GHz CPU and 16 GB RAM.

In our SFM pipeline, SIFT feature points [23] are used for detection and matching. The relative fundamental matrices are estimated for each pair of images with sufficient corresponding points, which are used to estimate relative camera pose and 3D structure. Next, the relative parameters are used to generate the global initial

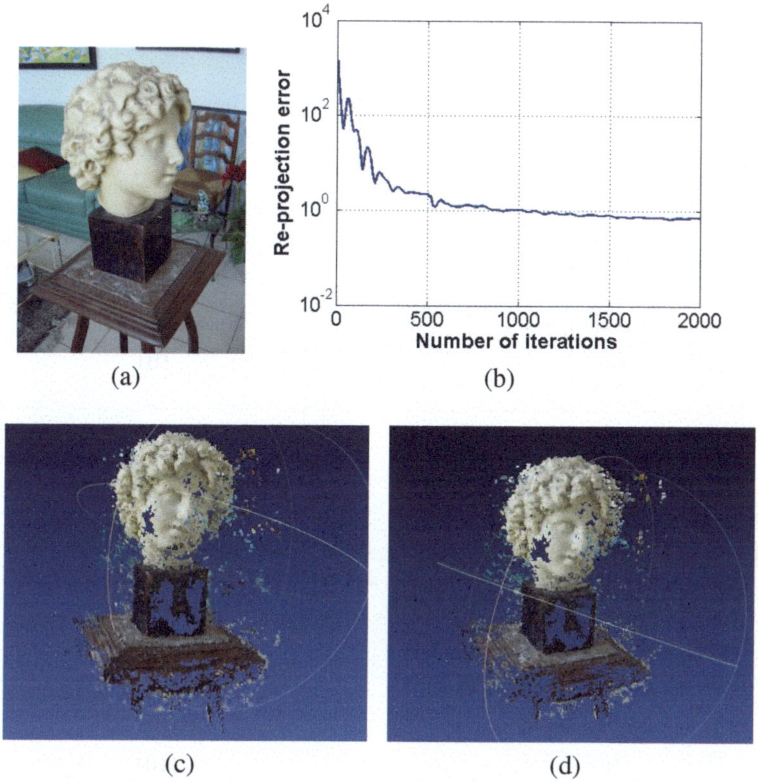

Fig. 6 (**a**) Original 2D image, (**b**) re-projection error for D-ADMM BA, (**c**) dense 3D point cloud estimated with LM BA (mean re-projection error = 0.93), (**d**) dense 3D point cloud estimated using D-ADMM BA (mean re-projection error = 0.67)

values for BA. The datasets were downloaded from the Princeton Vision Group and the EPFL Computer Vision Lab [33].

Since there are no ground truth 3D structures available for the real datasets, we compare the dense reconstruction results obtained using the method of [41]. The first dataset has five images and a sample image is shown in Fig. 6a. After keypoint detection and matching, centralized LM BA and D-ADMM BA are given the same input. There are a total of 104 world points and 252 observations. The final re-projection errors of LM and D-ADMM are 0.93 and 0.67, respectively. Figure 6c, d shows that the dense reconstruction quality of LM and the D-ADMM is similar. Figure 6b shows the convergence of re-projection error for the D-ADMM algorithm. Figure 7a shows the convergence of re-projection error for different values of $\rho = \rho_x = \rho_y$. Setting ρ to a high value accelerates convergence.

We also estimate camera parameters and scene points, applying the approach of Sect. 4.2.5 to the same dataset. Figure 7b shows that as the number of scene points per iteration increases, the runtime decreases, with 32 scene points per iteration

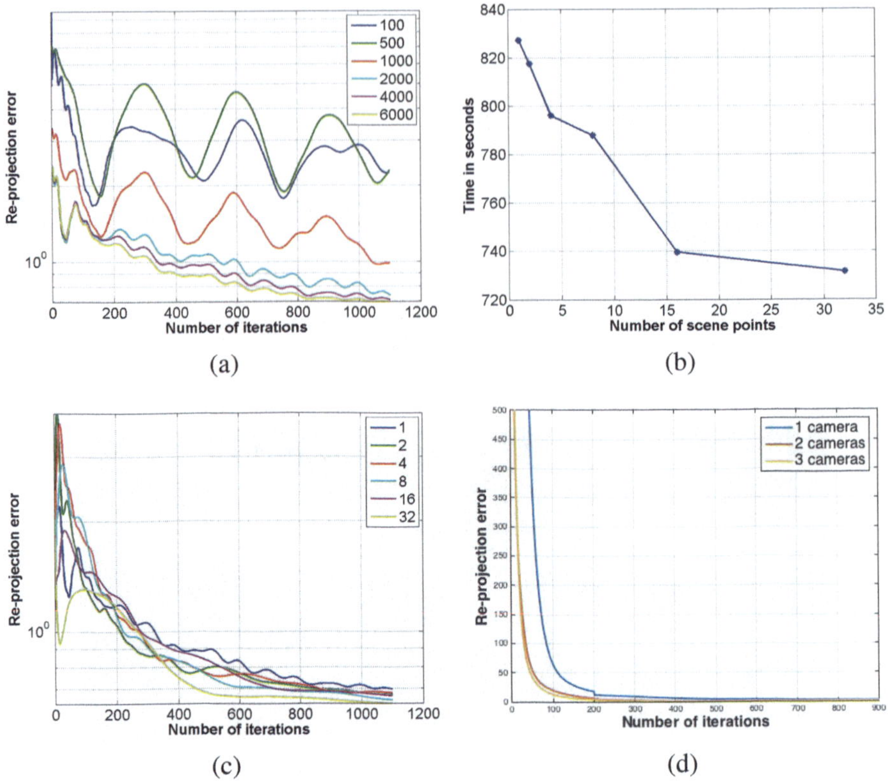

Fig. 7 (**a**) The re-projection error for different values of ρ, using generalized distribution approach, (**b**) runtime of D-ADMM BA, (**c**) re-projection errors with increasing number of scene points, (**d**) re-projection errors for multiple cameras per estimation vector

Fig. 8 (**a**) Original 2D image, (**b**) dense 3D point cloud estimated with D-ADMM BA (mean re-projection error = 0.76)

Fig. 9 Three reconstructed fountain-P11 views (11 images, 1346 world points, 3859 observations, mean re-projection error = 0.5)

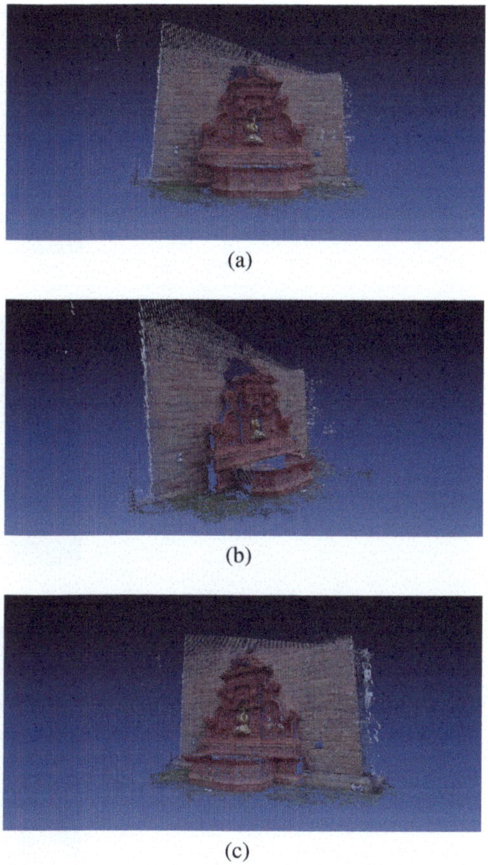

giving the fastest convergence, see Fig. 7c. Figure 7d compares re-projection errors with different number of cameras in each iteration. Initial values are the same as in the castle-P30 experiment (Fig. 12), and the number of scene points in each iteration is 64. Re-projection errors decrease faster as the number of cameras in each iteration increases.

We perform distributed BA on the Herz-Jesu dataset provided in [33] using the approach in Sect. 4.2.5. This dataset has seven images, 1140 world points, and 2993 observations. In this experiment, the LM BA algorithm using the same setting as in previous experiments does not converge and has the final re-projection error about 2500. Therefore, the dense reconstruction result is not presented. D-ADMM BA with eight scene points in each update step has a final re-projection error of 0.76. Figure 8b shows the dense 3D point cloud estimated with D-ADMM BA.

Additional results on other datasets (fountain-P11, entry-P10, Herz-Jesu-P25, and castle-P30) are presented in Table 1, Figs. 9, 10, 11, and 12. σ is mean re-projection error. Figures 9, 10, 11, and 12 present different perspectives of the dense reconstruction results to show the robustness of 3D parameter estimations.

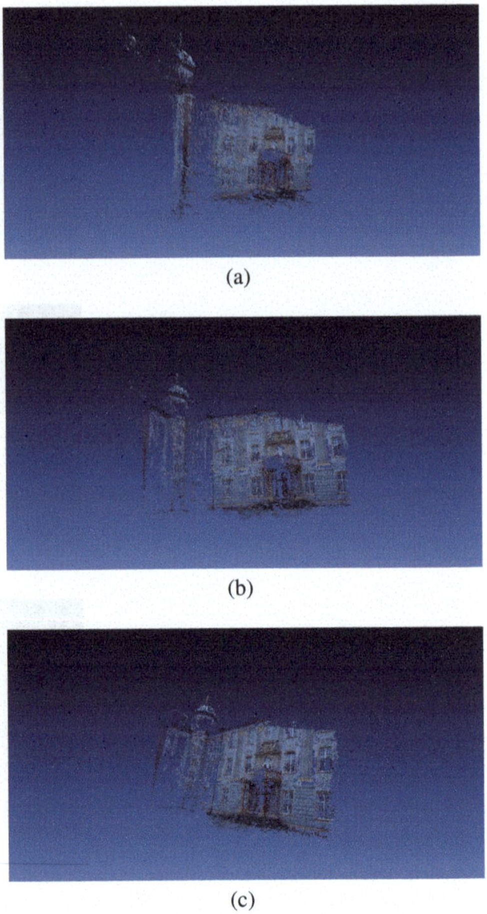

Fig. 10 Three reconstructed entry-P10 views (10 images, 1382 world points, 3687 observations, re-projection error = 0.7)

Table 1 The dataset information and experiment results

Dataset	Images	Scene pts	Obs	σ
Fountain-P11	11	1346	3859	0.5
Entry-P10	10	1382	3687	0.7
Herz-Jesu-P25	25	2161	5571	0.87
Castle-P30	30	2383	6453	0.84

Settings are fixed across experiments, and the maximum iteration counter is set to 1600. The experiments on fountain-P11 and Herz-Jesu-P25 dataset (Figs. 9 and 11) have better dense reconstruction results since there are more images covering the same regions. The real data experiments show D-ADMM BA achieves similar objective values (mean re-projection error < 1) as the number of observations increases; it is not necessary to increase the number of iterations as the size of the data increases. D-ADMM BA scales linearly with the number of observations and can be parallelized on GPU clusters.

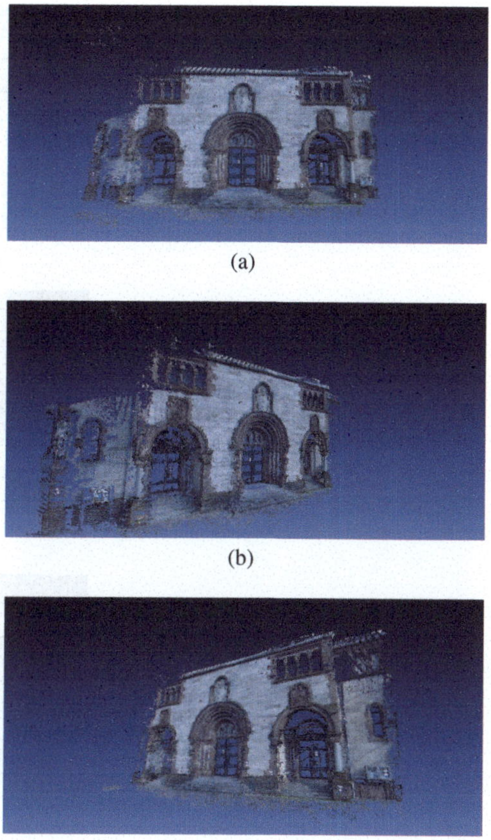

(a) (b) (c)

Fig. 11 Three reconstructed Herz-Jesu-P25 views (25 images, 2161 world points, 5571 observations, mean re-projection error = 0.87)

5 Conclusions

Although video analytics has been available for many years, the use of moving cameras has experienced a tremendous growth in recent years due to advances in mobile hardware and software. This phenomenon opens up new possibilities for mobile camera analytics and allows video analytics to expand into areas that were previously reserved for static cameras. That said, it also introduces technical challenges due to the vast amounts of data and metadata being generated.

We first presented a novel approach to summarize and visualize long videos from mobile cameras. Our method generates efficient representations of videos based on moving object detection and tracking while remaining robust in the face of abrupt camera motion. Experimental results on the challenging VIRAT aerial dataset demonstrated that our method can summarize the entire video by registering all the tracks in the 18,575 frames on a single panorama. This demonstrates that our system can accomplish a 10,000-fold data reduction without significant loss of events of interest.

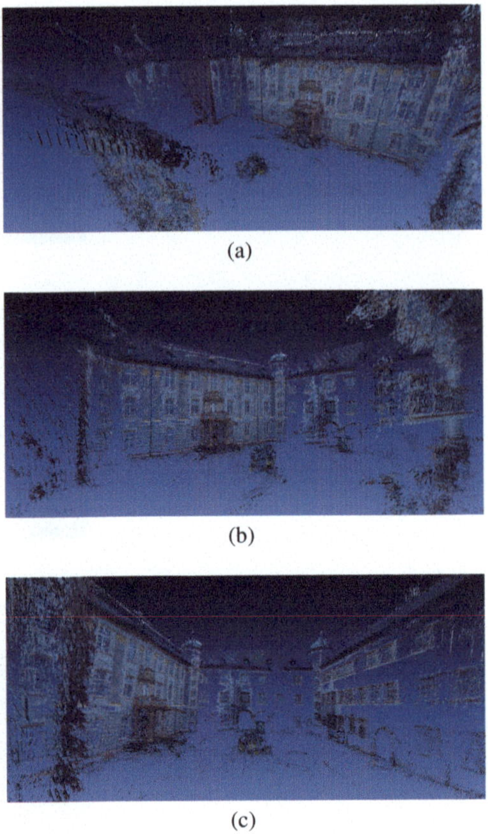

Fig. 12 Three reconstructed castle-P30 views (30 images, 2383 world points, 6453 observations, mean re-projection error = 0.84)

We then presented a new distribution algorithm for bundle adjustment, D-ADMM BA, which compares well to centralized approaches in terms of performance and scales well for SfM. Experimental results demonstrated the importance of robust formulations for improved convergence in the distributed setting. Even when there are no outliers in the initial data, robust losses are helpful because estimates of processors working with limited information can stray far from the aggregate estimates, see Fig. 4. Formulation design for distributed optimization may yield further improvements; this is an interesting direction for future work.

The results obtained with D-ADMM BA are comparable to those obtained with state-of-the-art centralized LM BA, and D-ADMM BA scales linearly in runtime with respect to the number of observations. Our approach is well-suited for use in a networked UAV system, where distributed computation is an essential requirement.

With more mobile video analytics capabilities, we can enhance applications in both individual and enterprise domains to save costs and labor by providing holistic situational awareness summaries. The implementation of our system proved that mobile devices are definitely suitable even for complex analytics applications such as the ones we described. Other domains that could benefit from this type of analytics include: shelf-monitoring, map and road update, parking spot management, inventory accuracy, large warehouse management, cart localization or supply, and spill and fall detection.

References

1. Agarwal S, Snavely N, Seitz SM, Szeliski R (2010) Bundle adjustment in the large. In: Computer Vision–ECCV 2010, Springer, pp 29–42
2. Aravkin A, Styer M, Moratto Z, Nefian A, Broxton M (2012) Student's t robust bundle adjustment algorithm. In: Image Processing (ICIP), 2012 19th IEEE International Conference on, IEEE, pp 1757–1760
3. Bhattacharyya A (1943) On a measure of divergence between two statistical populations defined by their probability distributions. Bull Calcutta Math Soc 35:99–109
4. Botchen RP, Bachthaler S, Schick F, Chen M, Mori G, Weiskopf D, Ertl T (2008) Action-based multifield video visualization. IEEE Transactions on Visualization and Computer Graphics 14(4):885–899
5. Boyd S, Parikh N, Chu E, Peleato B, Eckstein J (2011) Distributed optimization and statistical learning via the alternating direction method of multipliers. Foundations and Trends in Machine Learning 3(1):1–122
6. Byröd M, Åström K (2010) Conjugate gradient bundle adjustment. In: Computer Vision–ECCV 2010, Springer, pp 114–127
7. Comaniciu D, Ramesh V, Meer P (2000) Real-time tracking of non-rigid objects using mean shift. In: Computer Vision and Pattern Recognition, 2000. Proceedings. IEEE Conference on, IEEE, vol 2, pp 142–149
8. Eriksson A, Bastian J, Chin TJ, Isaksson M (2015) A consensus-based framework for distributed bundle adjustment. In: Computer Vision and Pattern Recognition, 2015. CVPR 2015. IEEE Conference on, IEEE
9. Feng S, Lei Z, Yi D, Li SZ (2012) Online content-aware video condensation. In: Computer Vision and Pattern Recognition (CVPR), 2012 IEEE Conference on, IEEE, pp 2082–2087
10. Fischler MA, Bolles RC (1981) Random sample consensus: a paradigm for model fitting with applications to image analysis and automated cartography. Communications of the ACM 24(6):381–395
11. Fusiello A, Crosilla F (2015) Solving bundle block adjustment by generalized anisotropic procrustes analysis. ISPRS Journal of Photogrammetry and Remote Sensing 102:209–221
12. Goldman D, Curless B, Salesin D, Seitz S (2006) Schematic storyboards for video editing and visualization. In: Proc. ACM SIGGRAPH, vol 25, pp 862–871
13. Hartley R, Zisserman A (2003) Multiple view geometry in computer vision. Cambridge university press
14. Heinly J, Schonberger JL, Dunn E, Frahm JM (2015) Reconstructing the world* in six days*(as captured by the yahoo 100 million image dataset). In: Proceedings of the IEEE Conference on Computer Vision and Pattern Recognition, pp 3287–3295
15. Indelman V, Roberts R, Beall C, Dellaert F (2012) Incremental light bundle adjustment. In: Proceedings of the British Machine Vision Conference (BMVC 2012), pp 3–7
16. Irani M, Anandan P (1999) About direct methods. In: International Workshop on Vision Algorithms, Springer, pp 267–277
17. Irani M, Anandan P, Hsu S (1995) Mosaic based representations of video sequences and their applications. In: Computer Vision, 1995. Proceedings., Fifth International Conference on, IEEE, pp 605–611

18. Konolige K, Garage W (2010) Sparse sparse bundle adjustment. In: BMVC, Citeseer, pp 1–11
19. Kopf J, Cohen MF, Szeliski R (2014) First-person hyper-lapse videos. ACM Transactions on Graphics (TOG) 33(4):78
20. Lin CC, Pankanti S, Ashour G, Porat D, Smith JR (2015) Moving camera analytics: Emerging scenarios, challenges, and applications. IBM Journal of Research and Development 59(2/3): 5–1
21. Lions PL, Mercier B (1979) Splitting algorithms for the sum of two nonlinear operators. SIAM Journal on Numerical Analysis 16(6):964–979
22. Lourakis MI, Argyros AA (2009) SBA: A software package for generic sparse bundle adjustment. ACM Transactions on Mathematical Software (TOMS) 36(1):2
23. Lowe DG (1999) Object recognition from local scale-invariant features. In: Computer vision, 1999. The proceedings of the seventh IEEE international conference on, Ieee, vol 2, pp 1150–1157
24. Lowe DG (2004) Distinctive image features from scale-invariant keypoints. International journal of computer vision 60(2):91–110
25. Mitra K, Chellappa R (2008) A scalable projective bundle adjustment algorithm using the l infinity norm. In: Computer Vision, Graphics & Image Processing, 2008. ICVGIP'08. Sixth Indian Conference on, IEEE, pp 79–86
26. Moré JJ (1978) The Levenberg-Marquardt algorithm: implementation and theory. In: Numerical analysis, Springer, pp 105–116
27. Nocedal J, Wright S (1999) Numerical optimization. Springer Series in Operations Research, Springer
28. Oh S, Hoogs A, Perera A, Cuntoor N, Chen CC, Lee JT, Mukherjee S, Aggarwal J, Lee H, Davis L, et al (2011) A large-scale benchmark dataset for event recognition in surveillance video. In: Computer vision and pattern recognition (CVPR), 2011 IEEE conference on, IEEE, pp 3153–3160
29. Pritch Y, Rav-Acha A, Peleg S (2008) Nonchronological video synopsis and indexing. IEEE Transactions on Pattern Analysis & Machine Intelligence (11):1971–1984
30. Pritt MD (2014) Fast orthorectified mosaics of thousands of aerial photographs from small UAVs. In: Applied Imagery Pattern Recognition Workshop (AIPR), 2014 IEEE, IEEE, pp 1–8
31. Ramamurthy KN, Lin CC, Aravkin AY, Pankanti S, Viguier R (2017) Distributed bundle adjustment. In: ICCV Workshops, pp 2146–2154
32. Rockafellar RT, Wets RJB (2009) Variational analysis, vol 317. Springer Science & Business Media
33. Strecha C, von Hansen W, Gool LV, Fua P, Thoennessen U (2008) On benchmarking camera calibration and multi-view stereo for high resolution imagery. In: Computer Vision and Pattern Recognition, 2008. CVPR 2008. IEEE Conference on, Ieee, pp 1–8
34. Tomasi C, Kanade T (1991) Detection and tracking of point features. Tech. rep., International Journal of Computer Vision
35. Triggs B, McLauchlan PF, Hartley RI, Fitzgibbon AW (2000) Bundle adjustment - a modern synthesis. In: Vision algorithms: theory and practice, Springer, pp 298–372
36. Trinh H, Li J, Miyazawa S, Moreno J, Pankanti S (2012) Efficient UAV video event summarization. In: Pattern Recognition (ICPR), 2012 21st International Conference on, IEEE, pp 2226–2229
37. Tuytelaars T, Van Gool L (2004) Matching widely separated views based on affine invariant regions. International journal of computer vision 59(1):61–85
38. Wang Y, Yin W, Zeng J (2015) Global convergence of ADMM in nonconvex nonsmooth optimization. arXiv preprint arXiv:151106324
39. Wu C (2013) Towards linear-time incremental structure from motion. In: 3D Vision-3DV 2013, 2013 International Conference on, IEEE, pp 127–134
40. Wu C, Agarwal S, Curless B, Seitz SM (2011) Multicore bundle adjustment. In: Computer Vision and Pattern Recognition (CVPR), 2011 IEEE Conference on, IEEE, pp 3057–3064
41. Xiao J, Chen J, Yeung DY, Quan L (2008) Learning two-view stereo matching. In: Computer Vision–ECCV 2008, Springer, pp 15–27

42. Yilmaz A, Javed O, Shah M (2006) Object tracking: A survey. ACM computing surveys (CSUR) 38(4):13
43. Zamir AR, Dehghan A, Shah M (2012) GMCP-tracker: Global multi-object tracking using generalized minimum clique graphs. In: Computer Vision–ECCV 2012, Springer, pp 343–356
44. Zhang J, Boutin M, Aliaga DG (2006) Robust bundle adjustment for structure from motion. In: Image Processing, 2006 IEEE International Conference on, IEEE, pp 2185–2188

Code Compression for Embedded Systems

Chang Hong Lin, Wei-Jhih Wang, Jui-Chun Chen, and Che-Wei Lin

Abstract Embedded systems are constrained by the available memory, and code compression techniques address this issue by reducing the code size of application programs. The main challenge for the development of an effective code compression technique is to reduce the code size without affecting the overall system performance. Code compression traditionally works on fixed-sized blocks with its efficiency limited by their small size. A new methodology, branch block, which is a series of instructions between two consecutive possible branch targets, provides larger blocks for code compression. Moreover, dictionary-based code compression schemes are the most commonly used ones, because they can provide both good compression ratio and fast decompression. In this chapter, several branch-block based methods, as well as new dictionary-based code compression methods are presented. These methods can achieve a good compression ratio (CR) (the compressed code size divided by original code size), with little or no hardware overheads.

Preface

When Shuvra asked me to write a chapter in the festschrift to pay tribute to Marilyn's 60th birthday, the first topic came to my mind is "code compression." When I joined Marilyn's group at Princeton University in 2002, the first project she assigned me was "code compression," and the results of this project were published in several papers as well. Even though I moved to the Smart Camera project later on, and focused more in multimedia applications after I joined National Taiwan University of Science and Technology, "code compression" was still a topic my students and I worked on and off occasionally. This chapter would be the summary of the four journal papers [1–4] and three conference papers [5–7] we published under this topic in the past decade. Hopefully, it would show a tip of the iceberg of Marilyn's exceptional contributions in guiding her students.

C. H. Lin (✉) · W.-J. Wang · J.-C. Chen · C.-W. Lin
National Taiwan University of Science and Technology, Taipei City, Taiwan
e-mail: chlin@mail.ntust.edu.tw

S. S. Bhattacharyya et al. (eds.), *Embedded, Cyber-Physical, and IoT Systems*,
https://doi.org/10.1007/978-3-030-16949-7_6

1 Introduction

Embedded systems have become an essential part of everyday life in the past decade as almost all electronic devices contain them. The complexity and performance requirements for embedded systems grow rapidly as system-on-chip (SoC) architecture becomes the trend. Embedded systems are cost and power sensitive, and their memory systems often occupy a substantial portion of chip area and system cost. Program size tends to grow as applications become more and more complex, and even for the same application, program size grows as RISC (reduced instruction set computer), superscalar, or VLIW (very long instruction word) architectures are used. Pure software techniques, such as constant propagation, procedural abstraction, or cross jumping, may help us in reducing the program size; however, powerful compilers are needed and these techniques may have negative effect on performance. Code compression is proposed as another solution to reduce program size and in turn to reduce the memory usage in embedded systems. It refers to compressing program codes offline and decompressing them on-the-fly during execution. This idea was first proposed by Wolfe and Chanin in the early 1990s [8], and much research has been done to reduce the code size for RISC machines [9–14]. As instruction level parallelism (ILP) becomes common in modern SoC architectures, a high-bandwidth instruction-fetch mechanism is required to supply multiple instructions per cycle. Under these circumstances, reducing code size and providing fast decompression speed are both critical challenges when applying code compression to modern embedded machines.

Furthermore, code compression methods have to be lossless; otherwise, the decompressed instructions will not be the same as the original program. Since a decompression engine is needed to decompress code during runtime, the decompression overhead has to be tolerable. Unlike text compression, compressed programs have to ensure random accesses, since execution flow may be altered by branch, jump, or call instructions. The compressed blocks may not be byte aligned, so additional padding bits are needed after compressed blocks when bit addressable memory is not available. Existing code compression methods use small, equally sized blocks as basic compression units; each block can be decompressed independently with or without small amounts of information from others. When the execution flow changes, decompression can restart at the new position with or without little penalty.

Dictionary-based code compressions (DCC) [10] are commonly used in embedded systems, because they are quite effective and pose a relatively simple decoding hardware, and provide a higher decompression bandwidth than the code compression by applying lossless data compression methods. Thus, it is suitable for architectures with high-bandwidth instruction-fetch requirements, such as VLIW processors. Although several existing code compression algorithms have exhibited favorable compression performance, no single compression algorithm has efficiently worked for all kinds of benchmarks.

In the following sections, branch blocks are defined as the instructions between two consecutive possible branch targets and use them as the basic compression units. Moreover, a branch block may contain several basic blocks in the control flow graph representation. Since the size of branch blocks is much larger than blocks previously used, there would be more freedom in choosing compression algorithms. After that, various steps in the code compression process were combined into new algorithms to improve the compression performance with a smaller hardware overhead. Based on the BitMask code compression (BCC) algorithm [15, 16], an exclusive-or (XOR) operator unit is used to improve the variety of bit-masked code; a small separated dictionary is proposed to restrict the codeword length of high-frequency instructions; a dictionary selection algorithm is proposed to achieve more satisfactory instruction selection, which in turn may reduce the average CR. Furthermore, the fully separated dictionary architecture is proposed to improve the performance of the dictionary-based decompression engine.

2 Previous Work

Wolfe and Chanin were the first to apply code compression to embedded systems [8]. Their compressed code RISC processor (CCRP) uses Huffman coding to compress MIPS programs. A line access table (LAT) is used to map compressed block addresses, and is inherited by most of the later studies. Based on the same concept, IBM built a decompression core, called CodePack, for the PowerPC 400 series [11]. Compressed code is stored in the external memory, and CodePack is placed between the memory and cache. Liao et al. [9] replaced frequently used instruction groups as dictionary entries, which make compressed code easy to be decoded. Lekatsas and Wolf proposed SAMC [12], a statistical scheme based on arithmetic coding in combination with a pre-calculated Markov model. Netto et al. proposed a DCC using variable length indices [13]. Instructions in the decoding table are selected based on both static and dynamic profiling to decrease code size and increase performance. Benini et al. used a DCC method formed by using both static and dynamic entropy [14]. Their schemes are competitive with CodePack for footprint compression, and achieve superior bus traffic and energy reduction. All of these methods focused on RISC and CISC architecture code.

Another approach to decrease code size for RISC processors is to define a dense instruction set using a limited number of short instructions, and has been implemented in several commercial core processors, such as Thumb [17], MIPS16 [18], Tensilica Xtensa [19], and ARCompact [20]. The dense instruction sets often cause performance penalties due to lack of instructions, and require modifications to the processor core and software development tools. This approach is not suitable for embedded systems with hard intellectual-property (IP) processors.

Other than using hardware components to decompress code on-the-fly, several groups proposed pure software techniques to reduce program size and decompress instructions during execution. Cooper and McIntosh developed compiler techniques for code compression on RISC architectures [21]. They map isomorphic instruction sequences into abstract routine calls or cross jumping. Debray and Evans proposed a profile-guided code compression to apply Huffman coding to infrequently executed functions [22, 23]. Ozturk et al. proposed a control flow graph centric software approach to reduce memory space consumption [24]. Application binaries are on-the-fly compressed/decompressed using separated thread in their approach. Shogan and Childers implemented IBM's CodePack algorithms within the fetch step of software dynamic translator (SDT) in pure software infrastructure [25]. Such approach provides a flexible decompressor and can be applied to multiple platforms.

Lefurgy et al. [10] proposed the first DCC algorithm, which replaced frequently executed instructions as dictionary indices. Gorjiara et al. [26] used DCC with a multi-dictionary for a no instruction set computer (NISC) architecture. Ros and Sutton [27] proposed improved DCC methods by considering Hamming distances and mismatches. Based on the DCC, Thuresson and Stenstrom [28] combined dynamic instruction stream editing and BitMask methods to compress instruction sequences. Seong and Mishra [15, 16] used several bits as a mask for increasing the instruction coverage rate, and they proposed a novel dictionary selection method to improve the CR. Qin et al. [29] combined the BCC and run-length coding with an improved dictionary selection method for field-programmable gate array bitstreams. Murthy and Mishra [30] used a skip map with a multi-dictionary for the NISC architecture. Bonny and Henkel [31] used dictionary-based and canonical Huffman coding to re-encode the codewords compressed by Huffman coding in embedded processors. Both instructions and lookup tables (LUTs) are compressed to achieve an optimal CR. Based on the same method, Ranjith et al. [32] applied the code compression in a delta-sigma control-system processor to reduce the memory cost and optimize power consumption in the processor. Based on the BCC, Chen et al. [33] used dictionary-entry replacement algorithm to reduce the power consumption of the systems. Azevedo Dias et al. [34] used Huffman coding to compress two adjacent instruction sequences and then used the same method to compress single instructions, which is called compressed code using Huffman-based multilevel dictionary. They also design a one instruction per cycle decompression engine.

Recent research in code compression has focused on two directions: (1) applying existing compression methods to various architectures for optimization and (2) combining several approaches to improve the performance, including CR. Seong and Mishra [15, 16] and Wang and Lin [7] observed that no single compression algorithm operated efficiently for all the benchmarks. Thus, the following sections integrate several approaches to form a new algorithm with smaller hardware overhead. New dictionary architecture is used to improve the decompression engine performance.

3 Code Compression with a Self-Generating Dictionary

Even though control flow can change due to the existence of branch instructions or function calls, and the destination addresses may be calculated during execution time, not all instructions will be the destinations of these flow changes. The possible branch targets (or function calls and returns) are determined once the program is compiled. A program can be broken into procedure calls, and the compiler expresses each procedure as a control flow graph (CFG), as illustrated in Fig. 1a. Each node in the CFG represents a *basic block*, which is a straight-line piece of code without any branch instructions or branch targets. Each basic block starts at a branch-target instruction, and ends at a branch or jump instruction. A CFG is a static representation of the program, and represents all the alternatives of possible execution paths.

Branch blocks, constructed from basic blocks in CFGs, are used as the basic components in code compression to ensure random accesses to all branch blocks. A basic block is considered as a branch target if there is at least one entry with address change; on the other hand, a basic block is not a branch target if it can only be visited through sequential execution. A branch block is defined as a series of basic blocks in a CFG with the only branch target located at its first basic block. The construction of branch blocks depends not only on the CFG, but also on the memory allocation of each basic block. As shown in Fig. 1b, only basic blocks *B7* and *B8* can be combined into a larger branch block, while other branch blocks contain only one

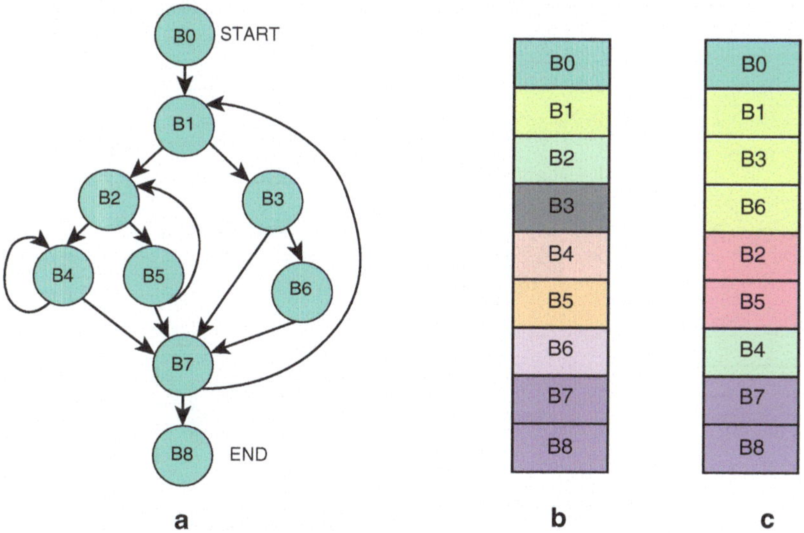

Fig. 1 (**a**) An example control flow graph (CFG) fragment; (**b**) simple and (**c**) optimized memory allocation for basic blocks. Adjacent basic blocks that can be combined together as a single branch block are marked with the same color

basic block in this simple memory allocation, and it results in eight total branch blocks. An optimized memory allocation is shown in Fig. 1c, and the basic blocks can be combined into five larger branch blocks. Various compiler techniques can be applied to optimize different control flow criteria, such as number or size of branch blocks, based on global data flow analysis.

Compressed blocks will not have the same address as their original ones. When execution flow changes, the target block address will be the uncompressed one, which does not correspond to the same location in the compressed code. Wolfe and Chanin's idea is borrowed to use a line address table (LAT) to map original program addresses into compressed program addresses [8]. Instead of storing the addresses of all cache lines as in most previous work, only addresses of branch targets are needed for the proposed approach, which can produce a much smaller LAT.

Source programs are compressed offline and stored in memory systems, either ROM or hard drives, for embedded systems using code compression. The codes are then decompressed on-the-fly when the branch blocks are needed. The dictionary used for the proposed approaches is self-generated during runtime, either compression or decompression, and is stored in memory systems. The dictionaries are reset when branch targets are met during execution. As illustrated in Fig. 2, the decompression engine can be placed in two possible configurations, *pre-cache* or *post-cache*. For pre-cache structure, the timing overhead for decompression can be hidden behind cache miss penalty, while post-cache has more area and power savings. When more than one level of caches is used, the closer the decompression unit to the processor, the larger the power and area saving for the memory systems; however, it also means the more critical impact to the system performance the decompression core has. The proposed methods can work on both pre- and

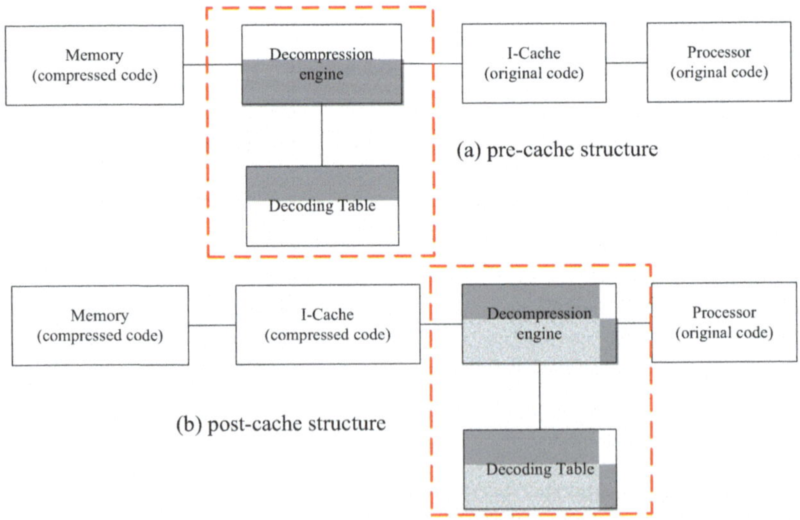

Fig. 2 (**a**) Pre-cache and (**b**) post-cache decompression architectures

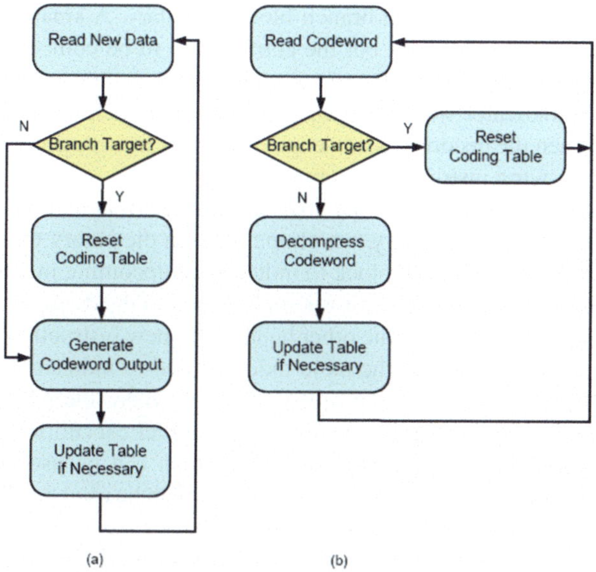

Fig. 3 Flow chart of the proposed approach: (**a**) compression and (**b**) decompression

post-cache structures: post-cache structure would get more benefit when parallel decompression is used due to larger decompression bandwidth, while pre-cache architecture fits better for sequential decompression.

The proposed code compression schemes use self and runtime generated dictionaries that can be used for both compression and decompression, so there would be no additional space needed to store dictionaries along with the compressed code. Figure 3 illustrates the flowchart of the proposed approaches. In both compression and decompression phases, the coding dictionary is reset if the incoming instruction or codeword represents a branch target; otherwise, the compression and decompression engines keep on using the existing dictionary to generate codeword or instruction outputs and update the dictionary when necessary.

Compression ratio (CR) is often used as a merit to measure the efficiency of code compression schemes, and is defined as

$$\mathrm{CR} = \frac{\text{Compressed Code Size} + \text{Stored Dictionary Size}}{\text{Original Code Size}}.$$

The stored dictionary size includes both coding dictionaries and LATs. For the proposed approaches, the dictionaries are not included in the compression ratio because they are self-generated during runtime and not stored in the memory systems. Moreover, the size of LATs is inversely proportional to the average size of basic blocks, so the use of branch blocks would reduce the size of LATs as well.

When programs are running, there will be no problem to identify branch targets if execution flow changes. However, a way to distinguish branch targets from regular instructions is needed when incrementing the program counter (PC) causes the execution to cross a branch block boundary. A small list of all the branch targets may be maintained, but the entries have to be compared every time an instruction is executed. The other way is to use a codeword as a branch-target indicator. An indicator would be generated before branch targets during compression phase. When the decompression engine sees a branch-target indicator, it will know that the following instruction is a branch target.

The first example that takes advantages of branch blocks is the DCC with a self-generating dictionary, which maintains a dictionary of previously seen instructions in the same branch block to indicate the upcoming instructions. If a new instruction is not inside the table, it is simply added to the dictionary, and the instruction is left uncompressed. On the other hand, if the new instruction is already in the dictionary, an escape tag plus the table index is used to indicate the instruction. For example, there is a 6-bit unused prefix (*1100sp*, where the s-bit determines the destination register file and p-bit serves as the parallel execution indicator) in the 32-bit wide TMS320C6x instruction set [35], which can be used to identify the compressed dictionary indices. This simple scheme can be used on all the processor architectures with unused instruction prefixes.

The compression process starts from the beginning of a source program file, and deals with the instructions sequentially with an empty initial dictionary. After the compression engine fetches a new instruction from the source file, the engine first compares it with all the entries in the dictionary. If the instruction is already in the dictionary, the engine generates the escape tag along with the dictionary index that contains the instruction as the output. Otherwise, the instruction is left uncompressed and is added as a new dictionary entry. Once the coding table is full, there will be two possible actions: (1) the dictionary can be simply left unchanged once it is full, or (2) the entries are replaced using round-robin, least recently used, or other complicated replacement algorithms. According to the experiments, the difference in the compression ratio is <1% for different replacement methods when a 1024-entry dictionary is used for TMS320C6x benchmarks.

During the decompression phase, the decompression engine reads in an instruction width from the main memory, and checks if there is a match with the escape tag. If it does not match, the instruction is bypassed to the decompression core output and added to the next available dictionary entry, and then the next instruction is fetched from the memory. On the other hand, if the escape tag exists, the decompressor would first check if the index matches the exception code for branch targets. If it matches, the decompression core will clear the dictionary, shift out the buffer along with the padded bits, and read the next instruction from the memory. Otherwise, the decompression engine will output the instruction stored in the dictionary indicated by the index, shift out the escape tag and index, and read the next instruction. During runtime, there are two ways to encounter branch targets. When branch targets are met sequentially, the decompression core will fetch the exception index and shift out the padding bits, while an execution flow change implies the existence of a branch target. In both cases, the decompression engine can clear the decoding dictionary and restart decompression at the byte-aligned location without error.

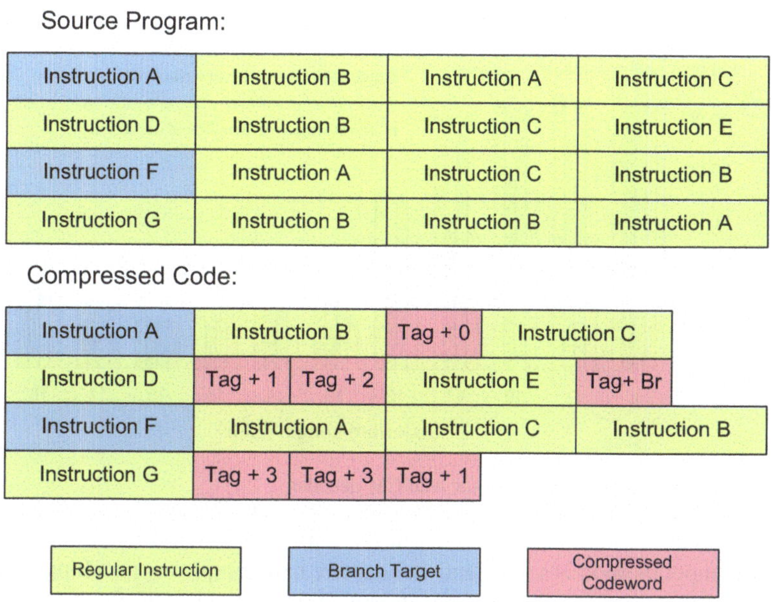

Fig. 4 Example of dictionary-based code compression

Figure 4 shows an example of DCC. The dictionary is initially empty, and is updated when a new instruction is encountered. Instructions A and B are put in entries 0 and 1, respectively, and when instruction A is met again, the compression engine sends out the escape tag with index 0. After eight instructions, the dictionary contains instructions A, B, C, D, and E. Since the next instruction input is a branch target, the engine generates the exception code (tag + branch indicator) along with the padding bits. After that, the engine restarts the compression procedure with an empty dictionary.

Figure 5 illustrates the impacts on compression ratio for different codeword lengths, replacement policies, and branch-target identification approaches using an ADPCM (adaptive differential pulse code modulation) decoder as the code compression example. Replacing the entries using the round-robin policy outperforms fixed dictionaries until saturation; however, the difference is <2% for codewords longer than 14-bit (dictionaries with more than 256 entries). The compression ratio using branch-target indicators is around 1% worse than using branch list (including the list) with better decompression performance. Codeword length determines the size of the dictionary exponentially. When the dictionary is too small, there is little or no benefit using code compression. When the number of entries is more than the number of identical instructions, increasing codeword length will increase the compressed code size. DCC achieves a better compression ratio for dictionaries with size around the number of identical instructions (codeword 14–16 bits wide). Since a byte-aligned codeword simplifies the decompression core implementation, 16-bit codeword is the best choice for DCC.

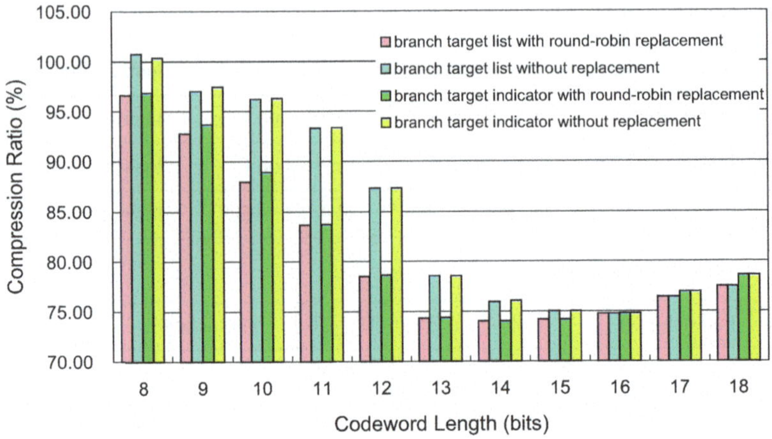

Fig. 5 Dictionary-based compression for ADPCM decoder

Ziv–Lempel method is a well-known adaptive text compression technique which uses previously seen data to compress the upcoming one [36]. The dictionary need not to be stored along with the compressed file, and can be regenerated on-the-fly during decompression. The LZ family is used to be considered as a method without random accessibility and with poor performance when dealing with small blocks of data. The use of larger branch blocks makes it possible to take advantage of the well-compressed and fast-decompressing LZW method [37], and apply it to code compression. Lempel–Ziv–Welch (LZW) compression was modified from Ziv–Lempel 78 [36] by Welch in 1984 [37], which replaces a series of basic elements with a single codeword. The LZW dictionary reserves all the possible elements as its initial codewords, and generates a new codeword to represent phrases (series of elements) during runtime. During the compression phase, the compression engine will search for the longest phrase already in the dictionary that matches the input stream, output the codeword that represents the phrase, and add the phrase along with the next element in the input stream as a new dictionary entry.

To apply LZW to code compression, *byte* can be used as the basic element. Since the compressed output of the LZW method only contains compressed codeword, and all the possible elements have to be included in the entire initial dictionary, the codeword length has to be long enough to contain the initial dictionary. In short, the chosen codeword length should be at least 9-bit wide, and the initial dictionary should contain 256 entries. A new entry will be generated per iteration till the dictionary becomes full. During the compression phase, the compressor will find the longest phrase in the dictionary that matches the code stream, send the codeword to the output, and add the phrase with the next byte in the code stream as a new entry. Once the dictionary is full, the compressor will keep on using the existing table to compress the upcoming program stream. When a codeword is read from the memory, the decompression core first checks if it is a branch target. If yes, the

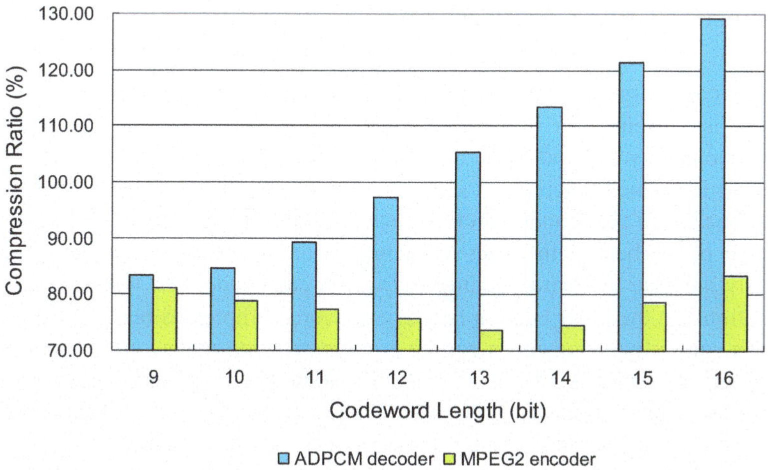

Fig. 6 LZW-based compression for ADPCM decoder and MPEG2 encoder

engine will shift out the padding bits from the buffer, reset the dictionary, and restart decompression at a byte-aligned position. Otherwise, the decompression core will get a codeword, look it up, output the content, and add the previous phrase with the first element of the current phrase as a new entry.

Codeword length and decoding bandwidth are two important parameters in our methods. LZW-based compression is a variable-to-fixed method. A fixed length codeword is used to represent variable length phrases. The codeword has to be at least 9-bit long, and determines the dictionary size exponentially. The larger the dictionary, the more phrases can be represented as codewords, which yields better compression result. However, a longer codeword is used to represent 8-bit basic elements. In the worst case, if none or only a few repeated phrases occur within a branch block, one-element codewords may be used most of the time. This often happens when the source file is not big enough. The other adjustable feature is the decompression bandwidth. From the simulation results, the compression ratio differs within 1% for widths from 8 to 20 bytes. Since the size of the dictionary is linearly dependent on the decoding bandwidth, 8-byte wide decoding table will be the desired choice. Figure 6 shows the compression ratio of LZW-based code compression by using different codeword length on two example benchmarks. For smaller benchmarks such as an ADPCM decoder, the longer the codeword, the worse the compression ratio. The benefit for a larger coding cable can be seen on larger benchmarks such as an MPEG2 encoder.

4 Selective Code Compression

Selective compression, a modified algorithm based on LZW-based code compression, is presented in this section, which may have a better CR and even higher decompression bandwidth. The dictionaries generated by different codeword lengths for the same branch block share exactly the same entries in the front parts of the dictionaries. If the branch block is too small to fill the smallest 9-bit dictionary, there will be no benefit in using more bits to encode this block. The more bits used only increase the extra padding in front of each codeword without additional information. On the other hand, a longer codeword compresses better for the larger blocks, as more byte-sequences can be compressed into codewords. According to the benchmarks for TMS320C6x, only 12.8% of the branch blocks can use up all the entries in the 9-bit dictionary, while only 1% of them can fill up the 12-bit dictionary. Statistically, more than 84% of the branch blocks in the benchmarks are <256 bytes; and more than 60% of the blocks are <100 bytes. This gives the inspiration to apply different compression methods to different branch blocks.

Two selective code compression schemes based on the LZW-based code compression are proposed. The basic idea of selective compression is to apply different compression methods to different branch blocks according to the block profile. As shown in Fig. 7, block sizes, instruction and execution frequency, and other information of each branch block are collected during profiling phase. Then the compression method for each branch block is determined based on the profile. Each branch block may choose a different code compression method.

The first selective compression method proposed is *minimum table-usage selective compression* (MTUSC), whose basic concept is to minimize the size of dictionary used for each branch block. Since no single codeword length can provide good compression results in all the branch blocks, the number of generated phrases is calculated for each branch block in the profiling phase. The shortest codeword length for a certain branch block is selected such that all the phrases generated by it can fit into its dictionary. For example, when the branch block generates <256 phrases, only 9-bit LZW compression is needed; 10-bit codeword is used for more than 256 but not exceeding 767 phrases; and so on. As a longer codeword is used, the size of the dictionary grows exponentially. Experiments show that only a few branch blocks really need larger dictionary and the number of branch blocks decreases as larger dictionary is used. So 12-bit is chosen as the maximum codeword length for selective code compression schemes.

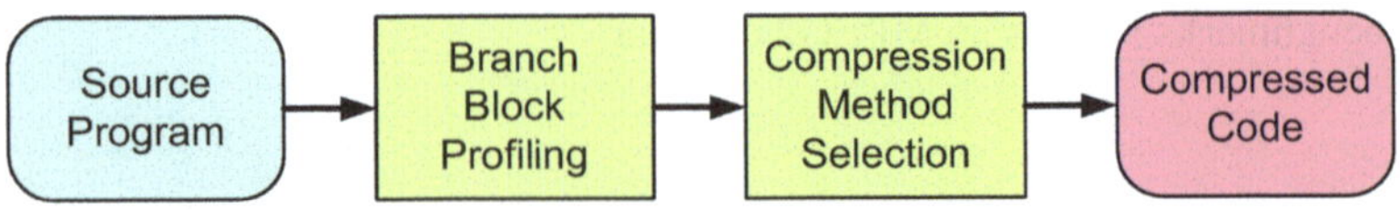

Fig. 7 Block diagram of selective code compression

One major drawback for Lempel–Ziv compression family is its poor performance on small blocks where only a few repeated phrases exist. Sometimes, the size of the compressed block is even larger than the original code when MTUSC is used. *Minimum code-size selective compression* (MCSSC) is proposed to alleviate this problem. Each branch block is first compressed using LZW-based code compression in different codeword lengths in the profiling phase, and the smallest compressed code (including the uncompressed original code) is then chosen for the block. By doing so, it is ensured that each branch block has the minimum code size when the LZW-based code compression is applied.

When the block is left uncompressed, there is no codeword to indicate the existence of branch targets. Experimental results show that only blocks with size 32, 64, or 96 bytes may be left uncompressed in MCSSC. These three situations are encoded along with the compression methods into a 3-bit header in front of each branch block. The average compression ratio for MCSSC is 76.8%, which is about 6.3% better than the 9-bit LZW-based code compression.

Although four different codeword lengths may be used in selective compression, only a single 12-bit LZW decompression core with dispatching logic is enough for all the codeword lengths. And the unused memory banks can be turned off when a shorter codeword is used. When a branch indicator is met, the coding table will be reinitialized, and the dispatching logic will reconfigure to the upcoming method based on the selection header. Otherwise, the decompression engine just performs its normal operations. For a codeword <12 bits, zeros would be padded in the front of the codewords. The decompression core will use those padded codewords to address the coding table. It also works the same way when dynamic LZW is applied. The only difference is the dispatching logic that has to count the number of incoming codewords, and change the padding when necessary for dynamic LZW compression. On the other hand, the dispatching logic will bypass the instructions directly when the branch block contains only uncompressed instructions for MCSSC.

In the following paragraphs, several experimental results on benchmarks for Texas Instruments' TMS320C6x VLIW processors are presented. The benchmarks are collected from *Texas Instruments* and *MediaBench*, which are general embedded system applications with strong digital signal processing components. The benchmarks are compiled using *Code Composer Studio IDE* from Texas Instruments.

Figure 8 shows the CR for all the benchmarks using 14–18-bit DCC, with round-robin as the replacement policy. For smaller benchmarks with limited instruction reoccurrence, the longer the codeword length, the worse the CR. For larger benchmarks, using medium length codeword (15 or 16 bits) results in better CR. Even though a longer codeword means more instructions can be represented as compressed codewords, not all branch blocks can take advantage of the larger dictionary. For small benchmarks and small branch blocks in large benchmarks, longer codewords only represent extra bits to represent the same instructions. Only larger branch blocks in large benchmarks have more repeated instructions and take advantage of the larger dictionary. These two effects compensate each other and result in poor performance for a longer codeword in small benchmarks, and larger

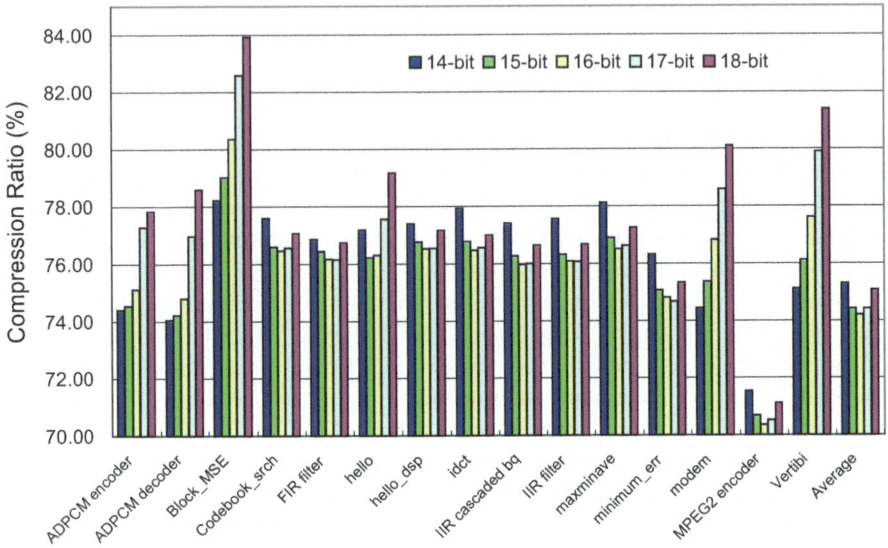

Fig. 8 CR for 14–18-bit DCC

benchmarks have better performance when a medium length codeword is used. A 16-bit codeword not only has the best average CR at 74.2%, but also has the simplest decompression engine due to byte-aligned codewords.

Figure 9 illustrates the CR for all benchmarks using 9–12-bit LZW-based code compression. A longer codeword performs worse in most of the benchmarks. However, for the huge files such as MPEG2 encoder, 12-bit LZW has better CR. Unlike DCC that guarantees smaller compressed programs, LZW-based compression may result in compressed code larger than the original program. Since LZW-based code compression generates only one new codeword per iteration, the front part of dictionaries contains the same phrases for different codeword length. For benchmarks that cannot fill up a small dictionary, longer codeword won't benefit. The effect of a large dictionary can only be seen on huge benchmarks. On average, the CRs for 9–12-bit LZW-based code compression are 83.4, 83.3, 84.8, and 87.9%, respectively.

Figure 10 summarizes the CR using selective code compression (MTUSC and MCSSC), both with and without dynamic LZW. Since MCSSC chooses the compression method with the smallest code size for each branch block, while MTUSC uses the smallest dictionary for each block, it is clear that MCSSC can achieve a better CR than MTUSC. Dynamic LZW can reduce the compressed code size for both selective methods. Among all four selective schemes, MTUSC is always the worst, and dynamic MCSSC is always the best in terms of CR. On average, dynamic MCSSC can achieve compression ratio at 75.6%.

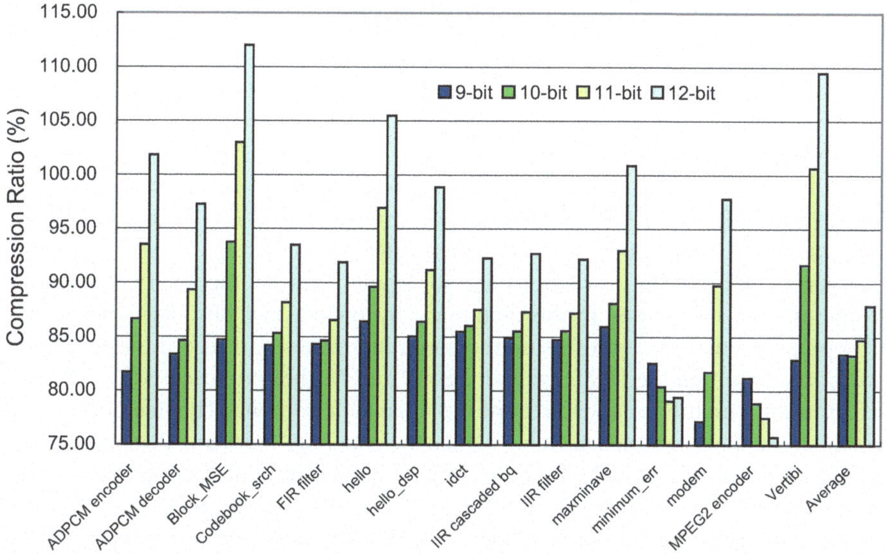

Fig. 9 CR for 9–12-bit LZW-based compression

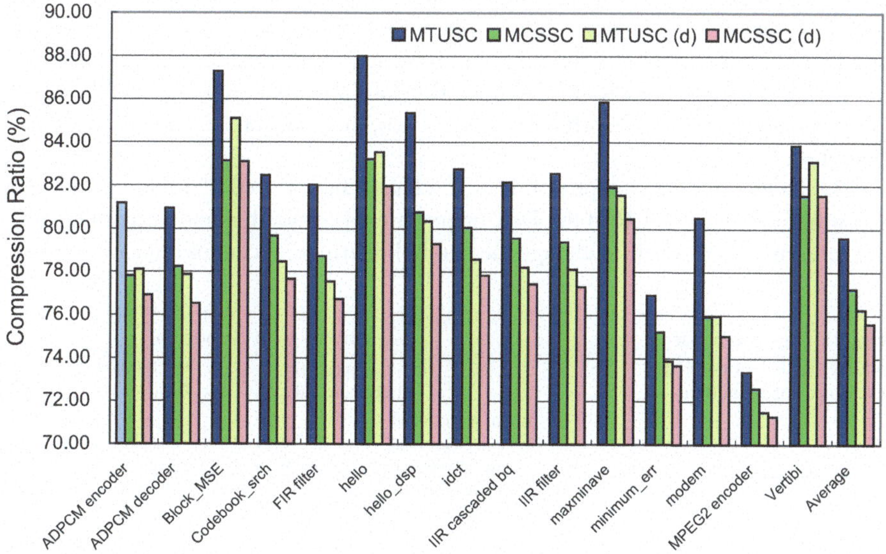

Fig. 10 CR for selective compression

5 XOR-Referenced Code Compression

Lefurgy et al. [2] proposed the first dictionary-based method in 1999. Although the dictionary-based methods result in simpler decompression engines, their CRs are usually less efficient than those of other entropy-based compression algorithms. Thus, there have been many modified versions of DCC methods proposed to improve the CR. The XOR-referenced code compression is one of these methods, which is a modified version of the so-called BitMask methods.

Even though certain instructions are not available in the dictionary, they differ from some of the entries in the dictionary by only a few bits. Ros and Sutton [27] proposed the concept of using the hamming distance to record the mismatch positions. Based on the DCC method, Seong and Mishra proposed BCC [15, 16], which used a multibit mask to record the mismatch portions. This method is able to match more instructions, and also reduce the encoding overhead, thus, improving the CR. An example of BCC is shown in Fig. 11. An additional tag bit is used for identifying whether an instruction has been compressed using only the dictionary, or using the BitMask compression approach. When an instruction is compressed by the BCC, the encoding codeword must contain the mask value and its position. The position of the mask starts from the left and move to the right. According to the mask value and its position, the original instruction is decompressed by performing an XOR operation on the dictionary entry. In the decompression engine design, these methods can be implemented into a similar architecture. First is the codeword fetch and identify codeword type stage. Second is the codeword decoding stage where the dictionary is accessed without or with the BitMask unit operation. The BitMask unit contains the program and shift buffer unit. Then, the final stage is the instruction(s) output stage.

The XOR-referenced approach tries to change the symbol distribution in order to further enhance the CR of the DCC algorithms. It is worth mentioning that using the reference XOR itself to change the symbol distribution cannot improve the CR.

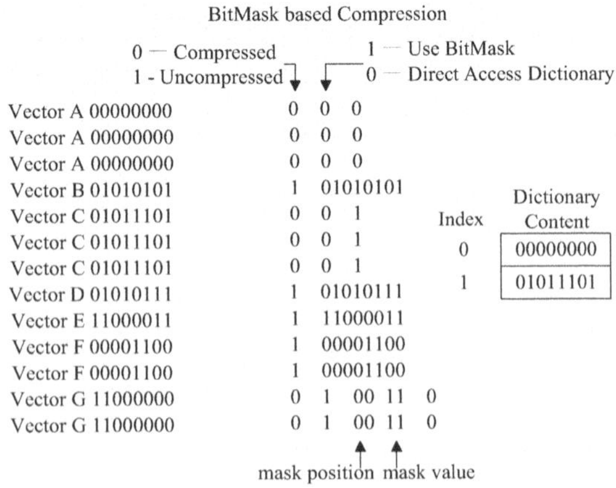

Fig. 11 An example of Seong and Mishra's [16] BitMask-based method

Format of Uncompressed Code

00	Uncompressed data (32 bits)

Format of Compressed Code with XOR reference

01	# of mask patterns	Mask type	Location	Mask pattern	…	Mask type	Location	Mask pattern	Dictionary Index

Format of Compressed Code with non-XOR reference

1	# of mask patterns	Mask type	Location	Mask pattern	…	Mask type	Location	Mask pattern	Dictionary Index

Fig. 12 An example coding format for modified BCC

It has to cooperate with other compression schemes in order to improve the CR, and the compressed code must indicate whether or not the instructions have to perform XOR operations with reference data. Therefore, the original encoding format has to be modified. Either an extra bit can be added to indicate whether the instructions need to perform XOR operations with reference data or a new encoding format can be used. A modified encoding format based on the BCC is shown in Fig. 12. An extra bit is used to indicate whether an instruction has been compressed or not, and another extra bit is used to mark the instructions to indicate whether an XOR operation with reference data is needed. Because DCC methods choose the most frequently used instructions as the dictionary entries, the size of compressed code dominates the size of the ones left uncompressed. The proposed modified format has to leave the frequent case unchanged, and use an extra bit to represent the uncompressed instructions and the ones using XOR references.

For traditional DCC, the dictionary is built based on the code compression method to encode the program. For the XOR references, the modified flowchart is shown in Fig. 13 that the reference data is sought out after the dictionary has been built and is used to change the bits for non-dictionary entries. Finally, the modified program (with some instructions XORed with the reference data) is encoded using the dictionary.

Different DCC schemes have different criterions in dictionary selection. In order to achieve the minimum CR, different situations have to be considered in XOR reference selection to maximize the matching rate for different algorithms. Two example algorithms are provided here to generate the reference data used for the reference XOR operation. The frequency-based reference data selection algorithm can find the optimal reference data for the standard DCC, while the XOR-referenced selection algorithm performs better when BCC is used.

For the standard DCC, the optimal algorithm to choose the reference is shown in Fig. 14. The standard dictionary is first generated based on the instruction frequency. Then XOR operations are performed on the non-entry instructions with every dictionary entry. The results of the XOR operations are stored as a histogram. The peak in the histogram represents the optimal reference data that can match

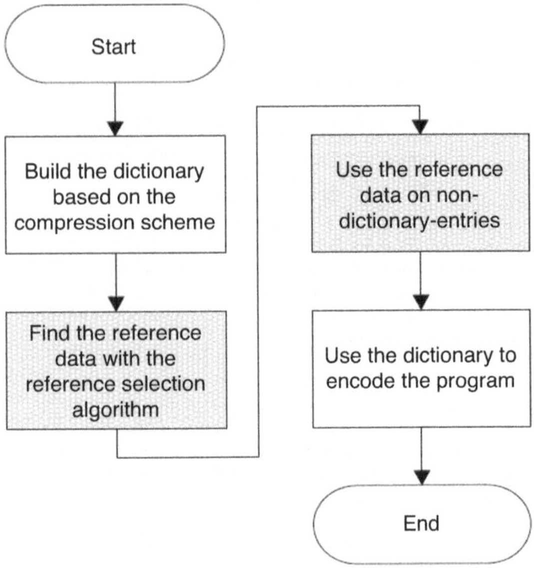

Fig. 13 Flowchart of XOR-referenced code compression

the most non-dictionary-entry instructions. Figure 15 shows an example histogram using the frequency-based reference data selection algorithm with more than ten matches for the *hello* benchmark. From the figure, it is clear that using the reference value *0x00000001* can reduce the most CR for the standard DCC.

To find the XOR reference data with the minimized CR, the basic idea is to increase the number of bit-changes in the reference data by one per iteration. The reference data are initially set to all-zero, and the number of ones in the reference data is gradually increased. For each iteration, the code compression with XOR reference is performed at most 32 times for the 32-bit instructions. If there are reference data with better CR compared with the previous iteration, another round of reference selection is needed. The XOR reference is used to generate more matches to the dictionary entries. When a match occurs, it has to conform to the following equation:

$$I \oplus D \oplus R = 0$$

The XOR of I (instruction), D (dictionary entry), and R (reference data) equals to a zero vector (indictor to 0) for a match. For BCC, an additional term, mask (M), has to be included. If more 1 s in the reference cannot generate a smaller CR, it indicates that more 1 s will reduce the number of zero vectors. So the more 1 s in reference will generate more non-zero vectors to reduce the matching rate. Therefore, the algorithm can be terminated immediately after a smaller CR is not obtained. The experimental results show that the frequency-based reference data selection algorithm can find the optimal reference data for the standard DCC, while the heuristic can only find a suboptimal estimation for the BCC. However, the XOR

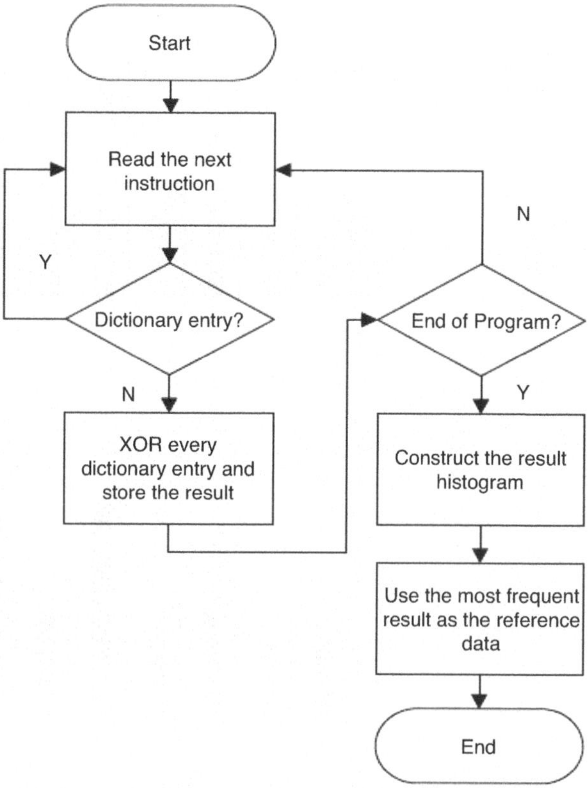

Fig. 14 Optimal reference selection for standard DCC

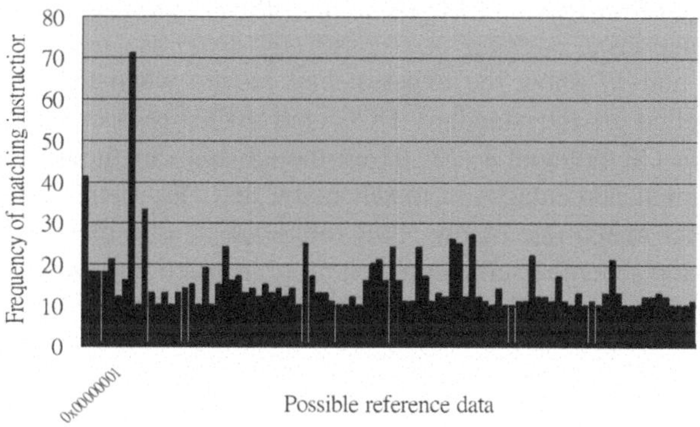

Fig. 15 An example histogram of the frequency-based reference data selection

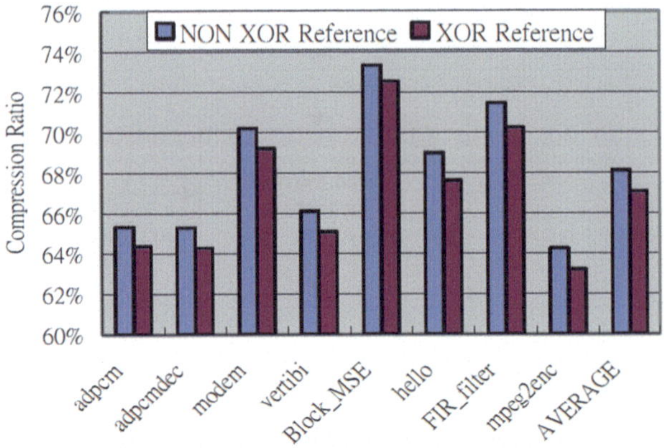

Fig. 16 CR for different benchmarks with standard DCC

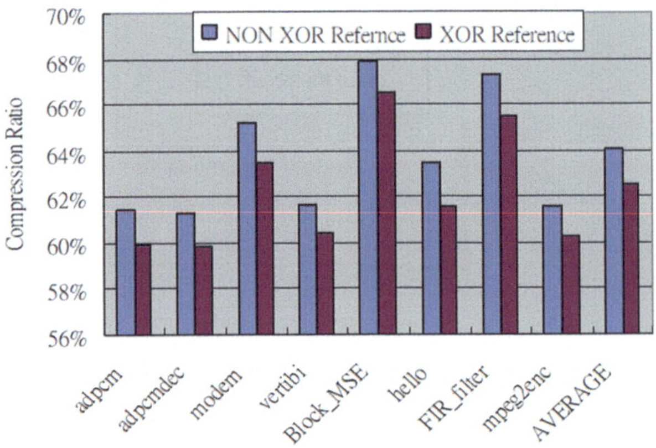

Fig. 17 CR for different benchmarks with BCC

reference can improve more for the BCC algorithms, since they can provide more matching instructions.

Figures 16 and 17 show the experimental results when the XOR-referenced method is applied to the standard DCC and BCC, respectively. The average improvement in CR is about 1–2%. Even though both methods have around the same improvement percentages in match rates, BCC has better CR improvement due to the higher match rate before XOR reference. The results suggest that if the code compression scheme itself can match more instructions, the proposed method can get a larger CR improvement.

6 Code Compression Using Separated Dictionaries

In this section, a separate dictionary was used to reduce the codeword length of high-frequency instructions. Variable mask numbers are used to eliminate the encoding redundancy. The combination of these methods is referred to as the codeword-length constrained BitMask code compression (CLCBCC). Moreover, a mixed-bit saving dictionary selection algorithm is used to select an improved instruction combination for the dictionary, and a fully separated dictionary architecture is proposed to reduce the access latency of the dictionary.

In certain cases, such as in low code-density architecture [26], which contains a high number of unique instructions, or because of algorithmic characteristics, a large LUT may be required to compress the programs. A large LUT has several disadvantages: it requires a large chip area, additional power consumption, a long LUT latency, and a long codeword length. Thus, it is desirable to minimize the dictionary size.

The static frequency distribution of the instructions was analyzed from the set of benchmarks [7] on TI C62xx processors; the results demonstrated that only a small set of instructions consistently exhibited extremely high frequencies. Figures 18 and 19 show the frequency distribution of dictionary entries from two benchmarks: *fft*, a smaller benchmark with 512 entries, and *susan*, a larger benchmark with 1024 entries. Both distributions were generated by using the frequency-based

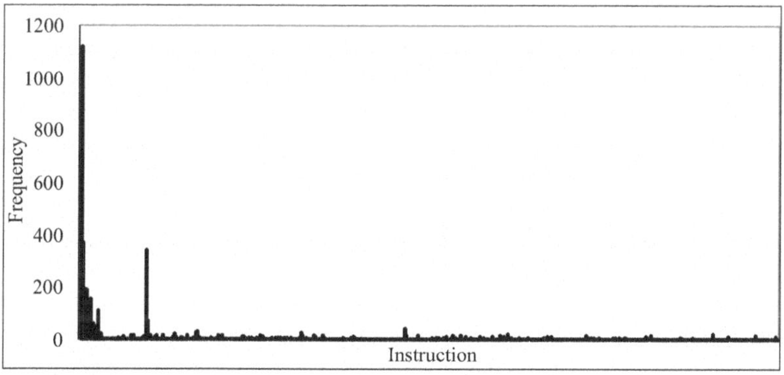

Fig. 18 *fft*: Frequency distribution of 512 dictionary entries

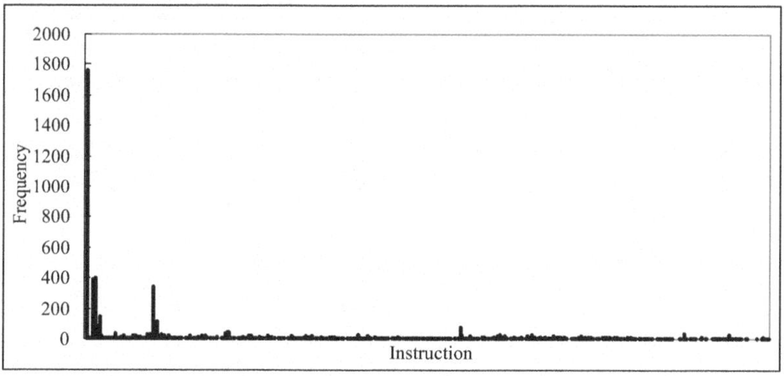

Fig. 19 *susan*: Frequency distribution of 1024 dictionary entries

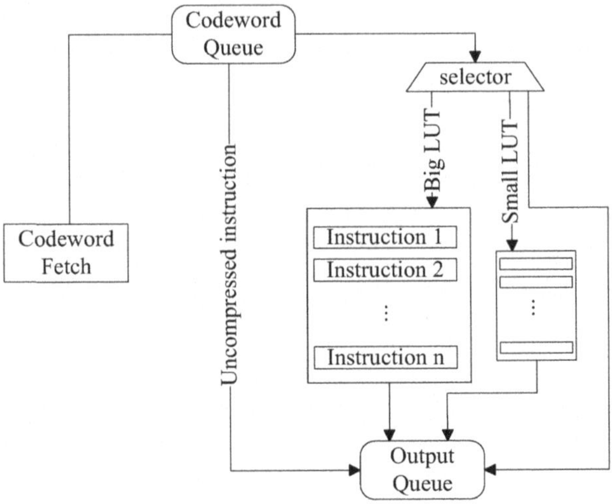

Fig. 20 Specific architecture for CLCBCC

dictionary selection algorithm. The frequency distributions were similar for all the benchmarks. Compressing these high-frequency instructions with the same codeword length as other low-frequency instructions would result in inefficient compression. In order to overcome this problem, these high-frequency instructions are separated into another small dictionary to obtain shorter codeword lengths. Two LUTs are used for the BitMask approach. A large LUT is used to compress single instructions, and a small LUT is used to compress the extremely high-frequency instructions. The small LUT was modifiable for storing either single instructions or instruction sequences [7]. The specific dictionary architecture for the CLCBCC is shown in Fig. 20.

Following suggestions from some prior arts [7, 16], using a 4-bit fixed and a 2-bit fixed masks in addition to a single 4-bit fixed mask achieves better compression results for the benchmarks. Although the maximum mask overhead was 13 bits (4 bits for 4-bit mask, 3 bits to record the position of the 4-bit fixed mask, 2 bits for 2-bit mask, and 4 bits to record the position of the 2-bit fixed mask), it was determined that approximately 50% of the instructions were compressed by using only the 4-bit fixed mask in the benchmarks. Thus, in this study, the 4f mask and the 4f–2f masks are combined, and 1-bit is used to identify whether the codeword uses one or two masks. The encoding format is shown in Fig. 21, which contains four situations, such as uncompressed, matched with small dictionary, matched with large dictionary, and matched using a variable number of masks.

Frequency-based dictionary selection cannot achieve an optimal CR in BCC, because it cannot guarantee that the matched rate of high-frequency instructions is maximized. The proposed dictionary selection algorithm is based on the graph representation, where instructions are transformed into nodes, and an edge between two nodes indicates that these two instructions have been matched to each other

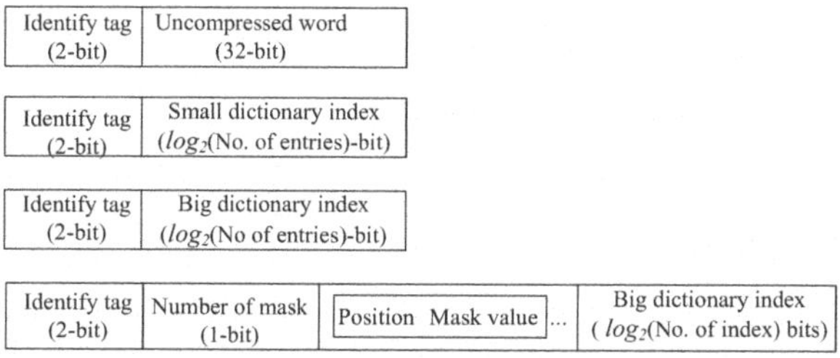

Fig. 21 Encoding format for the proposed approach

using the BCC approach. Generally, the nodes can be classified into five cases according to the frequency and connection pattern.

Case 1: a high-frequency node mostly connects to high-frequency nodes.
Case 2: a high-frequency node mostly connects to low-frequency nodes.
Case 3: a low-frequency node mostly connects to high-frequency nodes.
Case 4: a low-frequency node mostly connects to low-frequency nodes.
Case 5: a low-frequency node with a few connections.

Cases 1, 2, and 4 are better choices for the CR improvement, and Case 2 nodes can achieve the most savings. Because the high-frequency nodes are usually selected into the dictionary, the benefits for nodes in Case 3 are limited. Thus, they are unsuitable for the dictionary. The nodes in Case 5 would never be selected in the algorithm because their low frequency and few connections result in low savings.

Algorithm 1: Mixed-Bit Saving Dictionary Selection
Inputs:

1. 32-bit unique instruction vectors
2. Dictionary size
3. Mask types

Output: Optimized dictionary
Begin
Step 1: Transform every unique instruction to a graph, $G = (V, E)$.
If two nodes can be matched by using BitMask, use directional edges to connect them.
Step 2: Allocate bit savings to the nodes and edges.
$\forall$ node i,
Node saving (N_i) = (original instruction size − compressed codeword size) × frequency of the instruction–32 bits overhead.
$\forall$ edge between nodes i and k,
Edge saving (W_{ik}) = (original instruction size − compressed codeword size) × frequency of the matched instruction.
Total bit saving $(S_i) = N_i + \sum_{k=1}^{n} W_{ik}$
Step 3: Calculate the total bit saving distribution of all nodes.
Step 4: Select the most profitable node $i_* = \arg\max f(S_i)$.

Step 5: Remove the most profitable node i from G and insert it into the dictionary.
Step 6: ∀ nodes connect to i, construct the neighboring node set Nb(i).
Step 7: Delete all edges with at least one end in Nb(i).
Step 8: Update node savings of all the nodes in Nb(i), $S_i = N_k - W_{ik}$.
Step 9: Repeat Steps 3–8 until the dictionary is full.
Step 10: Return dictionary.
End

A mixed-bit saving dictionary selection (MBSDS) algorithm is proposed in Algorithm 1, which first transforms every unique instruction into a single node, and two-directional edges between two nodes indicate that these two instructions were matched to each other using the BCC. The proposed algorithm then calculates the bit savings of all nodes, and inserts the most profitable node into the dictionary. The most profitable node is then removed from the graph. Since all the neighboring nodes of the most profitable node can be covered by the most profitable node, the node saving of each neighboring node should subtract the edge saving from the edge with the most profitable node. Furthermore, all the edges of the neighboring nodes are removed. These steps are repeated until the dictionary is full.

The most profitable node achieves the savings from the combination of its own node saving and the edge savings of other nodes. However, connected instructions cannot be easily inserted into the dictionary. Whether these connected instructions should be selected into the dictionary in the following rounds is solely determined by their frequency values. Figure 22 shows an example of selection using MBSDS. All symbols in this example are 32-bit wide, the dictionary contained 1024 entries, only one 2-bit mask was used, and the overhead of the identification tag is 2-bit. After each symbol was transferred into the nodes, every node contained its frequency value. When two nodes were matched to each other using BCC, the algorithm will create two-directional edges to connect them: the direction pointed to the instruction, and the weight corresponding to the actual edge saving when the connected node was compressed by the matched node.

The node saving for every node equaled to 32 − 12 (codeword length 10 + tag width 2), which was multiplied by its frequency. The edge saving equaled to 32 minus 18 (codeword length 10 + tag width 2 + bitmask 6), which was multiplied by the frequency of the matched node. Node A was clearly the most profitable node; the bit saving value of SA was better than that of other nodes. In MBSDS, after the Node A was selected and inserted into the dictionary, all edges of Nodes B, C, and E were deleted. For Nodes B, C, and E which could be compressed by Node A using BCC, the edge saving from the edge of Node A was subtracted from their node saving. In other words, the node saving of Node B originally equaled 200. After Node A was selected and inserted into the dictionary, the node saving of Node B was updated into 60 (200 − 140). After all node savings were updated, the most profitable nodes were, in order, G, B, E, C, D, and F. For the bit-saving dictionary selection algorithm [16], a threshold value was used to determine whether the nodes were selected and inserted into the dictionary. Suppose a threshold value of 10 is used, Nodes A and B were selected and inserted into the dictionary, and Nodes C and E were deleted from

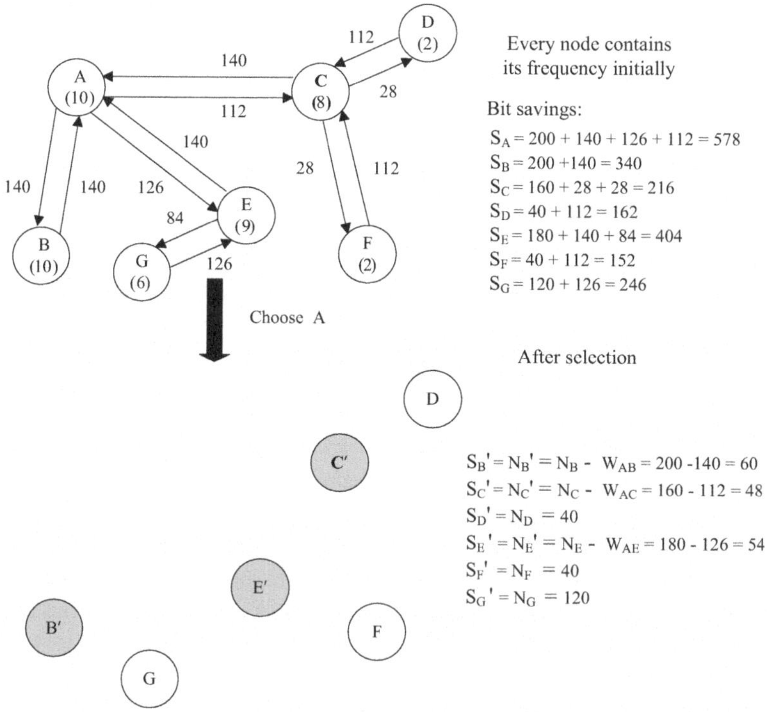

Fig. 22 An example of MBSDS

the graph. The rest of candidate nodes D and F were then selected in the following round. Nodes C and E (if G was not selected) were more efficient than B, D, or F. As this result demonstrated, an unsatisfactory threshold value reduced the efficiency of the bit-saving dictionary selection algorithm compared with the frequency-based dictionary selection algorithm. The CLCBCC with MBSDS is shown in Algorithm 2 and an example of CLCBCC with MBSDS is shown in Fig. 23.

Algorithm 2: CLCBCC with MBSDS
Inputs:

1. 32-bit instruction symbols
2. Small dictionary size
3. Big dictionary size
4. Mask types

Output: CR and Compressed codewords.
Begin
Step 1: Calculate the frequency distribution of all instruction symbols.
Step 2: Select the highest unique frequency symbols into the small dictionary based on the step 1.

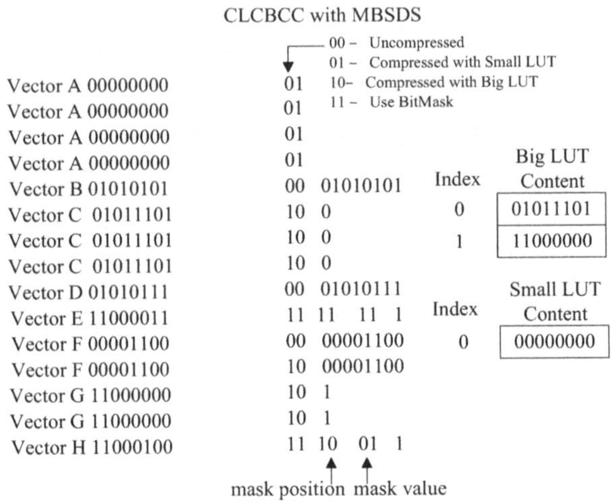

Fig. 23 An example of CLCBCC with MBSDS

Step 3: For every unique instruction symbols which are not selected into the small dictionary, use the MBSDS to construct the big dictionary.
Step 4: Use the BitMask based method to compress all instructions based on the current dictionaries and masks setting.
Step 5: Calculate the CR.
Step 6: Return the compressed codewords and CR.
End

The proposed decompression engine was implemented by using the Verilog Hardware Description Language (Verilog HDL), and synthesized using Synopsys' Design Compiler and a TSMC 0.13 μm cell library. The decompression engine, the logic diagram of which is shown in Fig. 24, consisted of a control unit, a demultiplexer, shift buffers, LUTs, and the BitMask unit. The control unit controls other units and assigns tasks to other units according to the control signals. The input queue initializes itself, collects compressed instructions from the storage space, and shifts the contents of the buffer after the decoding process is completed. The output queue stores the decompressed instructions and delivered them to the processor or cache. The large LUT and small LUT store the original binary instructions and synthesized by using flip-flop logic. The small LUT stores high-frequency instructions, enabling them to be quickly decoded with a shorter codeword length. The BitMask unit executes the shifting of masks and exclusive-or operations based on instructions from the large LUT to obtain the original instructions. The BitMask unit also accesses the dictionary and executes shift operations in parallel during decompression. The proposed decompression engine has a decompression bandwidth of 32 bits/cycle.

Figures 25, 26, and 27 show the probability distribution of all five possible codeword types when using the proposed approach with frequency-based dictionary selection (FDS) on the ARM and TI C6x architectures. Beginning with the longest,

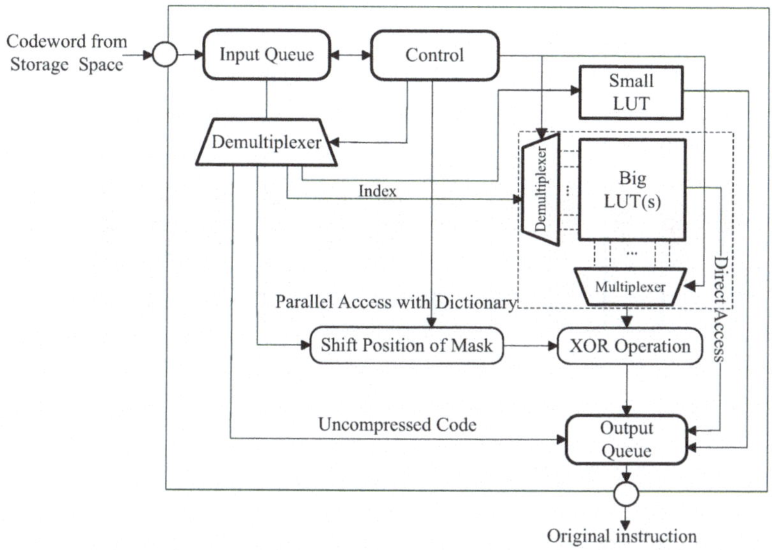

Fig. 24 Logic diagram of decompression engine

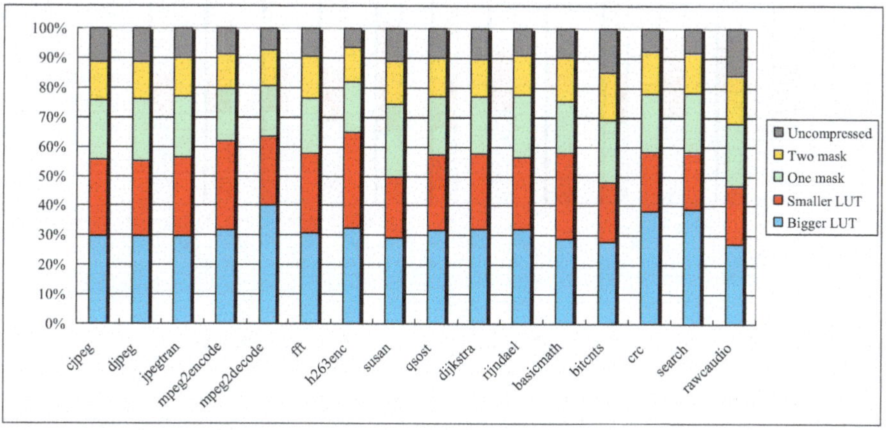

Fig. 25 Probability distribution of codeword types on ARM benchmarks

the sequence of the lengths of codewords was uncompressed, decompressed using two masks (4f, 2f), one mask (4f), the large LUT, and the small LUT. LUT with a size of 2048, and a small LUT with a size of 16, and codeword lengths of 34, 27, 21, 13, and 6, the introduction of a separated LUT clearly reduces the codeword sizes with 20–30% of the instructions in nearly all the benchmarks.

In Figs. 28, 29, and 30, the CRs of three different code compression algorithms were compared: DCC, BCC with fixed mask numbers (4f, 1 s), and CLCBCC by using a 2f and 4f masks, respectively. Furthermore, the saving rates of Thumb-2 [38]

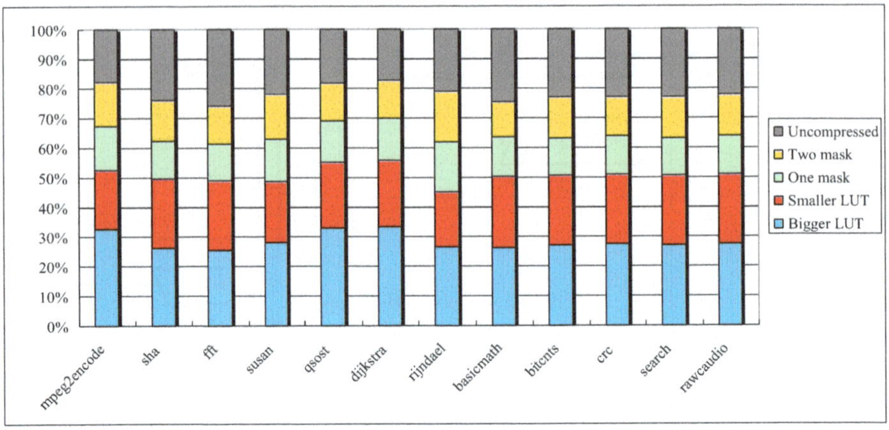

Fig. 26 Probability distribution of codeword types on TI C62xx benchmarks

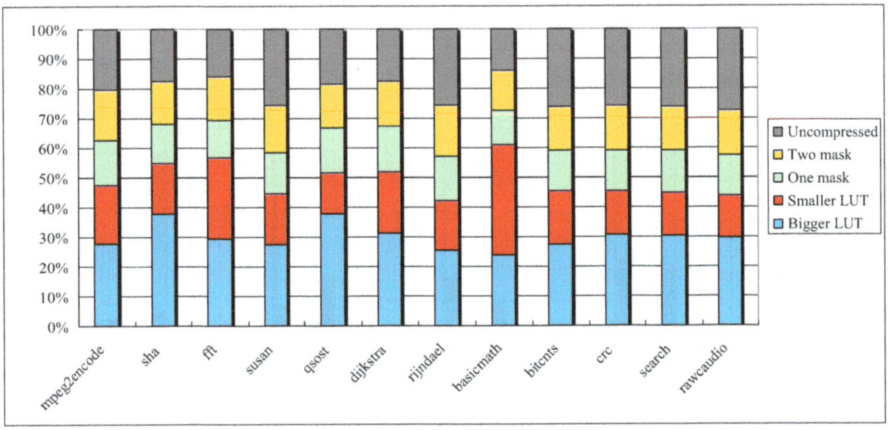

Fig. 27 Probability distribution of codeword types on TI C64xx benchmarks

for benchmarks on the ARM Cortex-A9 processor are included in Fig. 28 as well. A frequency-based dictionary selection algorithm was used to select LUT entries in all these approaches. For CLCBCC, a small LUT was first constructed, and then the remaining instructions were used to construct the large LUT. The results demonstrate that using a small LUT for storing high-frequency instructions can result in a CR improvement of 6% for the CLCBCC compared with BCC. Although using the debug configuration for the C6x series on the Code Composer Studio achieved a lower CR and greater improvements, the release configuration was used to simulate the experiment in this article.

In Figs. 31, 32, and 33, the CRs of four different dictionary selection algorithms are compared: frequency-based dictionary selection (FDS), bit-saving dictionary selection (BSDS) [16], decoding-aware dictionary selection (DADS) [29], and

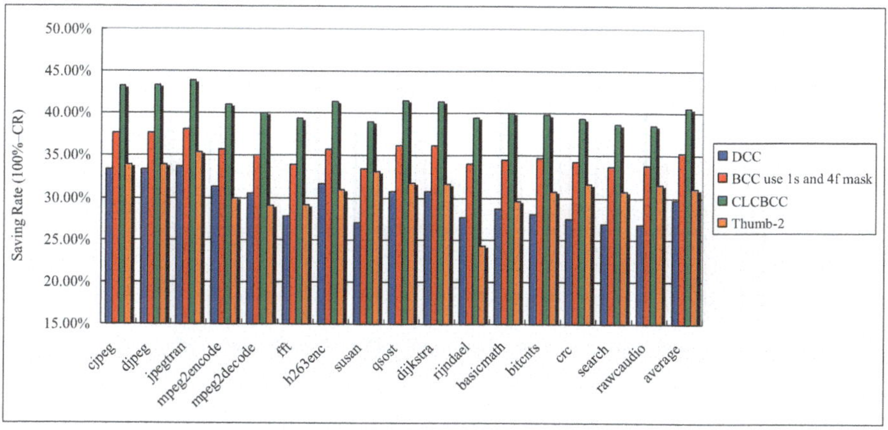

Fig. 28 Comparison of CR for benchmarks on ARM Cortex-A9 processors

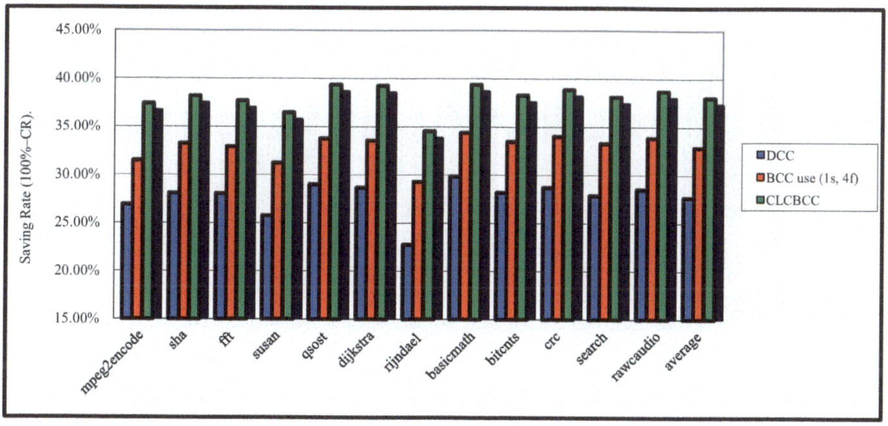

Fig. 29 Comparison of CR for benchmarks on C62xx processors

the proposed MBSDS. Compared with other selection algorithms, the proposed MBSDS outperforms the FDS and DADS for all the benchmarks. It is because FDS only took frequencies rather than BitMask matches into consideration. While DADS ensures higher instruction coverage, it also selects numerous unnecessary Case 3 instructions and several incorrect instructions, which results in a much larger LUT.

Seong and Mishra claimed that a threshold value between 5 and 15 is good for BSDS [16]; thus, 10 was set as the threshold value for BSDS in the simulations. The results showed that using the same threshold value can be unstable in different benchmarks, which is a disadvantage of BSDS. If there are n unique instructions, the time complexity for both BSDS (with a given threshold value) and MBSDS is $O(n^2)$. No studies have yet proposed an optimal method to obtain a superior threshold value. Thus, the threshold value can only be obtained by trial-and-error.

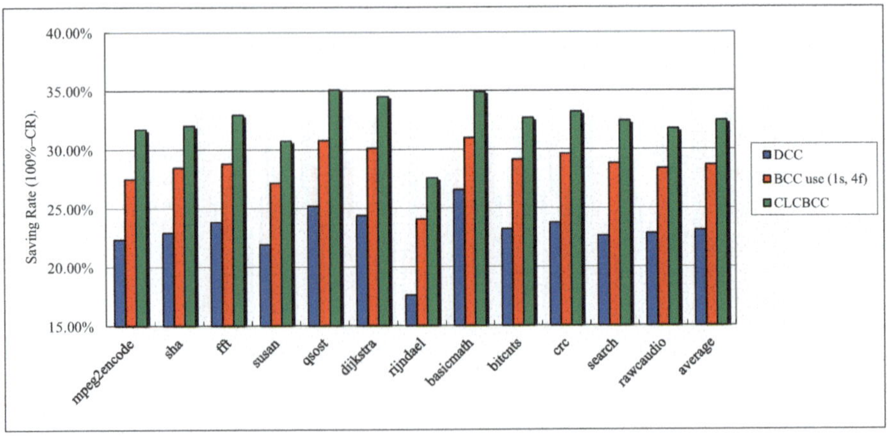

Fig. 30 Comparison of CR for benchmarks on C64xx processors

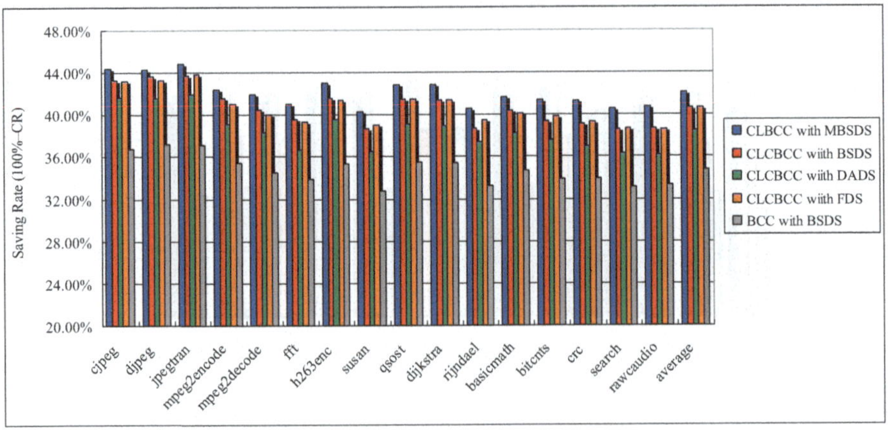

Fig. 31 Comparison of dictionary selection algorithms on ARM processors

The BSDS must be performed several times to determine the superior threshold value. Discovering the threshold value is time-consuming when the number of unique instructions is large. The proposed MBSDS algorithm thus offers another advantage: it does not require a threshold to decide whether a node can be selected, and can avoid the special case described previously. In other words, MBSDS not only improves the compression efficiency or CR, but also improves the performance of the algorithm.

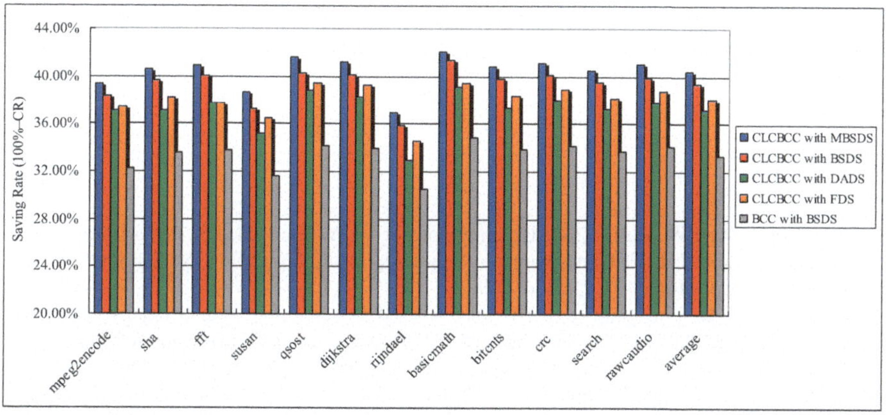

Fig. 32 Comparison of dictionary selection algorithms on C62xx processors

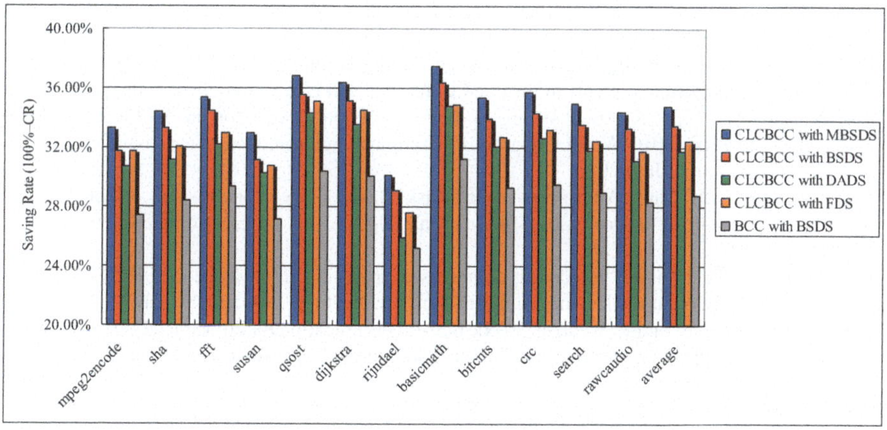

Fig. 33 Comparison of dictionary selection algorithms on C64xx processors

7 Conclusions

As mentioned in the preface, code compression was the first project I participated when I first joined Marilyn's group, and this chapter summarized the works we did in this field for the last decade. Several approaches have been proposed to reduce the memory usage in embedded systems, and may in turn reduce the cost and power consumption of embedded systems.

Multicore architecture has been a trend in modern embedded products, with requirements in higher communication bandwidths either between processors and the cache, or between cache and memory, than single-core systems. In future studies, the design and implementation of a general multilevel separated dictionary decompression engine with fully separated LUTs method and a parallel decompression engine will be investigated, for applying code compression to architectures with high-bandwidth requirements, such as multicore architectures.

References

1. C.H. Lin, Y. Xie, W. Wolf, "Code Compression for VLIW Embedded Systems Using a Self-Generating Table," IEEE Transactions on Very Large Scale Integration (VLSI) Systems. Vol. 15, Issue 10, pp. 1160-1171, October 2007.
2. J. C. Chen, C.H. Lin, "Improved Dictionary-based Code-compression Schemes with XOR Reference for RISC/VLIW Architecture," IEICE Transactions on Fundamentals of Electronics, Communications and Computer Sciences, Vol.E93-A, No. 12, pp. 2517-2523, Dec. 2010.
3. C.-W. Lin, C.H. Lin, W. J. Wang, "A Power-aware Code-compression Design for RISC/VLIW Architecture," Journal of Zhejiang University-SCIENCE C (Computers & Electronics), Vol.12, No. 8, pp. 629-637, Aug. 2011
4. W.J. Wang, C.H. Lin, "Code Compression for Embedded Systems Using Separated Dictionaries," IEEE Transactions on Very Large Scale Integration (VLSI) Systems, Vol. 24, No. 1, pp. 266-275, Jan. 2016
5. C.H. Lin, Y. Xie, W. Wolf, "LZW-Based Code Compression for VLIW Embedded Systems," in Design, Automation and Test in Europe Conference and Exposition., Feb. 2004, pp. 76–81.
6. C.W. Lin, C.H. Lin, "A Power-saving Code-compression Design for the VLIW Embedded Systems," in Proc. 2010 World Congress in Computer Science, Computer Engineering, and Applied Computing, July 2010
7. W.J. Wang, C.H. Lin, "An Improved BitMask Based Code Compression Algorithm for Embedded Systems," in Proc. 2011 International Symposium on Electronic System Design, December 2011
8. A. Wolfe and A. Chanin, "Executing compressed programs on an embedded RISC architecture," in Proc. International Symposium on Microarchitecture, 1992, pp. 81–91.
9. S. Liao, S. Devadas, and K. Keutzer, "Code density optimization for embedded DSP processors using data compression techniques," in Proc. Conference on Advanced Research in VLSI, 1995, pp. 272–285.
10. C. Lefurgy, P. Bird, I. Chen, and T. Mudge, "Improving code density using compression techniques," in Proc. International Symposium on Microarchitecture, 1997, pp. 194–203.
11. IBM, Armonk, NY, "PowerPC code compression utility user's manual, Version 3.0," 1998.
12. H. Lekatsas and W. Wolf, "SAMC: A code compression algorithm for embedded processors," IEEE Transaction on Computer-Aided Design Integrated Circuits System, vol. 18, no. 12, pp. 1689–1701, Dec. 1999.
13. E. W. Netto, R. Azevedo, P. Centoducatte, and G. Araujo, "Multi-profile based code compression," in Proc. Design Automation Conference, 2004, pp. 244–249.
14. L. Benini, F. Menichelli, and M. Olivieri, "A class of code compression schemes for reducing power consumption in embedded microprocessor systems," IEEE Transaction on Computer, vol. 53, no. 4, pp. 467–482, Apr. 2004.
15. S.-W. Seong and P. Mishra, "A bitmask-based code compression technique for embedded systems," in Proc. IEEE/ACM International Conference on Computer Aided Design, Nov. 2006, pp. 251–254.
16. S.-W. Seong and P. Mishra, "An efficient code compression technique using application-aware bitmask and dictionary selection methods," in Proc. Design, Automation and Test in Europe, 2007, pp. 1–6.
17. S. Segars, K. Clarke, and L. Goude, "Embedded control problems, thumb and the ARM7TDMI," IEEE Micro, vol. 15, no. 5, pp. 22–30, Oct. 1995.

18. K. Kissel, "MIPS16: High-density MIPS for the embedded market," Silicon Graphics MIPS Group, Sunnyvale, CA, 1997.
19. R. Gonzalez, "Xtensa: A configurable and extensible processor," IEEE Micro, vol. 20, no. 2, pp. 60–70, Mar./Apr. 2000.
20. ARC International, San Jose, CA, "ARCompact-ISA programmer's reference manual," 2005.
21. K. D. Cooper and N. McIntosh, "Enhanced code compression for embedded RISC processors," in Proc. SIGPLAN Conference on Programming language design and implementation, 1999, pp. 139–149.
22. S. Debray and W. Evans, "Profile-guided code compression," in Proc. SIGPLAN Conference on Programming language design and implementation, 2002, pp. 95–105.
23. S. Debray and W. Evans, "Cold code decompression at runtime," Commun. ACM, vol. 46, no. 8, pp. 54–60, Aug. 2003.
24. O. Ozturk, H. Saputra, M. Kandemir, and I. Kolcu, "Access pattern based code compression for memory-constrained embedded systems," in *Proc.* Design, Automation and Test in Europe., 2005, pp. 882–887.
25. S. Shogan and B. R. Childers, "Compact binaries with code compression in a software dynamic translator," in *Proc.* Design, Automation and Test in Europe., 2004, pp. 1052–1057.
26. B. Gorjiara, M. Reshadi, and D. Gajski, "Merged dictionary code compression for FPGA implementation of custom microcoded PEs," ACM Transaction on Reconfigurable Technology. Systems, vol. 1, no. 2, Jun. 2008.
27. M. Ros and P. Sutton, "A Hamming distance based VLIW/EPIC code compression technique," in Proc. International Conference on Compilers, Architecture, and Synthesis for Embedded Systems, 2004, pp. 132–139.
28. M. Thuresson and P. Stenstrom, "Evaluation of extended dictionary based static code compression schemes," in Proc. 2nd Conference on Computer Frontiers, 2005, pp. 77–86.
29. X. Qin, C. Muthry, and P. Mishra, "Decoding-aware compression of FPGA bitstreams," IEEE Transaction on Very Large Scale Integrated (VLSI) Systems, vol. 19, no. 3, pp. 411–419, Mar. 2011.
30. C. Murthy and P. Mishra, "Bitmask-based control word compression for NISC architectures," in Proc. 19th ACM Great Lakes Symposium on VLSI, 2009, pp. 321–326.
31. T. Bonny and J. Henkel, "Efficient code compression for embedded processors," IEEE Transaction on Very Large Scale Integrated (VLSI) Systems, vol. 16, no. 12, pp. 1696–1707, Dec. 2008.
32. J. Ranjith, N. J. R. Muniraj, and G. Renganayahi, "VLSI implementation of single bit control system processor with efficient code density," in Proc. IEEE International Conference on Communication Control and Computing Technologies (ICCCCT), Oct. 2010, pp. 103–108.
33. P.-Y. Chen, C.-C. Wu, and Y.-J. Jiang, "Bitmask-based code compression methods for balancing power consumption and code size for hard realtime embedded systems," Microprocessors Microsystems, vol. 36, no. 3, pp. 267–279, May 2012.
34. W. R. Azevedo Dias, E. D. Moreno, and I. Nattan Palmeira, "A new code compression algorithm and its decompressor in FPGA-based hardware," in Proc. 26th Symposium on Integrated Circuits System Design (SBCCI), Sep. 2013, pp. 1–6.
35. Texas Instruments Incorporated, Dallas, "TMS320C62xx CPU and instruction set: Reference guide," 1997.
36. T. Bell, J. Cleary, and I. Witten, Text Compression. Englewood Cliffs, NJ: Prentice-Hall, 1990.
37. T. Welch, "A technique for high performance data compression," IEEE Transactions on Computers, vol. 16, no. 6, pp. 8–19, Jun. 1984.
38. ARM Architecture Reference Manual ARMv7-A and ARMv7-R Edition, ARM, Cambridge, U.K., 2012.

Wearable Sensor Applications: Processing of Egocentric Videos and Inertial Measurement Unit Data

Yantao Lu and Senem Velipasalar

Abstract There has been a proliferation of smartphones, smart watches, and wearable sensors, making them ubiquitous in our daily lives. Mobile sensors have found widespread use due to their ever-decreasing cost, ease of deployment, and ability to provide continuous monitoring as opposed to sensors installed at fixed locations. Various techniques have been proposed for fall detection, gait analysis, activity monitoring, and heart rate and sleep sensing by wearable sensors and mobile phones. Compared to works that use inertial measurement unit (IMU) data or static cameras installed in the environment, there has been relatively less work using egocentric videos, meaning providing the first-person view from wearable cameras. Moreover, most of the existing studies on egocentric videos are based on only one sensor modality, namely the camera. There have been even fewer approaches that combine egocentric video data with IMU data. In this chapter, we will describe three different applications using wearable cameras together with IMU data. First, we will present an overview of a fall detection system using wearable devices, e.g., smartphones and tablets, equipped with cameras and accelerometers. Since the portable device is worn by the subject, monitoring is not limited to confined areas, and extends to wherever the subject may travel, as opposed to static sensors installed in certain rooms. Second, we will present an autonomous and robust method for counting footsteps, and tracking and calculating stride length by using both accelerometer and camera data from smartphones or Google™ glass. To provide higher precision, instead of using a preset stride length, the proposed method calculates the distance traveled with each step by using the camera data. This method is compared with the commercially available accelerometer-based step counter apps. The results show that the proposed method provides a significant increase in accuracy, and has the lowest average error rate both in number of steps taken and the distance traveled. Finally, we will provide an overview of a robust

Y. Lu · S. Velipasalar (✉)
Syracuse University, 4-206 Center for Science and Technology, Syracuse, NY, USA
e-mail: ylu25@syr.edu; svelipas@syr.edu

S. S. Bhattacharyya et al. (eds.), *Embedded, Cyber-Physical, and IoT Systems*,
https://doi.org/10.1007/978-3-030-16949-7_7

and autonomous method to detect activities with more details and context by using accelerometer and egocentric video data obtained from a smartphone.

1 Autonomous Fall Detection with Wearable Cameras and IMU Data

1.1 Introduction

Activity monitoring systems have been introduced as part of elderly care in recent years, especially for elderly people living independently. Fall detection is a crucial part of elderly activity monitoring systems, since falls are considered to be the eighth leading cause of death in the USA [1], and fall injuries can result in serious complications [2, 3]. Almost 20% of all falls require immediate medical attention and about 10% of the falls result in fractures [4]. It has been well established that timely response to a fall has a significant influence on lowering the morbidity–mortality rate [5, 6].

Even though several user-activated commercial devices are available, they have limited benefits, especially in situations where the user loses consciousness. In response to growing needs, researchers have been working on autonomous fall detection via dedicated signal processing devices. Noury et al. [7] and Mubashir et al. [8] have provided well-organized surveys on the principles and approaches involved in fall detection. According to Noury et al. [7], although there are some common characteristics among falls, different scenarios must be considered for different kinds of falls. For example, falls from a standing position are much easier to detect as compared to falls from sitting or lying down positions. An autonomous system should provide real-time detection of falls with tolerable number of, ideally zero, false positives.

The fall detection and activity monitoring systems can be broadly categorized into two main classes: (a) systems using non-vision sensors and (b) systems using vision-based sensors. Acoustic, vibrational, and other ambience sensor-based methods use characteristic vibration patterns to detect different events. However, with these systems, the monitoring is limited to only those areas, where the sensors are installed. Moreover, it is usually assumed that there is only one subject performing the activities. Accelerometer-based fall detection systems [9–12] are simple and cost effective. Yet, even with multiple sensors, these systems are prone to creating false positives, especially when people are exposed to acceleration, e.g., due to being in an elevator. The limitations of using just the accelerometer are also discussed by Wu et al. [13]. Thus, due to shortcomings of systems that rely only on acceleration and gyroscope data, more robust methods are needed to differentiate between falls and other regular daily activities. A comprehensive survey on activity detection and classification using wearable sensors can be found in [14].

Vision-based systems have been introduced as an alternative to employing non-vision sensors. Compared to non-vision sensors, cameras provide a much richer set of data including contextual information about surroundings, which allows the analysis of a wider variety of activities. Vision-based methods involve processing images from one or more cameras monitoring a subject [15]. Most approaches use raw video data, while others address privacy concerns [16] by using infrared or contrast-detection cameras [17, 18]. Stereoscopic vision and 3D scene reconstruction are other variations that aim to increase system accuracy [19, 20]. There have also been implementations of static camera-based algorithms on embedded platforms. Belbachir et al. [21] presented a dynamic visual sensor (DVS)-based system consisting of two optical sensors with 304×240 event-driven pixels and an FPGA for the processing. Fleck et al. [22] presented a distributed camera network for assisted living using FPGA and PowerPC-based smart cameras.

In most of the existing vision-based systems, cameras are static, and installed at fixed locations. Similar to the approaches using acoustic and vibrational sensors, using cameras installed at static locations confines the monitoring environment only to those regions. Furthermore, in many existing systems, videos captured by the cameras are transferred to a central location for processing, which requires extensive communication. In addition, subjects, continuously being monitored by these cameras, often raise privacy concerns [16].

In this section, we present an overview of our work on fall detection by using wearable camera and IMU data. In our work [23–27], we proposed a completely different approach compared to existing vision-based activity monitoring systems. In our system, the camera is worn by the subject providing an egocentric view, in contrast to static sensors installed only at certain locations. Therefore, monitoring extends to wherever the subject may travel including indoors and outdoors. Moreover, compared to the static cameras watching the subject, the privacy concerns are alleviated, if not eliminated, since the captured images are not of the subject. Instead, the captured images are egocentric images showing what the subject sees. Furthermore, the images are not saved or transmitted to a central processor, but are processed onboard locally. Based on the current trends, it is expected that *wearable cameras* will be increasingly employed to understand lifestyle behaviors for health purposes [28].

1.2 Proposed Approach

In our earlier work [25], we presented a fall detection algorithm that employed histograms of edge orientations and strengths, and proposed an optical flow-based method for activity classification. Similar to histograms of oriented gradients (HOG) [29], an image is divided into blocks and then each block is divided into n cells. Different from the original HOG algorithm, for every cell, two separate histograms are built for the edge orientation (EO) and edge strength

(ES). During a fall, edge orientations change significantly, which is reflected in the gradient orientation histograms. Also, since falls are relatively fast events, the edges in images get blurred as seen in Fig. 4. This is captured by the change in the gradient strength histograms, and the edge strength values decrease during a fall.

Only one block is used to lighten the processing load. In addition, the number of cells in the block is changed adaptively so that the cells that do not contribute to overall edge information are autonomously removed according to their content. The details of the fall detection algorithm can be found in [25]. For the falls starting from a standing up position, an average detection rate of 87.84% was achieved with pre-recorded videos from eight subjects.

In a following work [26], we implemented this algorithm, which uses edge orientation and edge strength histograms, on a Samsung Galaxy S®4 phone with Android™ OS. We also implemented an algorithm to detect falls from the accelerometer data. The fusion method is inspired by the sum rule of two normalized classifiers, since it gives the least detection error rate [30]. When a fall is detected, the result is displayed on the screen of the phone. The algorithm runs at 15 fps on the smartphone. We performed experiments with 10 people carrying this phone. The experimental setup and example images captured from the phone camera, during a fall from standing up position, can be seen in Fig. 1. We compared sensitivity values and the number of false positives for each modality alone and also for when they are fused. The average detection rates obtained when using only the accelerometer, only the camera, and when fusing accelerometer and camera modalities were 65.66%, 74.33%, and 91%, respectively. As shown in [26], fusing accelerometer and camera data provides much higher sensitivity, and helps to eliminate false positives caused by using only camera or only accelerometer as the fall detection sensor.

More recently, we presented an improved algorithm [27] to autonomously compute an optimal threshold for fall detection from training data, by employing the relative entropy approach from the class of Ali-Silvey distance measures. In our previous work mentioned above [25, 26], we had used an empirically determined threshold. In this work, we used edge orientation histograms together

(a) (b) (c) (d)

Fig. 1 (**a**) An Android™ smartphone attached to the waist; (**b–d**) Example images captured during a fall from standing up position [26] with the gradient local binary patterns (GLBP), which are more descriptive and discriminative than histograms of oriented gradients (HOG) [31].

In order to evaluate the performances and perform comparison, we obtained the receiver operating characteristic (ROC) curves for indoor and outdoor experiments for varying threshold values [27]. Experiments were performed with ten different people and a total of around 300 associated fall events indoors and outdoors. We compared the proposed method with original HOG [29], our earlier work [25] (referred to as modified HOG in figures), and using only the GLBP features. For the indoor dataset, as can be seen in Fig. 2, the proposed method operates closest to the upper left-hand corner, and provides a better performance compared to other approaches. The operating threshold point, which was obtained by the autonomous threshold computation described in detail in [27], is marked on the ROC curve. The operating point corresponds to the location (0.0755, 0.9378) in Fig. 2. We also obtained the sensitivity/specificity curves, for the indoor dataset and varying threshold values, which are shown in Fig. 3. As can be seen, the proposed method outperforms the others, and operates closest to the upper right-hand corner. The proposed method, with the autonomously computed threshold value, operates at the point of (0.938, 0.876).

As mentioned above, wearable sensors allow the monitoring of people wherever they may travel including indoors and outdoors. We also tested the proposed method on outdoor scenarios. Example frames captured by the body-worn camera during the course of a fall event can be seen in Fig. 4, which also shows the significant change in the scene, going from building and trees to wide open

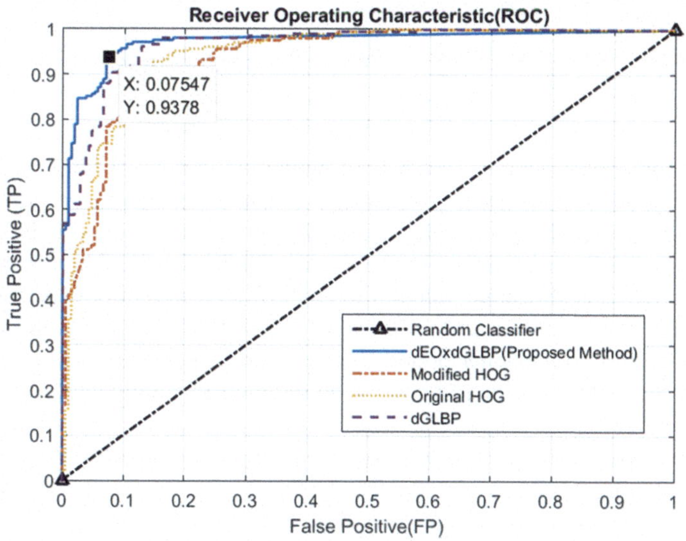

Fig. 2 Receiver operating characteristic (ROC) curves, for varying threshold values, obtained from the indoor dataset containing falls and non-fall activities [27]

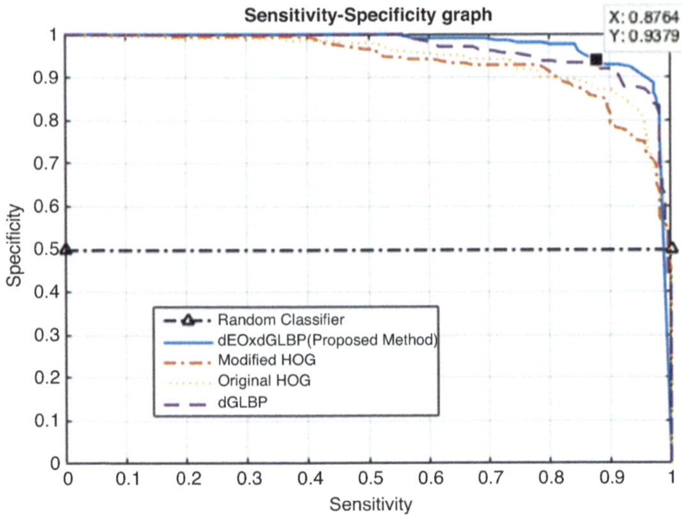

Fig. 3 Sensitivity–specificity curves, for varying threshold values, obtained from the indoor experiment dataset containing falls and non-fall activities [27]

Fig. 4 Example images captured from the body-worn camera during a fall from standing up position in an outdoor environment [27]

skies. The threshold calculated from the training set is $\tau_c = 0.4602$. For outdoor scenarios, the change in scenery is usually much more significant compared to indoor scenarios resulting in higher dissimilarity distances between frames. The proposed approach for autonomous threshold computation is able to capture this resulting in a higher threshold value of 0.4602. Similar to indoor experiments, we obtained the ROC curves and the sensitivity–specificity graph for the proposed method as well as three other approaches as seen in Figs. 5 and 6, respectively. The operating point corresponding to the autonomously computed threshold value is (0.0755, 0.898).

In summary, with this approach, since the sensors or the smartphone are worn by the subject, the monitoring can continue wherever the subject may travel including indoors and outdoors. Moreover, wearable cameras alleviate, if not eliminate, privacy concerns of users, since the captured images are not of the subjects but the surroundings. Also, with smartphone implementation, images are processed locally *on the device*, and they are not saved or transmitted anywhere. Moreover,

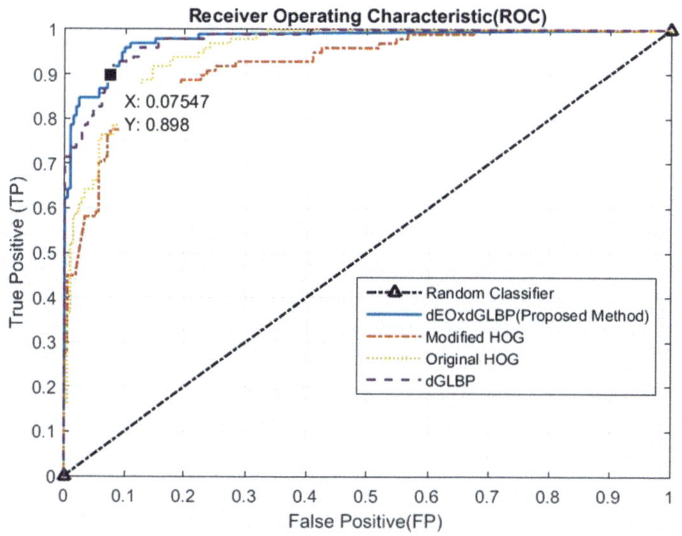

Fig. 5 Receiver operating characteristic (ROC) curve, for varying threshold values, obtained from the outdoor dataset [27]

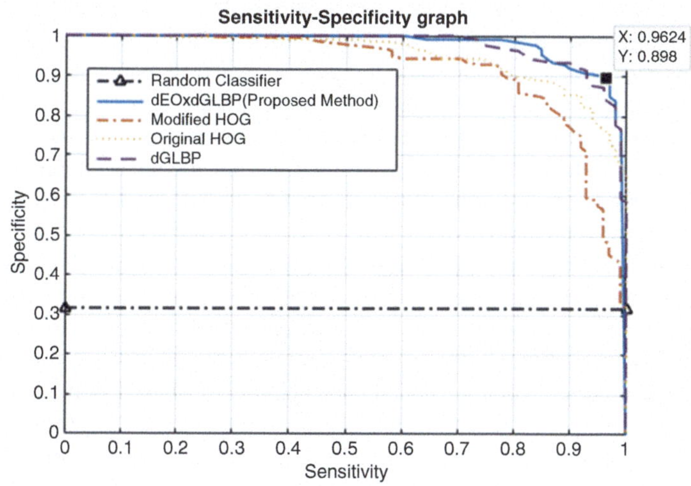

Fig. 6 Sensitivity–specificity curves, for varying threshold values, obtained from the outdoor dataset [27]

experimental results have shown that combining two different sensor modalities provides much higher sensitivity and a significant decrease in the number of false positives during daily activities, compared to accelerometer-only and camera-only methods.

2 Robust and Reliable Step Counting by Mobile Phone Cameras

2.1 Introduction

Step counting is being increasingly used as an activity-level measure, which is evidenced by different types of commercially available wristbands, pedometers, and apps developed for smartphones and smart watches. In addition to measuring daily activity levels and keeping logs for health monitoring, an accurate and reliable count of footsteps can be used for motion estimation, calculating traveled distance and indoor navigation. Yet, most of the available devices and approaches for step counting rely only on accelerometer data, and thus are prone to over-counting. Moreover, most existing devices calculate the traveled distance based on the counted number of steps and a preset stride length, or rely on GPS data, which might not be suitable for GPS-denied areas and indoor environments.

There has been significant amount of research on algorithms using the accelerometer data from smartphones. Park et al. [32] presented an accelerometer-based activity tracker on smartphones to provide high accuracy in location estimation for indoor environments. Pan and Lin [33] proposed an accelerometer-based step counting algorithm for smartphone users, which does not require the user to have the smartphone attached to the body while walking. Another smartphone application uses accelerometer data to analyze walking patterns and compute distance traveled for clinical purposes when the phone is attached on top of a belt around the waist [34]. Brajdic and Harle [35] tested accelerometer-based step counter algorithms while trying different locations for the smartphone. They have found that certain locations such as back pocket of trousers degrade the performance of the algorithms significantly.

In general, accelerometer-based applications on smart watches or smartphones are prone to over-counting, since they are sensitive to even small motions, and can count other routine movements as steps. Moreover, when people are exposed to acceleration, e.g., inside a vehicle or an elevator, accelerometer-based step counters usually keep counting. When we tested an accelerometer-based step counter application running on Apple™ iPhone 6, it was observed that the application increments the number of steps when the user is traveling in a car or taking the elevator. Also, it has been observed that when users walk really slowly, or when they stop and start walking again, the accelerometer-based counting becomes unreliable. Marschollek et al. [36] compared several accelerometer-based step counting algorithms on both healthy subjects and geriatric patients with mobility impairments. It is stated that [36] none of the algorithms worked very well, and that more research is needed to prove the validity of these algorithms for the elderly. Since accurate step detection is very important for indoor positioning systems, reliable and more precise alternatives are needed for step detection and counting.

Different from the aforementioned work, Aubeck et al. [37] use data from a camera sensor to detect steps. They use template matching based on the appearance

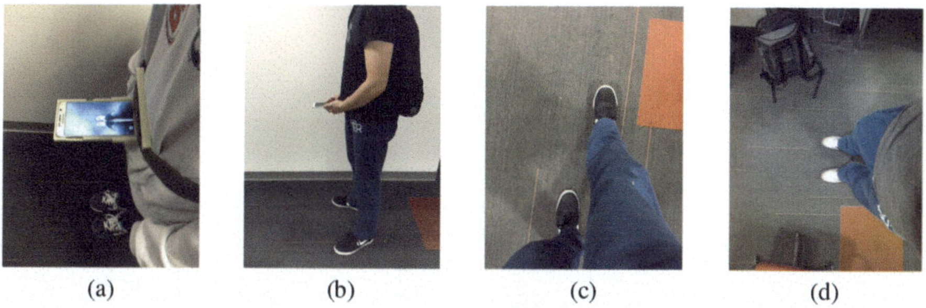

Fig. 7 Subjects (**a**) using the 3D-printed phone holder, (**b**) holding the phone, and example images obtained while walking from the camera of a (**c**) smartphone and (**d**) Google™ glass [41]

and disappearance of the forward section of the feet. It is stated that the method does not perform well for fast movements.

In addition to relying on only one sensor modality, most of the aforementioned works only focus on step counting and do not address the issue of calculating the total traveled distance. Most existing approaches and commercial devices calculate the traveled distance based on the number of counted steps and a preset stride length. This stride length is usually asked from the user, or is calculated based on the user's height. On the other hand, there is a body of work that relies on GPS data, and thus might not be suitable for GPS-denied areas and complex indoor environments [38–40], for which the GPS data is either unavailable or highly unreliable.

In this section, we present an overview of our work [41–43] on autonomous and robust footstep counting and tracking, and calculating stride length by using both accelerometer and camera data from smartphones or a Google™ glass. To provide higher precision, instead of using a preset stride length, the proposed method calculates the distance traveled with each step by incorporating data from different sensor modalities. In addition, if camera is tilted significantly, the angle data from the gravity sensor is used to account for camera geometry and increase the precision of the calculated step length. If a smartphone is being used, the subjects can hold the phone as they normally would when they read e-mails or browse pages on the web. As a proof of concept, we also designed and printed a holder, by using a 3D printer, for hands-free usage and longer walking experiments. The images captured by the rear-facing camera of the smartphone are used for the proposed method. Moreover, we used a Google™ glass as a different use scenario. Different usage styles and example images from a smartphone and Google™ glass can be seen in Fig. 7.

2.2 Proposed Method

The proposed method incorporates data from camera, accelerometer, and gravity sensors to accurately determine the start and end of a frame, and track and measure the length of *each step*. Data from the gravity sensor is used to detect if camera has been tilted significantly, and account for camera geometry. A flow diagram is shown in Fig. 8.

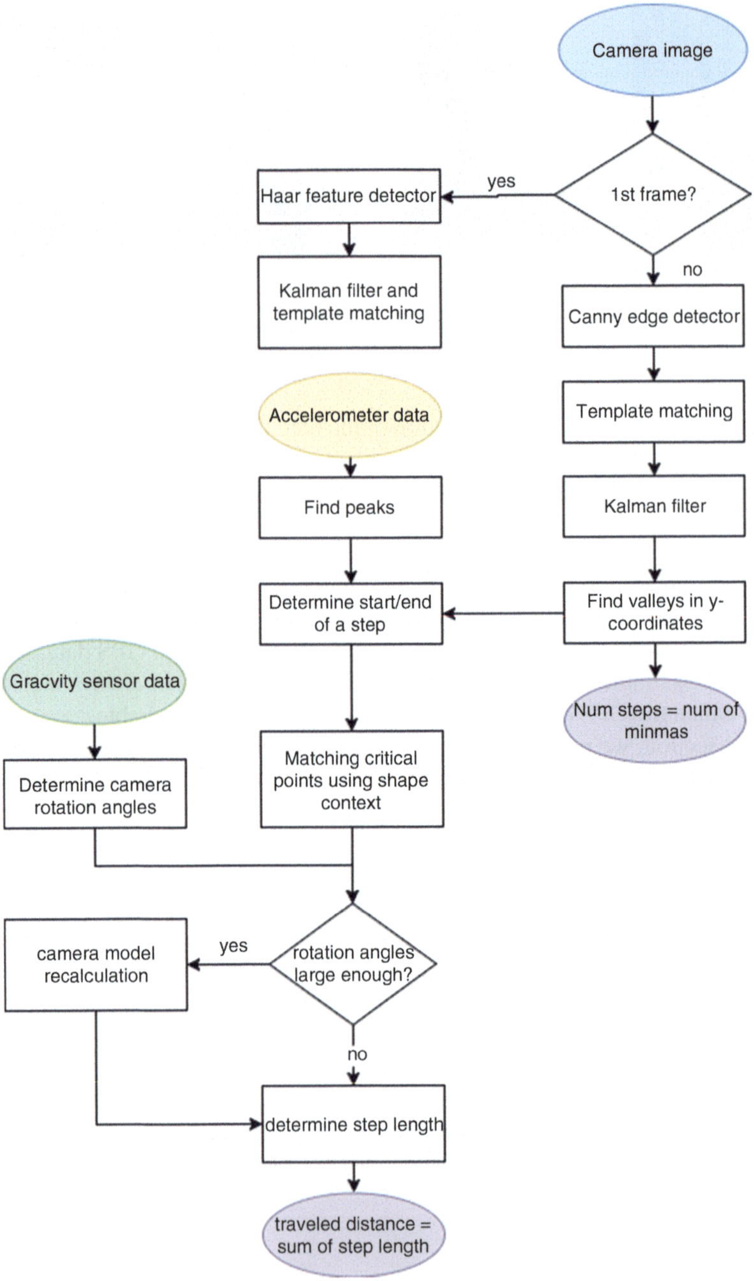

Fig. 8 Flow diagram [41]

Only for the first frame, Haar-based detection is used to detect the feet and obtain bounding boxes around them. The sizes and locations of these bounding boxes are used to initialize the matching template size, search region size, and the Kalman filter tracker. For subsequent images, edges are detected first by using the Canny edge detection algorithm. Several example images obtained after edge detection can be seen in Figs. 9b and 10b for a smartphone and Google glass. The blue boxes in these images indicate the search region. Then, the binary template shown in Fig. 11 is used to perform template matching only in these determined search regions. Since much smaller search regions are used instead of the entire image, the computational efficiency of the proposed method is increased. Figures 9c and 10c show the normalized value of correlation coefficient at each pixel location in the search regions, wherein brighter pixels correspond to higher values, and the two brightest spots are the locations of two detected feet. The detected locations are marked in green in Figs. 9b and 10b.

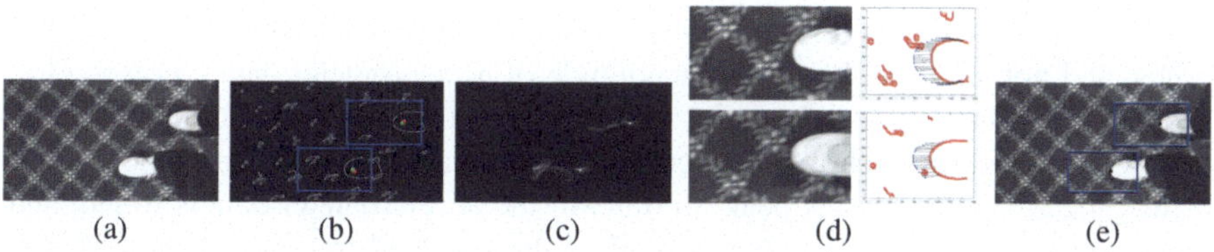

Fig. 9 (**a**) Gray scale image from input of a smartphone camera, (**b**) detected edges and search regions (blue boxes), (**c**) correlation coefficient values, (**d**) feet subimages and shape context analysis, (**e**) detected tip locations of feet [41]

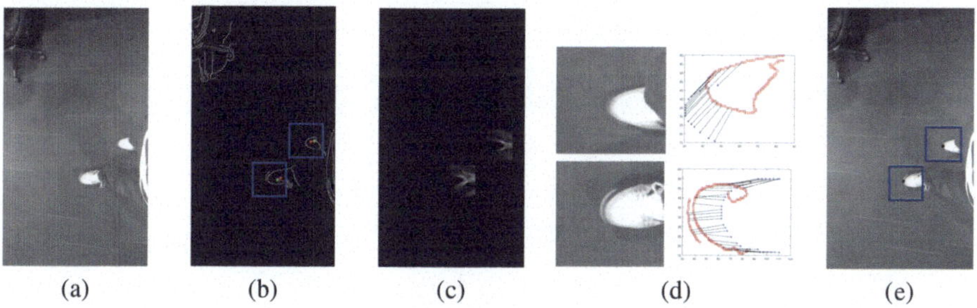

Fig. 10 (**a**) Gray scale image from input of a Google™ glass camera, (**b**) detected edges and search regions (blue boxes), (**c**) correlation coefficient values, (**d**) feet subimages and shape context analysis, (**e**) detected tip locations of feet [41]

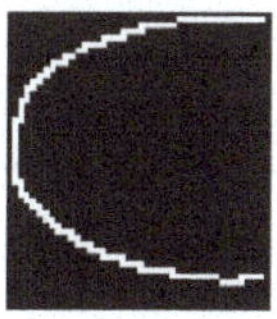

Fig. 11 The template used for matching [41]

A Kalman filter-based tracker is created in the first frame by using the bounding boxes around the feet detected by the Haar-based detector. The feet locations are then tracked in the subsequent frames. If feet locations cannot be detected by using template matching, then the locations are updated by using the predictions from the tracker. The detected and tracked locations for the feet are shown in green and red, respectively, in Figs. 9b and 10b. Then, the *y*-coordinates of the feet locations are saved. As seen in the flow diagram in Fig. 8, we then find the valleys in the *y*-coordinate values by using a peak detection algorithm by Yoder [44]. The number of steps taken is set to be the total number of detected valley points.

As mentioned above, instead of using preset stride lengths and assuming equal-length steps, we calculate the length of each step by incorporating data from the camera, accelerometer, and the gravity sensor. For better accuracy in calculating step length, it is very important to precisely determine the start/end frames of a step. The valley locations, detected as described above, may not exactly correspond to actual start/end of a step. In order to address this challenge, we incorporate accelerometer data. First, the peaks of the magnitude of accelerometer data are found. This, by itself, does not solve the problem either, since accelerometer data can be noisy at times, and not all the peaks in the magnitude of accelerometer data correspond to actual steps. Thus, we combine the valley information obtained from the camera data with the information from the accelerometer data. For a valley in *y*-coordinates of feet locations, the closest peak, in time, of the accelerometer data is found, and the corresponding frame number is used as the start/end frame of a step. This is illustrated in Fig. 12. This approach prevents the proposed method being affected by the noisy peaks in the accelerometer data.

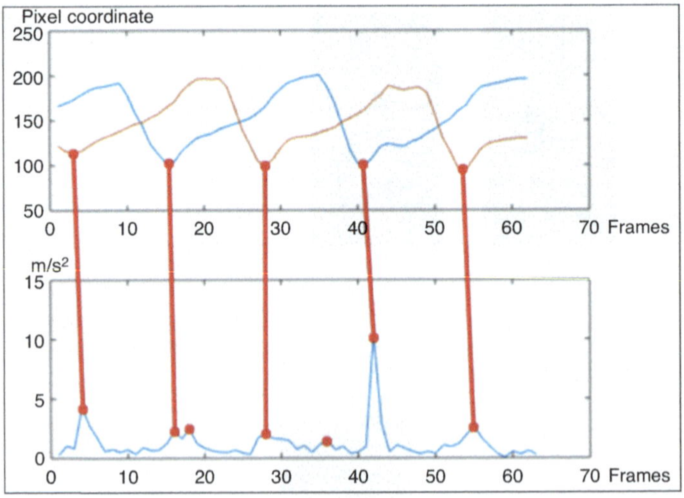

Fig. 12 Finding correspondences between the valleys of camera data, and the peaks of accelerometer data [41]

After the start/end frame of a step is determined, the next stage is to find the tips of each foot on those frames so that the step length can be calculated. For this, shape context [45] is applied to the output of the edge detection. The detected tip locations can be seen in Figs. 9d, e and 10d, e.

Before computing the step length, camera rotation angle obtained from the gravity sensor is checked. This angle is used to determine if a significant change happened in the camera orientation. More specifically, if the rotation angle is greater than 15°, which is an empirically determined threshold, then the camera geometry model is used to incorporate this change, and the image locations of extracted tip points are transformed to the world coordinate system as described below. This step is especially important when the camera is hold by hand.

If the angles obtained from the gravity sensor are θ' and ϕ', we use $\theta = \theta' - \frac{\pi}{2}$, $\phi = \phi' - \frac{\pi}{2}$ for the rotation of the camera along the x-axis and y-axis, respectively. Then, the image coordinates can be transformed to world coordinates by using the rotation matrix. Since the captured image is assumed to be of a ground plane, a 2D planar transformation is used. The details of how the step length is calculated can be found in [41].

2.3 Experimental Results

We have performed various experiments with a smartphone as well as a Google™ glass. We compared the proposed method with the accelerometer-based step counter apps in terms of both the counted steps and traveled distance. In different sets of experiments, subjects walked continuously, they stopped and started walking again, and also followed a zig-zag-like pattern. For comparison purposes, the subjects carried three mobile devices simultaneously at different body locations. More specifically, two smartphones were carried in a front pant pocket and in a backpack, and these phones ran readily available, accelerometer-based apps for step counting and distance calculation. The third mobile device is either a smartphone or a Google™ glass used to capture data for the proposed method. The captured image size is 1920×1080 pixels. The image size used for template matching is 240×160. Full-sized images are used for shape context calculation, and to improve the precision of the calculated traveled distance.

With smartphone experiments, 15–20 different subjects either carried the phone by hand or used the holder around the waist, and walked for various distances on a path involving several corridors. When 15 subjects held the phone by hand, the average error rates for the number of steps are 4.77%, 24.96%, and 24% for the proposed method, and the accelerometer-based apps running on the phone carried in the front pocket and in a backpack, respectively. The average error rates for the traveled distance are 3.95%, 53.31%, and 48.75% for the proposed method, and the accelerometer-based apps running on the phone carried in the front pocket and in a backpack, respectively. Thus, the proposed method provides the lowest average error rate for both the number of steps and the traveled distance compared to the accelerometer-based apps.

In addition, we performed smartphone experiments with 20 subjects by using the 3D-printed holder as a proof of concept. In these experiments, subjects walked a longer distance (2000 ft). Tables 1 and 2 summarize the results for the counted number of steps and the reported traveled distance, respectively, obtained by the proposed method and the accelerometer-based apps. The proposed method increases the accuracy significantly, and provides the lowest average error rate for both the number of steps and the traveled distance across 20 different subjects. The average error rates for the number of steps are 3.86%, 31.76%, and 30.98% for the proposed method, and the accelerometer-based apps running on the phone carried in the front pocket and in a backpack, respectively. The average error rates for the traveled distance are 3.92%, 67.65%, and 62.47% for the proposed method, and the accelerometer-based apps running on the phone carried in the front pocket and in a backpack, respectively. Table 2 also shows the improvement in accuracy obtained by incorporating the camera geometry. More specifically, by incorporating the camera geometry to the proposed method, the average error in the traveled distance decreases from 4.98% to 3.92%.

We also performed experiments with 10 subjects wearing a Google™ glass, and walking a distance of 2000 ft. The average error rates for the number of steps are 7.62%, 23.47%, and 25.17% for the proposed method, and the accelerometer-based apps running on the phone carried in the front pocket and in a backpack, respectively. The average error rates for the traveled distance are 7.78%, 77.44%, and 74.63% for the proposed method, and the accelerometer-based apps running on the phone carried in the front pocket and in a backpack, respectively. Experimental results show that the proposed method provides a significant increase in accuracy, and has the lowest average error rate both in number of steps taken and the distance traveled compared to commercially available, accelerometer-based step counters and apps.

For the experiments above, subjects were walking in a continuous manner, and only turning around the corners. Additionally, we performed experiments, with 10 subjects using the smartphone, during which subjects stopped and started walking again several times, and sometimes followed a zig-zag-like pattern while walking. For this experiment, the average error rates for the number of steps are 4.09%, 24.17%, and 25.1% for the proposed method, and the accelerometer-based apps running on the phone carried in the front pocket and in a backpack, respectively. The average error rates for the traveled distance are 5.56%, 48.1%, and 59.25% for the proposed method, and the accelerometer-based apps running on the phone carried in the front pocket and in a backpack, respectively.

In addition, we performed experiments with two subjects walking in low illumination conditions by using the built-in flashlight of the smartphone. If the highest brightness value in the image is smaller than 128, the flashlight is turned on. The proposed method still increases the accuracy significantly. More specifically, the accuracy rates for the number of steps are 96.3%, 77.23%, and 53.36% for the proposed method, and the accelerometer-based apps running on the phone carried

Table 1 Step counting results from the longer-distance experiments with a phone holder [41]

Subjects	Ground truth	Proposed method		Front pocket-accelerometer		Backpack-accelerometer	
	No. of steps	Counted	Error	Counted	Error	Counted	Error
Subject 1	1830	1749	4.43%	1165	36.34%	1373	24.97%
Subject 2	2023	1987	1.78%	1349	33.32%	996	50.77%
Subject 3	1996	1916	4.01%	1433	28.21%	1505	24.60%
Subject 4	2011	1936	3.73%	2679	33.22%	916	54.45%
Subject 5	1759	1707	2.96%	778	55.77%	1327	24.56%
Subject 6	1700	1635	3.82%	1521	10.53%	1299	23.59%
Subject 7	1861	1809	2.79%	2300	23.59%	1090	41.43%
Subject 8	1728	1631	5.61%	1517	12.21%	1444	16.44%
Subject 9	1838	1778	3.26%	996	45.81%	1919	4.41%
Subject 10	1830	1764	3.61%	1209	33.93%	1363	25.52%
Subject 11	1949	1880	3.54%	2596	33.20%	860	55.87%
Subject 12	2067	1953	5.52%	1247	39.67%	695	66.38%
Subject 13	1899	1807	4.84%	886	53.34%	2308	21.54%
Subject 14	1818	1734	4.62%	1541	15.24%	1404	22.77%
Subject 15	1800	1760	2.22%	1398	22.33%	2137	18.72%
Subject 16	2044	1959	4.16%	692	66.14%	1002	50.98%
Subject 17	1703	1608	5.58%	1416	16.85%	1923	12.92%
Subject 18	1895	1803	4.85%	1478	22.01%	1384	26.97%
Subject 19	1767	1700	3.79%	1503	14.94%	1993	12.79%
Subject 20	1906	1866	2.10%	2640	38.51%	2670	40.08%
			Avg. err 3.86%		Avg. err 31.76%		Avg. err 30.98%
			Std. dev. 0.011		Std. dev. 0.1489		Std. Dev. 0.166

Table 2 Traveled distance results from the longer-distance experiments with a phone holder [41]

Subjects	Ground truth		Proposed meth. w/ cam. geometry		Proposed Meth. w/o cam. geometry		Front pocket-accelerometer		Backpack-accelerometer	
	Distance (ft)	Distance (mile)	Distance (ft)	Error	Distance (ft)	Error	Distance (mile)	Error	Distance (mile)	Error
Subject 1	2000	0.38	1942.6	2.87%	1920.3	3.99%	0.48	25.13%	0.56	47.48%
Subject 2	2000	0.38	1978.0	1.10%	1951.1	2.45%	0.55	44.74%	0.41	7.89%
Subject 3	2000	0.38	1930.2	3.49%	1902.9	4.86%	0.58	53.92%	0.61	61.65%
Subject 4	2000	0.38	1986.8	0.66%	1970.2	1.49%	1.09	187.76%	0.37	1.61%
Subject 5	2000	0.38	1926.0	3.70%	1899.7	5.02%	0.32	16.43%	0.54	42.53%
Subject 6	2000	0.38	1894.1	5.30%	1878.7	6.07%	0.62	63.37%	0.53	39.53%
Subject 7	2000	0.38	1912.2	4.39%	1889.5	5.53%	0.94	147.05%	0.44	17.08%
Subject 8	2000	0.38	1862.5	6.88%	1849.3	7.54%	0.62	62.94%	0.59	55.10%
Subject 9	2000	0.38	1936.0	3.20%	1908.1	4.60%	0.41	6.98%	0.78	106.12%
Subject 10	2000	0.38	1940.3	2.99%	1919.0	4.05%	0.49	29.86%	0.56	46.40%
Subject 11	2000	0.38	1932.0	3.40%	1907.4	4.63%	1.06	178.84%	0.35	7.63%
Subject 12	2000	0.38	1910.7	4.47%	1891.5	5.43%	0.51	33.94%	0.28	25.35%
Subject 13	2000	0.38	1932.1	3.40%	1913.5	4.33%	0.36	4.83%	0.94	147.91%
Subject 14	2000	0.38	1927.9	3.61%	1917.9	4.11%	0.63	65.52%	0.57	50.81%
Subject 15	2000	0.38	1922.6	3.87%	1898.6	5.07%	0.57	50.16%	0.87	128.54%
Subject 16	2000	0.38	1948.2	2.59%	1930.2	3.49%	0.28	25.67%	0.41	7.63%
Subject 17	2000	0.38	1847.9	7.61%	1825.0	8.75%	0.58	52.09%	0.78	106.55%
Subject 18	2000	0.38	1887.8	5.61%	1867.8	6.61%	0.60	58.75%	0.56	48.66%
Subject 19	2000	0.38	1874.1	6.30%	1846.9	7.66%	0.61	61.44%	0.81	114.07%
Subject 20	2000	0.38	1939.4	3.03%	1921.5	3.93%	1.08	183.57%	1.09	186.79%
				Avg. err 3.92%		Avg. err 4.98%		Avg. err 67.65%		Avg. err 62.47%
				Std. dev. 0.017		Std. dev. 0.017		Std. dev. 0.5672		Std. dev. 0.5077

in the front pocket and in a backpack, respectively. The accuracy rates for the traveled distance are 95.28%, 64.34%, and 13.53% for the proposed method, and the accelerometer-based apps running on the phone carried in the front pocket and in a backpack, respectively.

In conclusion, we have proposed an autonomous and robust method for counting footsteps, and tracking and calculating stride length by using both accelerometer and camera data from smartphones or a Google™ glass. To provide higher precision, instead of using a preset step and/or stride length, the proposed method calculates the distance traveled with each step. The proposed method has been compared with the commonly used accelerometer-based step counter applications (apps). The results show that the proposed method provides a significant increase in accuracy, and has the lowest average error rate both in number of steps taken and the distance traveled compared to commercially available, accelerometer-based step counters and apps. The experimental results also show that the performance of the accelerometer-based methods varies significantly across different subjects with higher standard deviation values, depends on the walking pace of the subjects, and is sensitive to the location of the device.

3 Human Activity Classification from Egocentric Images and IMU Data

3.1 Introduction

Many methods have been proposed for human activity classification, which rely either on inertial measurement unit (IMU) data [46] or data from static cameras watching subjects (third-person video data) [47–51]. Systems relying only on IMU data are limited in the complexity of the activities that they can detect. For instance, we can detect a sitting event by using accelerometer data, but cannot determine whether the user has sat on a chair or sofa, or what type of environment the user is in.

There have been relatively less work using egocentric videos and even fewer approaches combining egocentric video and IMU data. Existing work using egocentric videos has focused on classifying the activities "seen" by the camera [52, 53] or the activity of the person wearing the camera [54–56, 58–60, 63]. Zhan et al. [61] use both IMU and camera sensors, in the form of reading glasses, to classify activities of elderly and disabled patients, such as walking, standing, lying down, drinking, writing, and sitting down. Windau and Itti [62] also use both IMU sensor and camera data and report 81.5% accuracy for 20 activities. Both of these methods still focus on activities that can be classified by accelerometer data. They do not perform object detection in the scene, and do not focus on activity types that cannot be classified by only accelerometer data. Spriggs et al. [57] explore first-person sensing through a wearable camera and IMUs for temporally segmenting human motion into actions and performing activity classification in the context of cooking and recipe preparation. They extract GIST features from camera data. A review of the egocentric vision systems, for the recognition of activities of daily living, is presented in the survey by Nguyen et al. [64], and it was concluded that the performance of current systems is far from satisfactory.

In this section, we present an overview of our work on activity classification using accelerometer and camera data from a smartphone [65]. In this work, we focus on the type of activities, which cannot be classified by just accelerometer data, and require detection of objects in the environment from camera data. Without loss of generality, we have performed experiments on differentiating between sitting on a sofa, sitting on a chair, and walking through doorways. Only accelerometer data can be enough to detect walking, but it is not enough to detect that the person is walking through a doorway, and thus changing rooms. Similarly, camera data becomes necessary to detect whether the person sat on a sofa or a chair.

3.2 Proposed Method

In our proposed approach, both camera and accelerometer data, obtained from a smartphone, are employed. Camera data is used to compute optical flow vectors, and perform object detection via an aggregate channel features (ACF)-based detector [66]. Different features are extracted from the accelerometer data to obtain a 14D vector. Then, a 3D vector is obtained based on the y component of the optical flow vectors, and concatenated to the 14D vector to obtain a 17D feature vector. A multi-class SVM is trained and employed to differentiate between motion-related activities of walking, sitting from standing position, standing from sitting position, and being immobile.

The ACF-based detector is trained to detect chairs, sofas, and doors/doorways. It provides bounding boxes around the detected objects. The features obtained from these bounding boxes together with the label describing the detected object type are saved as a vector. These vectors are used to train another multi-class SVM to detect approaching a sofa, chair, or door. Then, a hidden Markov model (HMM) is built to be used in the final stage to detect whether the person carrying the smartphone is sitting on a chair, sitting on a sofa, or walking through a door. More details about these steps are described in [65].

3.3 Experimental Results

For training the two multi-class SVMs and the HMM, a 30-min video is captured from a subject wearing the smartphone as seen in Fig. 13. The subject performs a total of 193 activities consisting of sitting on a chair, sitting on a sofa, and walking through doors.

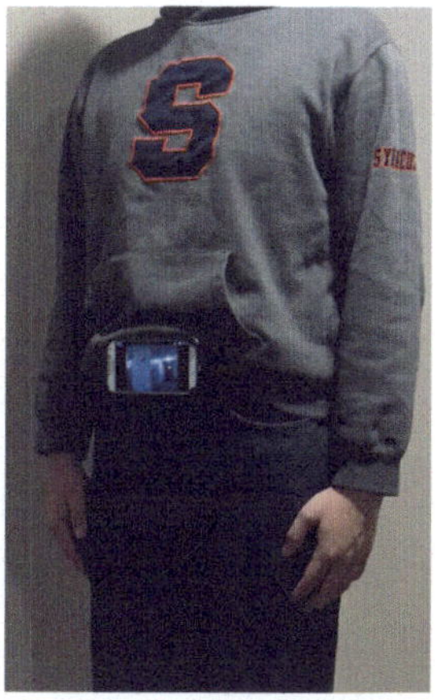

Fig. 13 Experimental setup [65]

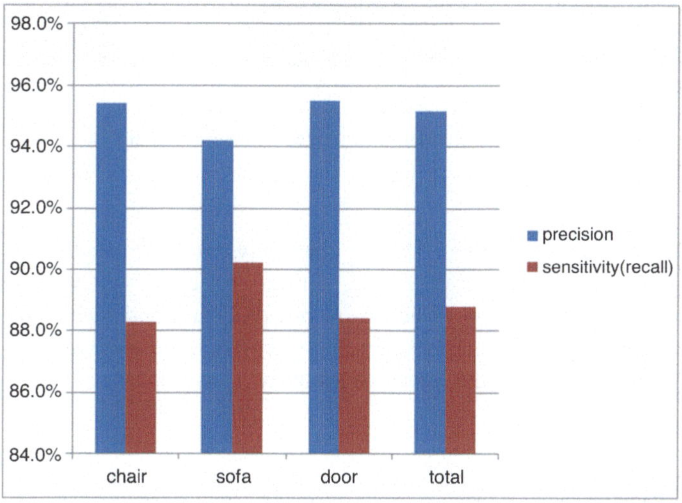

Fig. 14 The overall precision and recall values across all subjects [65]

Experiments have been performed with seven different subjects in an indoor environment. During a period of about 10 min, each subject sits on a chair, sits on a sofa, and walks through doors several times. More specifically, there are 100, 90, 80, 101, 88, 89, and 120 activities performed by the subjects 1 through 7, respectively. The smartphone is carried on a belt around the waist, and the camera faces forward as seen in Fig. 13. The captured image size is 640×480 pixels.

The overall precision and recall values, across all seven subjects, for all three activity types are plotted in Fig. 14. Example images from different test videos showing the detection of chairs, sofas, and doorways by the ACF-based detector are shown in Fig. 15.

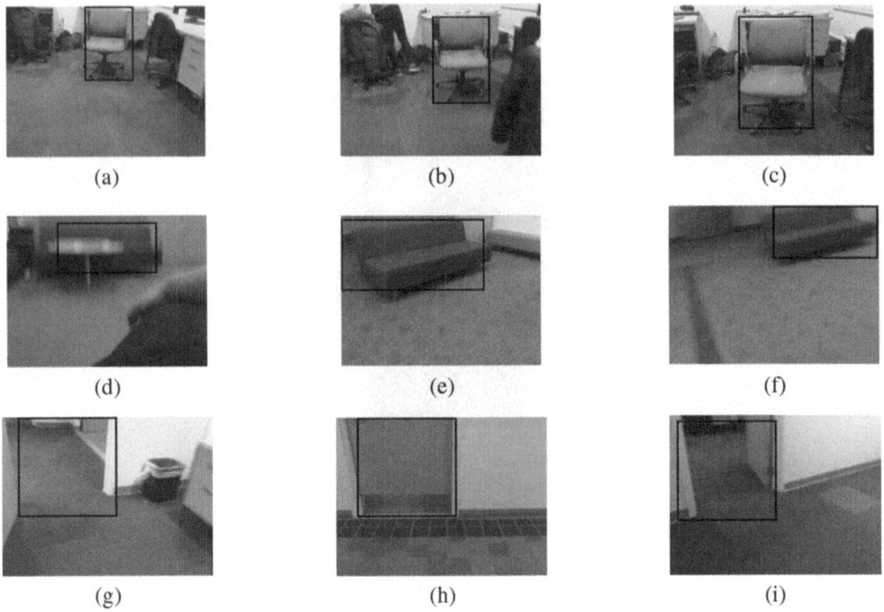

Fig. 15 Example images from a test video showing the detection of chairs (**a–c**), sofas (**d–f**), and doorways (**g–i**) [65]

As mentioned above, without loss of generality, we have presented results on differentiating between sitting on a sofa, sitting on a chair, and walking through a door. The proposed method achieved overall precision and recall rates of 95% and 89%, respectively. The number of activity types can be increased by detecting different types of objects in the scene, and modifying the HMM.

4 Conclusion

Wearable sensors, including cameras, provide wide-ranging and beneficial applications for the society. Thanks to the advances in wearable and mobile device technology, it has become feasible to employ them as stand-alone platforms and perform various tasks. Mobile devices now include various sensors, such as camera, accelerometer, gyroscope, and magnetometer, and provide high processing capability. Moreover, wearable sensors provide continuous monitoring as opposed to sensors installed at fixed locations.

In this chapter, we have presented our work on three different applications using egocentric videos from a wearable camera and also IMU data. First, we have provided an overview of our fall detection system. Second, we have presented an autonomous and robust method for counting footsteps, and tracking and calculating stride length. Third, we have provided an overview of our work on detecting activities with more details and context by using accelerometer and egocentric video data obtained from a smartphone. The proposed algorithms and solutions presented in this chapter can be utilized in activity monitoring for different applications including healthcare, monitoring elderly people leaving by themselves, and for indoor navigation.

Promising results have been obtained when wearable camera and IMU data is used together. Experimental results have shown that the proposed fall detection method, combining two different sensor modalities, not only increases the detection rate, but also decreases the number of false alarms during daily activities, compared to only accelerometer-only or camera-only methods.

For the footstep counting and traveled distance calculation, the proposed method has been compared with the commonly used accelerometer-based step counter applications (apps). It has been observed that the proposed method provides a significant increase in accuracy, and has the lowest average error rate both in number of steps taken and the distance traveled compared to commercially available, accelerometer-based step counters and apps. Also, experimental results have shown that the performance of the accelerometer-based methods varies significantly across different subjects with higher standard deviation values, depends on the walking pace of the subjects, and is sensitive to the location of the device.

As the third application, we focused on types of activities that cannot be differentiated by just accelerometer data, since they are more complex, and require context and detection of objects in the environment from camera data. Without loss of generality, we have presented results on differentiating between sitting on a sofa, sitting on a chair, and walking through a door. The activity types can be increased by detecting different types of objects in the scene, and modifying the HMM. The proposed method achieves overall precision and recall rates of 95% and 89%, respectively.

Acknowledgements The information, data, or work presented herein was funded in part by the National Science Foundation (NSF) under CAREER grant CNS-1206291, Grant CNS-1302559, Grant CNS-1739748, and by the Advanced Research Projects Agency-Energy (ARPA-E), U.S. Department of Energy, under Award Number DE-AR0000940. The views and opinions of authors expressed herein do not necessarily state or reflect those of the United States Government or any agency thereof.

References

1. Melonie Heron, "Deaths: Leading causes 2007," August 2011, number 8, pp. 17, 21–22.
2. Janet Shelfer, David Zapala, and Larry Lundy, "Fall risk, vestibular schwannoma, and anticoagulation therapy," United States, McLean, 2008, pp. 237–45, American Academy of Audiology.
3. R. C. O. Voshaar, S. Banerjee, M. Horan, R. Baldwin, N. Pendleton, R. Proctor, N. Tarrier, Y. Woodward, and A. Burns, "Predictors of incident depression after hip fracture surgery," *The American Journal of Geriatric Psychiatry*, vol. 15, no. 9, pp. 807–814, 2007.

4. LD Gillespie, WJ Gillespie, MC Robertson, SE Lamb, RG Cumming, and BH Rowe, "Interventions for preventing falls in elderly people," 2003, pp. 692–693.
5. S. Cagnoni, G. Matrella, M. Mordonini, F. Sassi, and L. Ascari, "Sensor fusion-oriented fall detection for assistive technologies applications," in *Intelligent Systems Design and Applications, 2009. ISDA '09. Ninth International Conference on*, Nov 2009, pp. 6730–678.
6. Miao Yu, A. Rhuma, S.M. Naqvi, Liang Wang, and J. Chambers, "A posture recognition based fall detection system for monitoring an elderly person in a smart home environment," *Information Technology in Biomedicine, IEEE Transactions on*, vol. 16, no. 6, pp. 1274–1286, Nov 2012.
7. N. Noury, T. Herve, V. Rialle, G. Virone, E. Mercier, G. Morey, A. Moro, and T. Porcheron, "Monitoring behavior in home using a smart fall sensor and position sensors," in *Microtechnologies in Medicine and Biology, 1st Annual International, Conference On. 2000*, 2000, pp. 607–610.
8. Muhammad Mubashir, Ling Shao, and Luke Seed, "A survey on fall detection: Principles and approaches," 2013, pp. 144–152, <ce:title>Special issue: Behaviours in video</ce:title>.
9. G.A. Koshmak, M. Linden, and A. Loutfi, "Evaluation of the android-based fall detection system with physiological data monitoring," in *Engineering in Medicine and Biology Society, 35th Annual Int'l Conf. of the IEEE*, July 2013, pp. 1164–1168.
10. Yabo Cao, Yujiu Yang, and Wenhuang Liu, "E-falld: A fall detection system using android based smartphone," in *Fuzzy Systems and Knowledge Discovery (FSKD), 2012 9th Int'l Conf. on*, May 2012, pp. 1509–1513.
11. Shih-Hau Fang, Yi-Chung Liang, and Kuan-Ming Chiu, "Developing a mobile phone-based fall detection system on android platform," in *Computing, Communications and Applications Conf. (ComComAp)*, Jan 2012, pp. 143–146.
12. Jiangpeng Dai, Xiaole Bai, Zhimin Yang, Zhaohui Shen, and Dong Xuan, "Perfalld: A pervasive fall detection system using mobile phones," in *Pervasive Computing and Communications Workshops, IEEE Int'l Conf. on*, March 2010, pp. 292–297.
13. WanminWu, Sanjoy Dasgupta, E. Ernesto Ramirez, Carlyn Peterson, and J. Gregory Norman, "Classification accuracies of physical activities using smartphone motion sensors," *J Med Internet Res*, vol. 14, no. 5, pp. e130, Oct 2012.
14. M. Cornacchia, K. Ozcan, Y. Zheng, and S. Velipasalar, "A survey on activity detection and classification using wearable sensors," *IEEE Sensors Journal*, vol. 17, pp. 386–403, Jan. 2017.
15. M. Yu, A. Rhuma, S. M. Naqvi, L. Wang, and J. Chambers, "A posture recognition-based fall detection system for monitoring an elderly person in a smart home environment," nov. 2012, pp. 1274–1286.
16. N. Noury, A. Galay, J. Pasquier, and M. Ballussaud, "Preliminary investigation into the use of autonomous fall detectors," in *Engineering in Medicine and Biology Society, 2008. EMBS 2008. 30th Annual International Conference of the IEEE*, aug. 2008, pp. 2828–2831.
17. Zhengming Fu, T. Delbruck, P. Lichtsteiner, and E. Culurciello, "An address-event fall detector for assisted living applications," 2008, pp. 88–96.
18. A. Sixsmith and N. Johnson, "A smart sensor to detect the falls of the elderly," 2004, pp. 42–47.
19. S. Zambanini, J. Machajdik, and M. Kampel, "Detecting falls at homes using a network of low-resolution cameras," in *Proc. 10th IEEE Int Information Technology and Applications in Biomedicine (ITAB) Conf*, 2010, pp. 1–4.
20. P. Siciliano, A. Leone, G. Diraco, C. Distante, M. Malfatti, L. Gonzo, M. Grassi, A. Lombardi, G. Rescio, and P. Malcovati, "A networked multisensor system for ambient assisted living application," in *Proc. 3rd Int. Workshop Advances in sensors and Interfaces IWASI 2009*, 2009, pp. 139–143.
21. A.N. Belbachir, M. Litzenberger, S. Schraml, M. Hofstatter, D. Bauer, P. Schon, M. Humenberger, C. Sulzbachner, T. Lunden, and M. Merne, "Care: A dynamic stereo vision sensor system for fall detection," in *Circuits and Systems (ISCAS), 2012 IEEE International Symposium on*, may 2012, pp. 731–734.

22. S. Fleck, R. Loy, C. Vollrath, F. Walter, and W. Strasser, "Smartclassysurv - a smart camera network for distributed tracking and activity recognition and its application to assisted living," in *Distributed Smart Cameras, 2007. ICDSC '07. First ACM/IEEE International Conference on*, sept. 2007, pp. 211–218.
23. K. Ozcan, A. Mahabalagiri, and S. Velipasalar, "Fall detection and activity classification using a wearable smart camera," in *Proc. of the IEEE International Conf. on Multimedia and Expo (ICME)*, July 2013.
24. Mauricio Casares, Koray Ozcan, Akhan Almagambetov, and Senem Velipasalar, "Automatic fall detection by a wearable embedded smart camera," in *Distributed Smart Cameras (ICDSC), 2012 Sixth International Conference on*, 2012, pp. 1–6.
25. K. Ozcan, A.K. Mahabalagiri, M. Casares, and S. Velipasalar, "Automatic fall detection and activity classification by a wearable embedded smart camera," *Emerging and Selected Topics in Circuits and Systems, IEEE Journal on*, vol. 3, no. 2, pp. 125–136, 2013.
26. K. Ozcan and S. Velipasalar, "Wearable camera- and accelerometer-based fall detection on portable devices," *IEEE Embedded Systems Letters*, vol. 8, no. 1, pp. 6–9, March 2016.
27. K. Ozcan, P. Varshney, and S. Velipasalar, "Autonomous fall detection with wearable cameras by using relative entropy distance measure," *IEEE Transactions on Human-Machine Systems*, vol. 47, pp. 31–39, Febr. 2017.
28. Aiden R. Doherty, Steve E. Hodges, Abby C. King, Alan F. Smeaton, Emma Berry, Chris J.A. Moulin, Sin Lindley, Paul Kelly, and Charlie Foster, "Wearable cameras in health: The state of the art and future possibilities," *American Journal of Preventive Medicine*, vol. 44, no. 3, pp. 320–323, 2013.
29. N. Dalal and B. Triggs, "Histograms of oriented gradients for human detection," in *Computer Vision and Pattern Recognition, 2005. CVPR 2005. IEEE Computer Society Conference on*, june 2005, pp. 886–893 vol. 1.
30. J. Kittler, M. Hatef, R.P.W. Duin, and J. Matas, "On combining classifiers," *Pattern Analysis and Machine Intelligence, IEEE Transactions on*, vol. 20, no. 3, pp. 226–239, Mar 1998.
31. N. Jiang, J. Xu, W. Yu, and S. Goto, "Gradient local binary patterns for human detection," in *Circuits and Systems (ISCAS), 2013 IEEE International Symposium on*, 2013, pp. 978–981.
32. Kwanghyo Park, Hyojeong Shin, and Hojung Cha, "Smartphone-based pedestrian tracking in indoor corridor environments," *Personal and Ubiquitous Computing*, vol. 17, no. 2, pp. 359–370, 2013.
33. Meng-Shiuan Pan and Hsueh-Wei Lin, "A step counting algorithm for smartphone users: Design and implementation," *Sensors Journal, IEEE*, vol. 15, no. 4, pp. 2296–2305, April 2015.
34. N.A. Capela, E.D. Lemaire, and N.C. Baddour, "A smartphone approach for the 2 and 6-minute walk test," in *Engineering in Medicine and Biology Society (EMBC), 2014 36th Annual International Conference of the IEEE*, Aug 2014, pp. 958–961.
35. Agata Brajdic and Robert Harle, "Walk detection and step counting on unconstrained smartphones," in *Proceedings of the 2013 ACM International Joint Conference on Pervasive and Ubiquitous Computing*, New York, NY, USA, 2013, UbiComp '13, pp. 225–234, ACM.
36. M. Marschollek, M. Goevercin, K.-H.Wolf, B. Song, M. Gietzelt, R. Haux, and E. Steinhagen-Thiessen, "A performance comparison of accelerometry-based step detection algorithms on a large, non-laboratory sample of healthy and mobility-impaired persons," in *Engineering in Medicine and Biology Society, 2008. EMBS 2008. 30th Annual International Conference of the IEEE*, Aug 2008, pp. 1319–1322.
37. F. Aubeck, C. Isert, and D. Gusenbauer, "Camera based step detection on mobile phones," in *Indoor Positioning and Indoor Navigation (IPIN), 2011 International Conference on*, Sept 2011, pp. 1–7.
38. Stephan Weiss, Davide Scaramuzza, and Roland Siegwart, "Monocular-slam–based navigation for autonomous micro helicopters in GPS-denied environments," *J. Field Robot.*, vol. 28, no. 6, pp. 854–874, Nov. 2011.

39. A. Bachrach, A. de Winter, Ruijie He, G. Hemann, S. Prentice, and N. Roy, "Range - robust autonomous navigation in GPS-denied environments," in *Robotics and Automation, 2010 IEEE Int'l Conf. on*, May 2010, pp. 1096–1097.
40. L. Ojeda and J. Borenstein, 'Personal dead-reckoning system for GPS-denied environments," in *Safety, Security and Rescue Robotics, 2007. SSRR 2007. IEEE International Workshop on*, Sept 2007, pp. 1–6.
41. Y. Lu and S. Velipasalar, "Autonomous footstep counting and traveled distance calculation by mobile devices incorporating camera and accelerometer data," *IEEE Sensors Journal*, vol. 17, no. 21, pp. 7157–7166, Nov 2017.
42. Y. Lu and S. Velipasalar, "Robust footstep counting and traveled distance calculation by mobile phones incorporating camera geometry," in *2016 IEEE International Conference on Image Processing (ICIP)*, Sept 2016, pp. 464–468.
43. K. Ozcan and S. Velipasalar, "Robust and reliable step counting by mobile phone cameras," in *Proceedings of the 9th International Conference on Distributed Smart Cameras*. 2015, ICDSC'15, pp. 164–169, ACM.
44. N. Yoder, "Peakfinder: Quickly finds local maxima (peaks) or minima (valleys) in a noisy signal," 2014.
45. S. Belongie, J. Malik, and J. Puzicha, "Shape matching and object recognition using shape contexts," *Pattern Analysis and Machine Intelligence, IEEE Transactions on*, vol. 24, no. 4, pp. 509–522, Apr 2002.
46. F. J. Ordez and D. Roggen, "Deep convolutional and LSTM recurrent neural networks for multimodal wearable activity recognition," *Sensors*, vol. 16, no. 1, 2016.
47. A. Karpathy, G. Toderici, S. Shetty, T. Leung, R. Sukthankar, and F. Li, "Large-scale video classification with convolutional neural networks," in *Computer Vision and Pattern Recognition*, 2014, pp. 1725–1732.
48. J. Donahue, L. A. Hendricks, S. Guadarrama, M. Rohrbach, S. Venugopalan, T. Darrell, and K. Saenko, "Long-term recurrent convolutional networks for visual recognition and description," in *Comp. Vision and Pattern Recogn.*, 2015, pp. 677–691.
49. D. Tran, L. Bourdev, R. Fergus, L. Torresani, and M. Paluri, "Learning spatiotemporal features with 3d convolutional networks," pp. 4489–4497, 2014.
50. A. Montes, A. Salvador, S. Pascual, and X. Giroinieto, "Temporal activity detection in untrimmed videos with recurrent neural networks," *arXiv:1608.08128*, 2017.
51. S. Buch, V. Escorcia, C. Shen, B. Ghanem, and J. C. Niebles, "SST: Single-stream temporal action proposals," in *IEEE Conf. on CVPR*, 2017.
52. Y. J. Lee, J. Ghosh, and K. Grauman, "Discovering important people and objects for egocentric video summarization," in *Computer Vision and Pattern Recognition*, 2012, pp. 1346–1353.
53. M. S. Ryoo and L. Matthies, "First-person activity recognition: What are they doing to me?," in *IEEE Conf. on CVPR*, 2013, pp. 2730–2737.
54. C. Li and K. M. Kitani, "Pixel-level hand detection in ego-centric videos," in *IEEE Conference on Computer Vision and Pattern Recognition*, 2013, pp. 3570–3577.
55. K. M. Kitani, T. Okabe, Y. Sato, and A. Sugimoto, "Fast unsupervised ego-action learning for first-person sports videos," in *Computer Vision and Pattern Recognition*, 2011, pp. 3241–3248.
56. Y. Li, A. Fathi, and J. M. Rehg, "Learning to predict gaze in egocentric video," in *IEEE International Conference on Computer Vision*, 2014, pp. 3216–3223.
57. E. H. Spriggs, F. De La Torre, and M. Hebert, "Temporal segmentation and activity classification from first-person sensing," in *Computer Vision and Pattern Recogn. Workshops. IEEE Computer Society Conf. on*, 2009, pp. 17–24.
58. A. Fathi, X. Ren, and J. M. Rehg, "Learning to recognize objects in egocentric activities," in *IEEE Conference on Computer Vision and Pattern Recognition*, 2011, pp. 3281–3288.
59. T. Mccandless and K. Grauman, "Object-centric spatio-temporal pyramids for egocentric activity recognition," in *British Machine Vision Conf.*, 2015, pp. 30.1–30.11.
60. M Moghimi, P Azagra, L Montesano, and A. C Murillo, "Experiments on an RGB-D wearable vision system for egocentric activity recognition," in *Computer Vision and Pattern Recognition Workshops*, 2014, pp. 611–617.

61. Kai Zhan, Steven Faux, and Fabio Ramos, "Multi-scale conditional random fields for first person activity recognition on elders and disabled patients," *Pervasive and Mobile Computing*, vol. 16, Part B, pp. 251–267, 2015, Selected Papers from the Twelfth Annual {IEEE} International Conference on Pervasive Computing and Communications (PerCom 2014).
62. Jens Windau and Laurent Itti, "Situation awareness via sensor-equipped eyeglasses," *IEEE Int'l Conf. on Intelligent Robots and Systems*, pp. 5674–5679, 2013.
63. Ekaterina H. Spriggs, Fernando De La Torre, and Martial Hebert, "Temporal segmentation and activity classification from first-person sensing," *2009 IEEE Conf. on Computer Vision and Pattern Recognition*, pp. 17–24, 2009.
64. T. H. Nguyen, J. C. Nebel, and F. Florezrevuelta, "Recognition of activities of daily living with egocentric vision: A review," *Sensors*, vol. 16, no. 1, pp. 72, 2016.
65. Y. Lu and S. Velipasalar, "Human activity classification from wearable devices with cameras," in *2017 51st Asilomar Conference on Signals, Systems, and Computers*, Oct 2017, pp. 183–187.
66. P. Dollár, R. Appel, S. Belongie, and P. Perona, "Fast feature pyramids for object detection," *IEEE Transactions on Pattern Analysis and Machine Intelligence*, vol. 36, no. 8, pp. 1532–1545, Aug 2014.

Nonvolatile Processor Architecture Exploration for Energy-Harvesting Application Scenarios

Kaisheng Ma, Shuangchen Li, Vijaykrishnan Narayanan, and Yuan Xie

Abstract Every shift in the way our devices are connected or powered brings with it a potential for revolution in the usage and capabilities of the systems built around them. Just as the transition from wired to wireless telephones led to unprecedented changes in our communications and the shift from wall-power to battery-power transformed our expectations for computational systems, the shift from battery-powered systems to self-powered systems promises to fuel the next revolution in the Internet of Things (IoT). The ability to utilize ambient, scavenged energy, such as solar energy, radio-frequency (RF) radiation, piezoelectric effect, thermal gradients, etc., can liberate IoT devices from the lifetime, deployment, and service limitations of a fixed battery. However, the power supplied by energy-harvesting sources is highly unreliable and dependent upon ambient environment factors. Hence, it is necessary to develop specialized IoT architectures and systems that are tolerant to this power variation, and also capable of making forward progress on the computation tasks. In this chapter, one of the potential solutions called nonvolatile processor is introduced, in which nonvolatility feature is designed within a processor to overcome the unstable power supply through distributed energy, time efficient backup, and recovery operations. The chapter provides insights on the design space of different architectures, different input power sources, and

---

K. Ma (✉)
Institute for Interdisciplinary Information Sciences, Tsinghua University, Beijing, China
e-mail: Kaisheng@mail.tsinghua.edu.cn

S. Li
Department of Electrical and Computer Engineering, University of California at Santa Barbara, Santa Barbara, CA, USA
e-mail: yuanxie@ece.ucsb.edu

V. Narayanan
Pennsylvania State University, State College, PA, USA

Y. Xie (✉)
Department of Electrical and Computer Engineering, University of California at Santa Barbara, Santa Barbara, CA, USA

S. S. Bhattacharyya et al. (eds.), *Embedded, Cyber-Physical, and IoT Systems*,
https://doi.org/10.1007/978-3-030-16949-7_8

policies for maximizing forward progress. Through exploration of the various factors involved in designing a battery-less energy-harvesting system, this chapter brings opportunities and accelerates the innovations of self-powered IoTs.

1 A Short Introduction

Improvements in electronics integration, miniaturization, and power efficiency continue to stretch the boundaries of what constitutes a sufficient power envelope within which to operate a computational platform. This trend is now fueling the emergence of self-powered systems that harness ambient energy sources without reliance on batteries. Such battery-less systems are attractive for ubiquitous deployment in the emerging Internet of Things paradigm by eliminating reliance on the presence of, and constant connection to, underlying power infrastructure and easing the restrictions imposed when provisioning batteries and enabling their recharge and/or replacement [1–5].

Some of such systems have already been successfully deployed, and some are very reachable in the near future. The applications that make use of this paradigm are diverse, including (1) area monitoring, e.g., the position of the enemy; (2) environmental monitoring; (3) industrial monitoring; (4) medical and healthcare monitoring; (5) traffic control systems; (6) underwater acoustic sensor networks; and (7) near-body wearable device networks.

There are, however, several drawbacks in relying on ambient sources of energy for such computing purposes. Most of these energy sources operate at relatively low conversion efficiencies since only a small fraction of the total transmitted power can be tapped. In addition, they are not reliable energy sources, since external factors could cause a disruption in supply.

The rest of the chapter proceeds as follows: Section 2 provides basic background on energy-harvesting platforms, including features of energy sources and comparison of various potential energy-harvesting solutions. Section 3 introduces a model from the system level. Section 4 explores the architectures from three different complexity: non-pipelined, N-stage pipelined, and out-of-order nonvolatile processors. Section 5 validates our proposed architectures with fabricated NVPs. Section 6 discusses the design guidelines for self-powered system designers. Section 7 concludes the chapter and discusses the future works.

2 Background of Energy Harvesting

2.1 Ambient Energy Sources: Weak and Unstable

Typical ambient energy sources that could be harvested to power an embedded system include solar energy, radio-frequency (RF) radiation, piezoelectric effect, and thermal gradients[6], as shown in Fig. 1. These sources can be classified

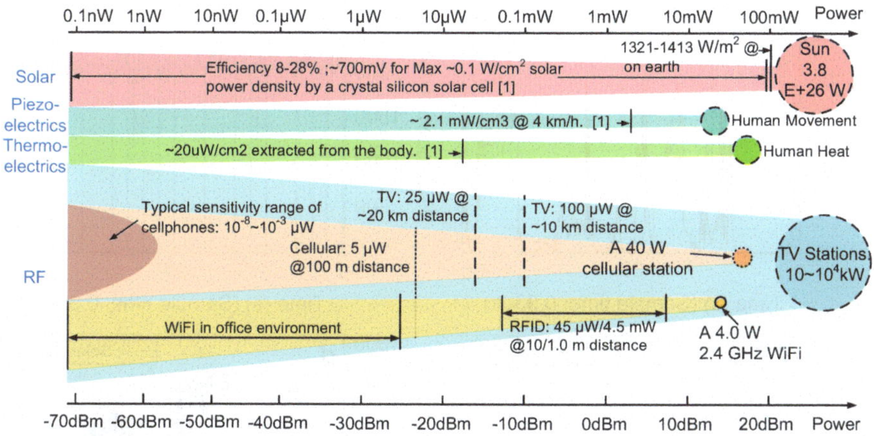

Fig. 1 Typical power ranges of ambient sources

according to three characteristics: signal magnitude, variability in signal strength, and granularity of variation/intermittency frequency. Research also indicates that implantable biofuel cells (BFCs) are able to generate electric power from sugars found in the body fluid of an insect [7]. A comprehensive comparison among these power sources can be found in references [8] and [9].

Figure 2a shows the power traces for four typical ambient energy sources. The RF energy is obtained by measuring the power of the frequency spectrum from a TV station, the piezo energy is measured through devices fixed on a bike, the thermal energy is generated from characterizations described in references [10–12], and the solar trace is obtained using data from MIDC [13]. It is observed that there is substantial variation in power, even over a few milliseconds for RF in Fig. 2a with the ratio between the maximum and minimum power over this period around 250× [6, 14, 15]. The piezo power is more stable than RF with just some short power loss in Fig. 2b. The thermal power, shown in Fig. 2c, is even more stable, due to the gradual nature of temperature variation. Variation in solar power seen in Fig. 2d is contingent on the weather conditions and orientation of the solar cell.

Another feature is the intermittency frequency that influences how soon the power drops below a given threshold as shown in Fig. 2a. The intermittency frequency decides the backup and recovery overheads. Sources with periodic behavior, like Fig. 2b, facilitate prediction of power loss and enable efficient scheduling of tasks.

2.1.1 Typical Energy-Harvesting System Structures

With the development of the Internet of Things (IoT), smart cities, and implantable and wearable applications, extremely low-power systems powered by ambient

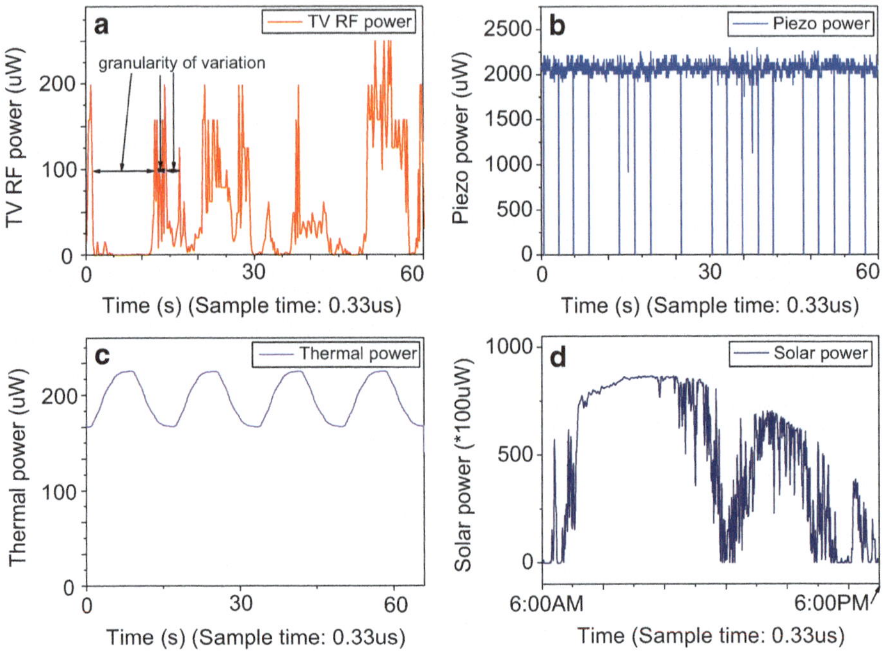

Fig. 2 Power traces (**a**)TV RF (**b**) Piezo (**c**) Thermal (**d**) Solar

energy sources are gaining popularity. Figure 3a shows an archetypical system structure consisting of (a) an energy harvesting and power management block, (b) a digital signal processor which is usually implemented using a microcontroller unit (MCU), and (c) the wired or wireless I/O interface. The capacity and implementation of the energy storage medium is also critical to the system design because it directly affects the trade-off between leakage and other overheads and the maximum stored power. In later sections, this trade-off will be discussed in more detail.

It is also noted that different energy sources require different energy harvesters for power conversion. For example, the output of a solar cell is a DC signal, while the RF signal and the output of piezoelectric-based systems are AC signals which require an extra AC–DC rectifier. When the input power is weak, the output voltage may also be low and potentially require an extra DC–DC voltage booster [6].

The baseline energy-harvesting block is illustrated in Fig. 3. Subsequent to the AC–DC or DC–DC conversion, an MPP tracking (MPPT) interface is employed to control the charging power for the highest power-conversion efficiency from the energy harvester.

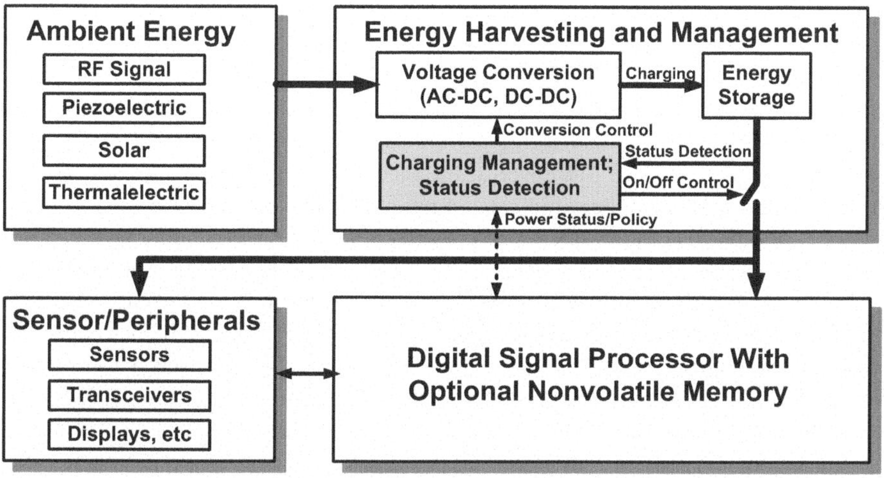

(a), Energy harvesting system diagram

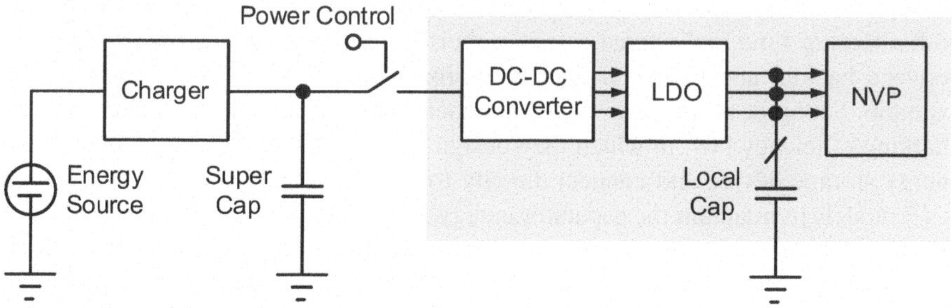

(b), Baseline front end circuits

Fig. 3 (**a**) Energy-harvesting system diagram. (**b**) Baseline front-end circuits

2.2 *Why Not Do Wait-Compute on A Volatile Processor with A Large Capacitor?*

On account of these limitations, most current energy-harvesting platforms tend to restrict themselves to applications that require relatively simple signal capturing mechanisms involving minimal computation and processing with a large capacitor for energy storage and start only when there are ample energy stored. Such traditional strategy in energy-harvesting systems is to employ a volatile low-power MCU or an MCU with checkpointing capability (e.g., FeRAM MSP430 is used in [16]) that waits before starting to execute while charging an energy storage device which must be large enough to store sufficient energy to complete an entire logical work unit, such as an image frame [16]. Systems operating on this paradigm will alternate between periods where they accumulate energy in the energy storage devices (ESD), and ones where powering the system drains the energy. While such a system is able to offer strong guarantees for execution once execution begins, this conventional solution has several limitations, including energy conversion efficiency overheads brought by frequently charging and discharging the capacitor, capacitor leakage [16, 17], minimum charging current (e.g., 20 μA for the GZ115 [17]), and slow charging curve [17]. Moreover, if the incoming unit of work is too large, the incoming power may not be sufficient compared to leakage in the ESD [16], or there may be long periods of complete power outage that drain the accumulated charge, and it may take arbitrarily long to reach the threshold for beginning execution.

An alternative execution paradigm is to utilize only a small on-chip capacitor, i.e., one sufficient for backup operations, and employ an NVP. This can reduce capacitor leakage and improve front-end conversion efficiencies. Its approach is to mitigate the overheads of moving charge into and out of a large energy storage device at the cost of additional system complexity for the NVP. More complicated guarantees on the granularity of work are accomplished once an execution period begins and the overheads imposed during more frequent backup and restore events during the execution of each logical unit of work. The two approaches can be seen as similar if the logical unit of work is at instruction or similar granularity, thereby minimizing both charging time and charge lost if a shortfall occurs during a charging period between backup and recovery. Hybrid approaches have also been proposed. For example, Sheng et al. propose a dual channel front-end solution to overcome low charging efficiency [18] in which they design another power channel to bypass the energy storage device and connect directly to the load, and Ma et al. extend prior NVP models to maintain the capacitor energy level [19] within a bounded range for charging efficiency during execution rather than greedily consuming energy. *Thus, the key energy trade-off between the two approaches is between the energy wasted on charging and discharging a capacitor with leakage and the backup and recovery overheads of NVP.*

It is observed through simulation and validation that the NVP-based execution approach can outperform the wait-compute scheme by 2.2X-5X.

2.3 Volatile with Checkpointing or Nonvolatile?

The wait-compute scheme suffers from low efficiency brought by the front-end circuits. One possible solution is that to build a direct path from source to the load. While this brings another challenge: unstable power supply. Figure 4 illustrates the difference in the behavior of a volatile processor with periodic checkpointing to an external nonvolatile memory and a completely nonvolatile processor when working under variable power source conditions. While both processors can only run when the input power exceeds a certain threshold, the volatile processor does not retain the instantaneous state of the system when the power drops below the threshold, resulting in a forced rollback from previously checkpointed state. This could limit the amount of forwarding progress from being made.

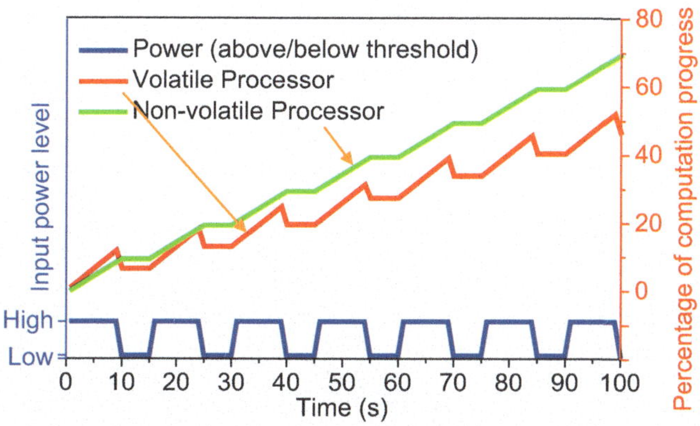

Fig. 4 VP with checkpointing v.s. NVP processing progress comparison

3 An Energy Model for Energy-Harvesting System

A key challenge in designing battery-less electronics with energy scavenged power sources is the erratic and unreliable power supply derived from ambient sources. One solution is to add an energy storage capacitor, such as a super-capacitor to smooth the power. Figure 6 shows an abstracted charging metaphor that highlights the key components of the systems considered in this paper and Fig. 5 illustrates an energy trace harnessed from an RF power source stored in a capacitor accounting for its leakage. Since the capacitor leaks, periods during which the power harnessed is less than the leakage power cause the overall stored energy to decrease. Conversely, leakage rates and size/weight restrictions bound the practical scale of capacitor volume, so the capacitor may saturate and be unable to store additional energy during periods of high input power.

Three key insights into the dynamics of energy storage in energy-harvesting systems are that, firstly, over the highly varying input power ranges that these systems must operate, they will frequently encounter both periods where the energy lost from capacitive storage is greater than that replaced by energy harvesting and periods where short term increases in input power provide more energy than a practical capacitor can store. Second, the effects of capacitive leakage, finite capacitor storage, charging losses, and other front-end power components are large enough to require co-optimization when considering processor or other compute-engine optimizations for these platforms. Third, as Fig. 5 depicts, policy management of the power demand at the processor, the load for this front end provides substantial leverage in mitigating both capacitive underflow and overflow by changing the slope of energy consumed. Changes in processor policies such as DVFS can also affect front-end efficiency: For example, since DC–DC conversion losses depend on the difference in voltage, actively depleting or restoring the energy

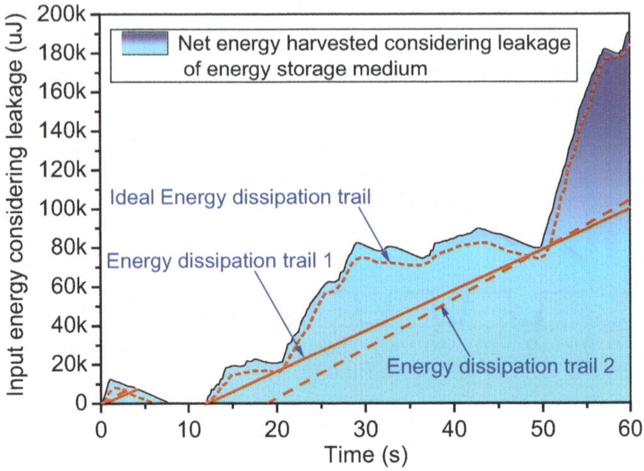

Fig. 5 A comparison of energy input and consumption trails for a TV RF power trace

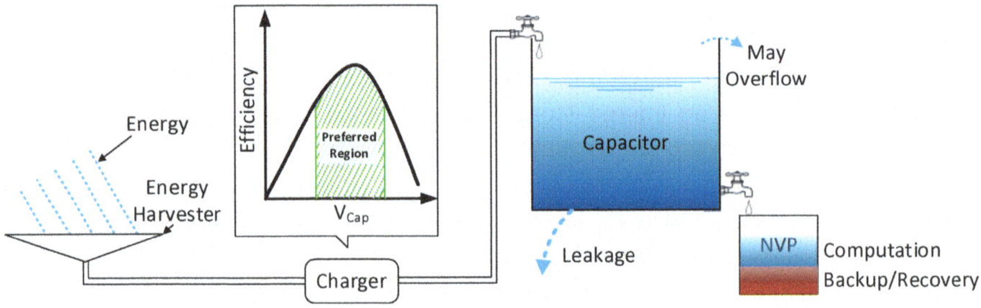

Fig. 6 Plumbing analogy for energy harvesting and consumption

storage capacitor to a particular voltage range will impact the efficiency of the system in harvesting future power (Fig. 6).

4 Architectural Exploration

This section focuses on determining which architectural configurations are best suited to optimally utilize the available power and energy by maximizing the processor performance under different energy constraints. Hence, depending on the energy that is harvested, various parameters are analyzed, such as the number of pipeline stages, the data to be backed up, and the frequency of backups.

Some common configuration assumptions for these structures include:

1. ISA MIPS
2. Clock frequency is 8 kHz for all configurations. The selection of clock frequency is driven by the limited strength of the WiFi signal used, rather than the limits of the microarchitectures.
3. Instruction Memory and ICache: The instruction memory is assumed to be ROM. The ICache can be SRAM, hybrid [20], or NVM [20, 21]. In this paper, ICache is designed using NVMs.
4. Data Memory and DCache: Data memory is assumed to be nonvolatile. An SRAM-based DCache employing a write-through strategy does not require any backing up policy. On the other hand, a write-back strategy necessitates writing the dirty data back to memory. Our system assumes a nonvolatile write-back DCache which preserves the dirty data even during periods of power down.

4.1 Non-pipelined Configuration (NP)

In the absence of any pipeline stages, the entire state of the processor can be characterized by a single instruction state. Hence it is sufficient to focus on the following structures for retrieving architectural state.

1. Program Counter (PC): The PC address relates to the instruction being executed and surely needs to be stored.
2. Register File (RegFile): Due to its frequent usage, the RegFile undergoes a large number of writes and hence volatile RegFile is more energy efficient than an NVM based RegFile.

   However, all the volatile RegFiles need to be moved to a nonvolatile memory on power failures to save state.

In addition to the architecture, there are also trade-offs between the energy consumed in backing up and recovering the data and the overall performance. These trade-offs are explored, by choosing which data to save, and when to save it, as demonstrated by the following policies:

Backup Every Cycle (BEC)
In spite of the significant energy penalty, this solution employs an NVM register file, or else both the contents of a volatile RegFile and its counterpart nonvolatile location need to update every cycle. As shown in Fig. 11, only the PC and a few registers are written into the RegFile every cycle. Some instructions such as Store Word and Jump do not require any further RegFile write. Consequently, the power increase due to the use of a power-hungry nonvolatile memory is moderate.

On Demand All Backup (ODAB)
This differs from the previous solution in that all RegFile entries need to be backed up only in the event of a reduced power state. We develop a control structure shown

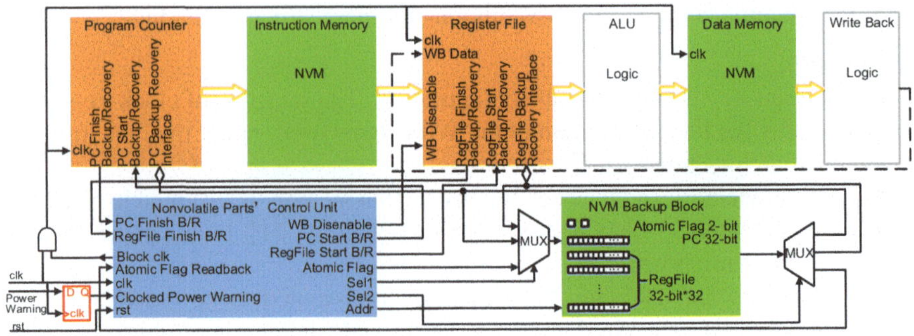

Fig. 7 NP on demand all backup structure

in Fig. 7, in which there is an individual NVM backup block to back up the PC and RegFiles. If the input power drops below a preset threshold, a power warning signal is activated. At that instant, the control unit starts to back up the PC and resets the atomic flag for PC to indicate that the PC has been successfully backed up. A similar procedure is carried out for the RegFile. When the power is available again, I first need to accumulate energy in the capacitor to ensure that there is enough energy for the next backup and recovery operation before continuing execution (Fig. 8).

On Demand Selective Backup (ODSB)
In order to reduce the backup time and energy penalty, I develop an *on demand selective backup* solution, as described below. A synchronous power warning signal is used, which may delay the power warning signal a little bit, but can guarantee that the current PC finishes executing and writing back (if necessary). To avoid re-executing the instruction corresponding to the current PC, $PC + 4$ is stored except in case of jump or branch instructions. This solution can save one clock cycle. Since the frequency of this system is very low, even a single clock cycle may be very significant if the power down happens frequently. In the volatile RegFile, I add a change flag to each register to identify if a register has been written into between two backup operations. If the register has not been changed during the interval, the control unit will know it from the change flag and would not need to generate addresses for the unchanged data, as shown in Fig. 9.

Simulation Results and Comparison
Figure 8 shows the area of each of the components for the schemes described above. It is observed that the total area is similar since the NVM cache and backup blocks are much larger than the logic components. The critical path delay shown in Fig. 10 indicates that the BEC has the lowest peak frequency due to the frequent backups. However, there are overheads in the other schemes which also prevent them from running at peak performance. These overheads are illustrated in Fig. 11. It shows the details of computing, backup, recovery, and off times for each scheme described above. BEC distributes the backup energy penalty to every cycle. Thus these penalties are the smallest for this case, as shown in Figs. 12 and 13. The

Fig. 8 Non-pipelined area

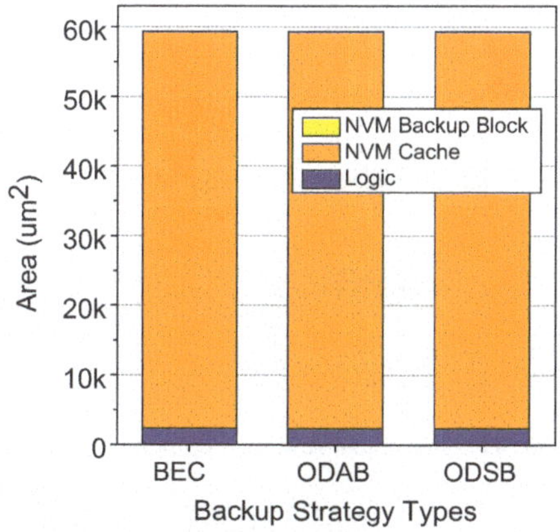

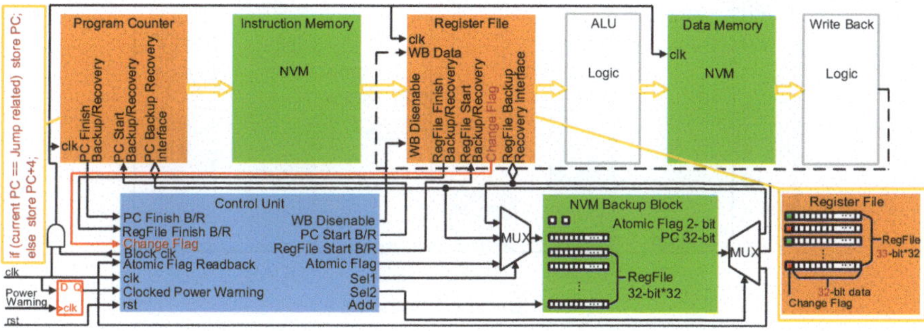

Fig. 9 Non-pipelined on demand selective backup structure

Fig. 10 Non-pipelined critical path delay (VDD = 0.95 V)

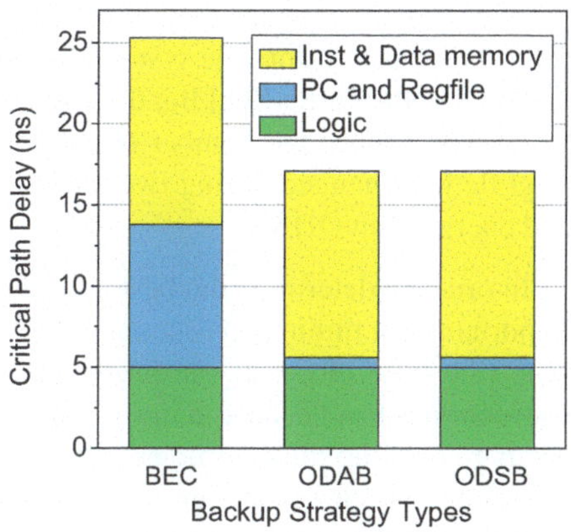

recovery time is defined as the time from the activation of the *Energy OK signal* to the time all backup operations are completed. The recovery times are similar across all schemes, but BEC does not need to accumulate energy for backup. Consequently,

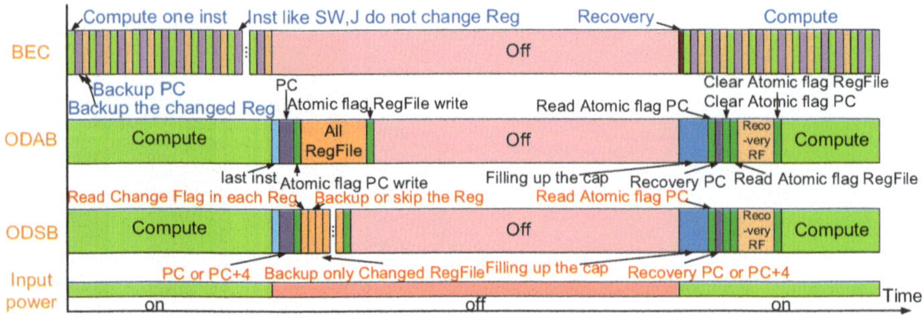

Fig. 11 Individual runtime components for BEC, ODAB, and ODSB schemes

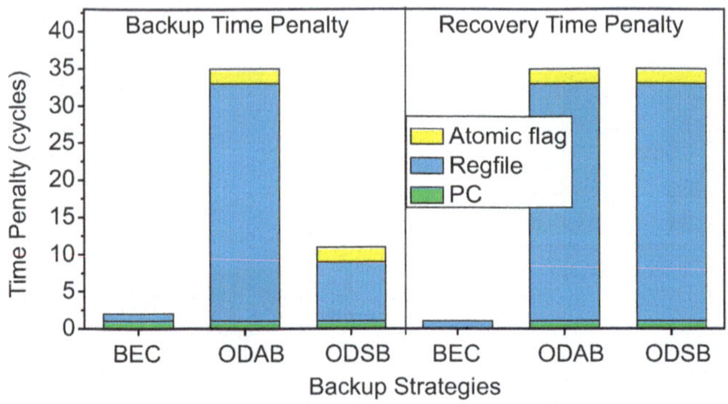

Fig. 12 NP time penalty comparison

this scheme can restore the system the fastest. The ODAB scheme needs to back up the PC and the entire RegFile; thus, the time and energy penalty is the largest.

ODSB reduces the number of RegFile entries to be copied, by detecting if the RegFile has changed during two backup intervals, thus requiring less backup time and energy than ODAB.

In order to determine the best NP scheme, optimizing power and energy is more important than timing, on account of the low clock frequency. In BEC, if the interval time (the time of power on during two power loss) is very short, the energy per instruction is low because at most only one RegFile entry is backed up, while ODAB needs to back up all RegFile entries. ODSB backs up only one entry at a time, but it is more complex in design. As the backup interval time is increased, ODAB and ODSB are more energy efficient, as observed in Fig. 13, on account of backing up only in the event of a power warning.

In order to avoid a large peak power which can result in system instability, I choose to back up and recover data serially. Although a parallel approach can reduce the backup and recovery time, it increases the peak power requirement. From this point of view, the ODSB is better than ODAB.

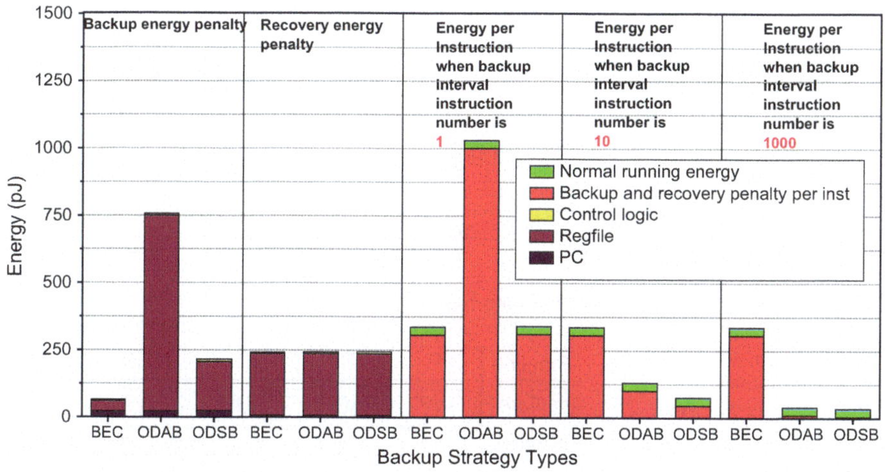

Fig. 13 Energy overheads for each NP scheme. For high frequency of backups, ODAB has the highest overhead, while BEC consumes maximum energy when the backup interval exceeds 10

- *ODSB is a most energy-efficient strategy when the source is relatively stable like solar energy. Compared to ODAB, ODSB can reduce the backup energy penalty by 69% with only 0.002% area overhead.*
- *While BEC is not the most energy efficient with very weak sources like WiFi, it does not require the time to accumulate energy in the capacitor to ensure sufficient backup energy is available, as shown in Fig. 11. Hence it is viable when the power failures are extremely frequent (less than 1 in 10 cycles), which rarely happens even in WiFi sources.*

4.2 N-Stage-Pipeline

In contrast to the MIPS non-pipelined case, a MIPS *N-stage pipeline* is traditionally used to improve the clock frequency. Due to the increase in circuit complexity and the activity factor of the processor, the power threshold of this design in energy-harvesting systems is higher than that of the non-pipelined case. In this subsection, we assume a five-stage-pipeline structure (5SP) and propose two backup schemes.

Shifted PC and Volatile Flip-Flops (SPC/VFF)
The main differences between NP and 5SP configurations are the pipelined data flow with bypass and forward and the complex control flows to handle hazards (Fig. 14). In the SPC/VFF scheme, a shifter buffer is designed to remember the PC value in each pipeline stage, as shown in Fig. 15. This means the PC no longer needs to pass through all pipeline stages to be stored. When the power is down, the clocked power warning signal can guarantee that the PC in the write-back stage will be

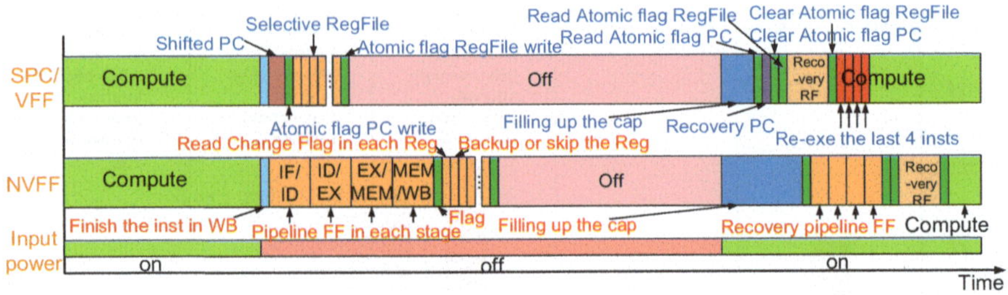

Fig. 14 Five-stage-pipeline NVM flip-flops backup

Fig. 15 Comparison of individual runtime components for SPC/VFF and NVFF

Pipeline	IF	RF	EX	MEM	WB
InstQue1	LW	ADD	SUB	SW	ADD
Shifter	PC	PC-4	PC-8	PC-12	
InstQue2	LW	J	SUB	SW	ADD
Shifter	PC2	PC1	PC1-4	PC1-8	

finished. The unfinished PC to be backed up would then be in the data memory stage. The reason that we use a shifter instead of simply rolling back the PC is that if some of the instructions in the pipeline are jump or branch instructions as shown in InstQue2 in Fig. 15, a different PC would need to be backed up. If the instruction in MEM stage is SW as shown in InstQue1&2 in Fig. 15, this SW instruction will be guaranteed finished by the clocked power warning signal. We can try to identify if the backed up instruction is SW, if yes, back up the PC in EX stage in the shifter instead of PC in MEM. Instead of dealing with the increased design complexity, we can just back up PC in MEM stage. Once the power is on again, the first instruction will be SW. In this case, we run SW actually twice: the first time is during the backup operation, another one is the first instruction after recovery in case the former one is not properly finished.

Nonvolatile Flip-Flops Solution (NVFF)
This solution involves the use of NVM flip-flops (Fig. 16). Here, the PC and the RegFile are automatically backed up through NVM flip-flops in the IF/ID pipeline stages.

Simulation Results and Comparison
SPC/VFF requires 11% less time and 57% less energy than NVFF in Fig. 17.

However, an extra 4 clock cycles are needed to re-execute the last 4 instructions that are lost from the latter pipeline stages after recovery, which we regard this as part of the recovery time penalty..

Counter to intuition, we show that SPC/VFF is more energy efficient than NVFF. Instead of backing up all the data in the pipeline latches, SPC/VFF only backs up one PC with a small shifter. Hence, a smaller backup capacitor with lower leakage

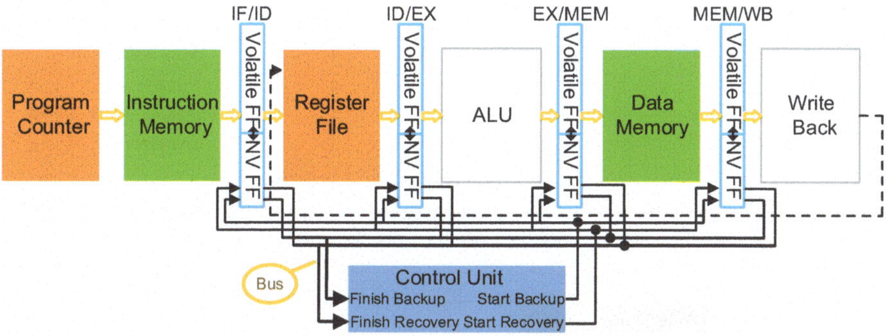

Fig. 16 Illustration of Shifted PC

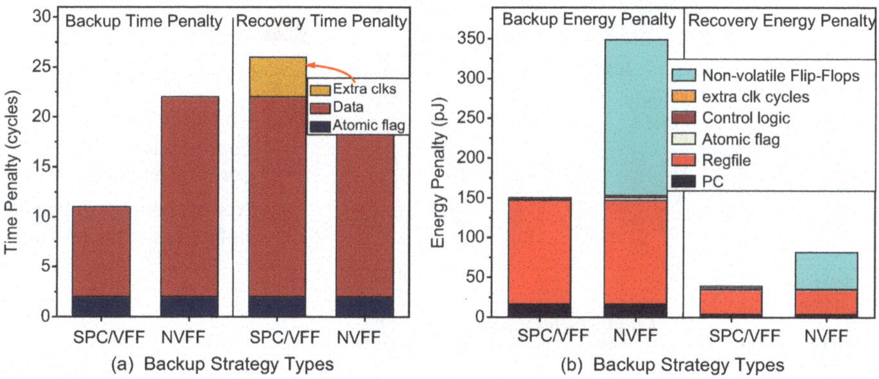

Fig. 17 Comparison of 5SP (**a**) time and (**b**) energy overheads

is sufficient for SPC/VFF, which will in turn affect the power threshold. In this case, SPC/VFF will also be able to outperform NVFF after several repeated instructions.

4.3 *Out-of-Order Processor (OoO)*

Compared to the MIPS 5SP configuration, our MIPS out-of-order (OoO) processor configuration, described in Table 1, is much more complex. Figure 18 indicates the key blocks we consider in our OoO processor model derived from FabScalar [22]. Conceptually, system state, unlike in the previous two examples is broadly distributed across several structures such as the PC, ROB, RegFile, Map Table, Issue Queue, Load Store Queue as well as the Branch History Table and Branch Target Buffer. Some of these structures are essential to maintain the integrity of the state of the system, while others contribute toward optimizing the performance and/or energy of execution in the presence of frequent backups and recoveries.

Table 1 Parameters for OoO processor

Parameter	OoO	Parameter	OoO
Fetch width	4	Map Table	32
Issue width	4	PRegFile	128
ROB size	32	Ready Table	128
IQ size	32	BHT/BTB	128
LSQ size	32/32	ARegFile	32
ICache/DCache	32 kB/32 kB	Free List	128

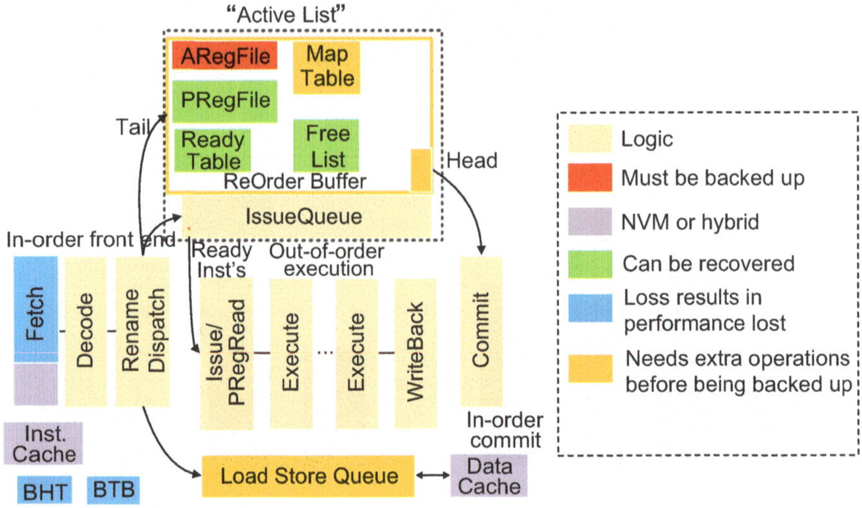

Fig. 18 OoO volatile/non-volatile structure

Fig. 19 Backup schemes for OoO configuration

OoO solutions	ROB	IQ	AReg File	Map Table	PReg File	Ready Table	Free List	BHT	BTB
MinR	★		⭐	☆					
LLB	⭐	⭐	⭐	⭐	⭐				
MLB	⭐	⭐	⭐	⭐	⭐	⭐	⭐		
MPL	⭐	⭐	⭐	⭐	⭐	⭐	⭐	⭐	⭐

⭐ Back up ★ Last uncommitted PC ☆ Pseudo-misprediction

Due to the relatively larger power requirements of an OoO processor, there are both fewer periods where the input power exceeds the minimum threshold, as compared to the previous cases, and more state to consider saving during power emergencies. Hence it is imperative to judiciously select the structures to be backed up, in order to ensure a comparable performance to the no-pipeline and n-stage pipeline designs.

We propose several resource selection strategies for this purpose, as illustrated in Fig. 19.

Minimum State Resource Backup Solution (MinR)
MinR backs up the minimal number of bits required to preserve functionality across power interruptions, as shown in Figs. 19 and 21. Fundamentally, this approach piggybacks on the branch misprediction mechanism to minimize the number of valid/relevant state bits prior to initiating backup, at the cost of some time and effort being required to enact the misprediction logic prior to checkpointing.

1. ROB and PC: to minimize state storage, we only back up the first uncommitted PC at the head of ROB. This means all the other instructions in the ROB will be abandoned regardless of status.
2. IQ: IQ does not need to be backed up because all the instructions in IQ are uncommitted.
3. ARegFile: We can either choose to backup ARegFile or PRegFile. The ARegFile is preferred since it is designed to be smaller.
4. Map Table:

   It is possible that uncommitted instructions that follow the head of the ROB could have modified the Map Table. However, since we need to restore the state to the instruction at the head of the ROB, the Map Table should also be correspondingly restored. In order to achieve this, we trigger an instruction flush identical to that following a branch misprediction on the ROB head. Since no actual branch prediction occurs, we term this operation *pseudo-misprediction.*
5. PRegFile, Ready Table, Free List, BHT, and BTB can be recovered.

Low-Latency Backup Solution (LLB)
While the MinR policy minimizes bits pushed to nonvolatile storage, it does so at the expense of requiring additional work before backup can begin. We next consider a backup solution that aims to minimize the number of bits to store if backup begins immediately. Rather than backing up only the first uncommitted PC, this solution backs up the entire ROB, IQ, ARegFile, Map Table, and PRegFile. Compared to the MinR policy, structures such as the Ready Table and Free List (Figs. 23 and 24) can be more easily reconstructed, resulting in a penalty of only a few recovery cycles. While LLB stores more state than MinR, it can sometimes nonetheless be more energy efficient, due to the extra work required of MinR on both backup and recovery.

Middle-Level Backup Solution (MLB)
Instead of using extra recovery time and energy to restore the Ready Table and Free List in the low-level backup solution, MLB backs up Ready Table and Free List as well (Fig. 19).

Min-State-Lost Backup Solution (MPL)
In this solution, all the structures are backed up including the BHT and BTB as shown in Fig. 19.

Integrated Flexible Atomic Backup Solution (IFA)
All of the previous solutions save and restore a fixed amount of state determined by the structures in question. However, one key feature of the backup process is that it

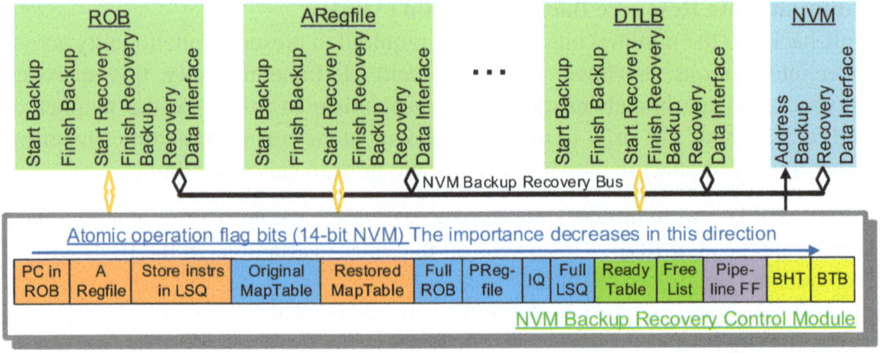

Fig. 20 OoO integrated flexible atomic backup solution

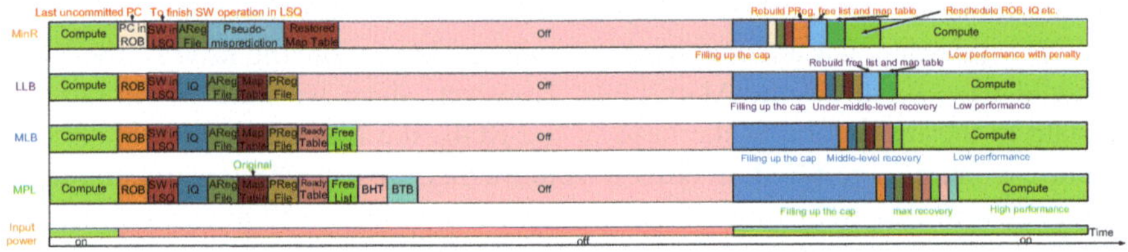

Fig. 21 Out-of-order structure backup design trade-off

must necessarily be triggered conservatively: The backup signal must be issued at a point where the processor can guarantee sufficient energy to complete the backup **even assuming zero additional input power during backup**. However, in practice, when a power emergency occurs in an energy-harvesting system, it is not usually because input power has dropped to zero, but because it has fallen below some threshold for some period of time. Thus, there may frequently be additional energy available during the backup period that, while insufficient to continue operation, would allow for the optional state, such as the BHT, to be subject to optimistic attempts at backup (Figs. 20 and 21).

We propose a flexible backup mechanism that integrates aspects of the four previous solutions together to exploit the conservative nature of the backup trigger. The key idea of the solution is to regard each backup operation as an atomic operation. A backup operation has only two states: success or failure. Figure 20 shows the systematic structure of this solution. Figure 22 shows how the power may be dropping at a different pace to zero and can execute more or less backup.

Simulation Results and Comparison For MinR, the pseudo-misprediction operation for the Map Table requires extra backup clock cycles as shown in Fig. 21. When recovering, we also need to pay extra clock cycles to restore the PRegFile, Ready Table, and Free List. Further, since we discard all instructions in the ROB following the head, we need to re-execute these instruction, resulting in the timing penalty in Fig. 23 as well as energy overheads, shown in Fig. 24.

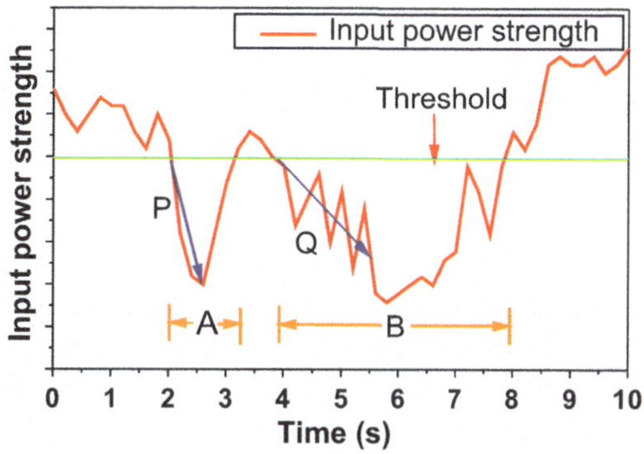

Fig. 22 Scenarios in which IFA can be applied

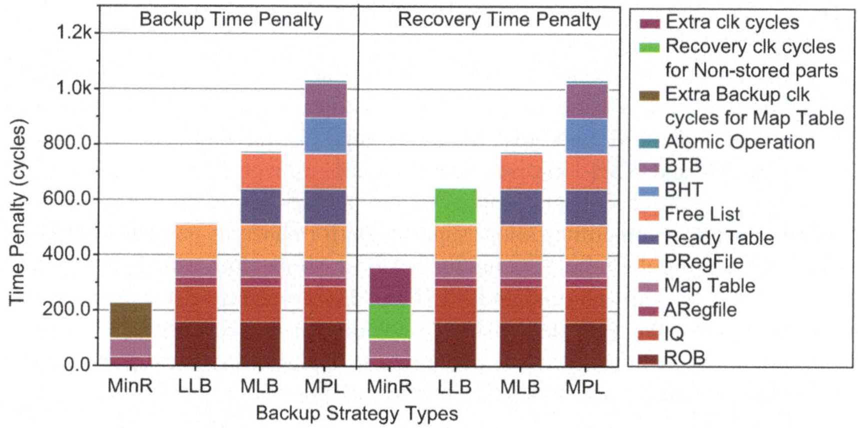

Fig. 23 OoO time penalty

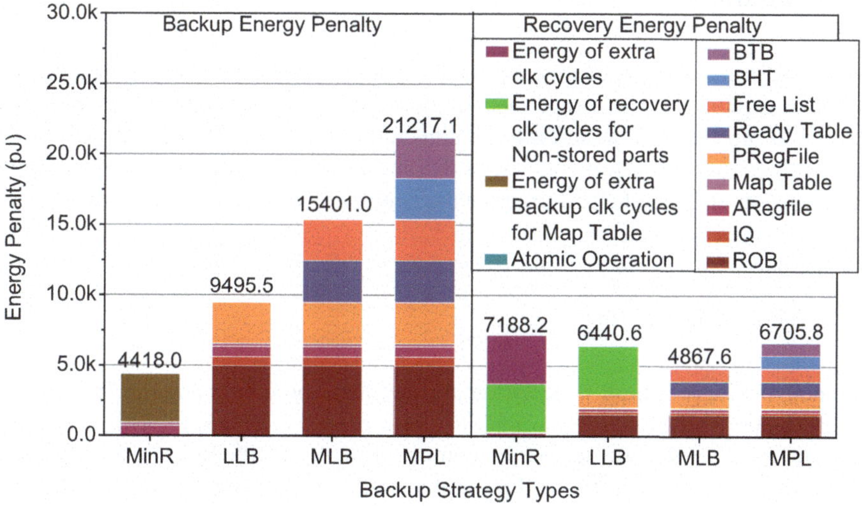

Fig. 24 OoO energy penalty

In the case of LLB, the ROB and PRegFile are relatively large and significantly increase the backup time and energy in Figs. 23 and 24. On the other hand, the recovery energy penalty is smaller than MinR, because all the instructions and their information in the ROB are backed up, eliminating the need to re-execute these instructions.

The backup time and energy penalty of MLB are larger than those of LLB, as shown in Figs. 23 and 24, but the recovery energy is lower.

The designer should use this MLB strategy when the system is optimizing the time to resume execution after a power failure.

In MPL, the backup and recovery time penalty and energy penalty are the largest in among all the solutions, but backing up all the additional structures incurs the minimum latency to return to peak performance after a power failure.

The results show the gain of 29 cycles for MinR, but not backing up the BHT and BTB negatively affects IPC.

On account of OoO being thought to be too complex for energy harvesting systems, prior work has not considered OoO platforms. Since OoO needs a much higher threshold than NP and NSP, the percentage of time OoO can run is much smaller than NP and NSP. However, OoO remains a favored option in several of the test scenarios in subsequent sections because the periods of sufficient power are common enough to sometimes allow superior performance to pay for lost cycles. In summary, storing the minimum number of bits, MinR, does not always provide the least energy backup solution, and MLB has the shortest time to execution after the power failure. We also demonstrate that, due to the conservative nature of backup initiation, there is a sizeable potential for opportunistic backup of optional, performance-enhancing bits with a flexible backup policy.

5 Validation

While the primary focus of this paper has been on a simulation-based exploration, we have explored the non-pipelined on-demand-backup strategy using an actual fabricated processor. In addition to demonstrating the execution of real workloads on the processor, this effort enabled us to gain insights to approximations in initial simulation models and helped to refine the simulation model used in this work.

5.1 System Overview

The nonvolatile processor is based on an Intel 8051 processor and the ISA choice for fabrication was driven by availability of a pre-existing design. This processor supports multi-cycle instructions as compared to the MIPS ISA used in the rest of the paper. Consequently, in this implementation, the saved state includes the state machine that captures the exact cycle in which the instruction was carried out currently. This is the most relevant difference from the perspective of the nonvolatile

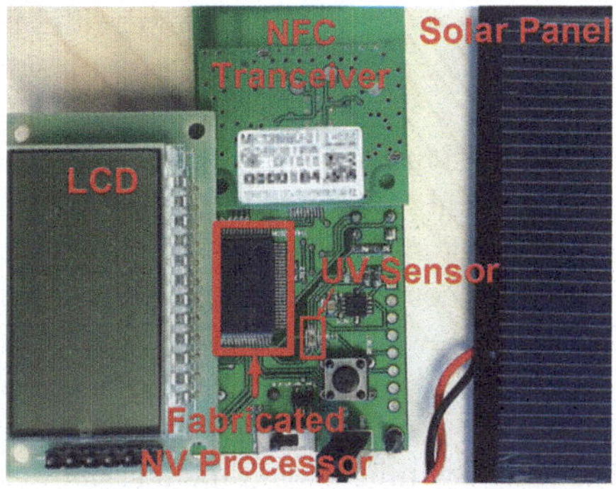

Fig. 25 System prototype

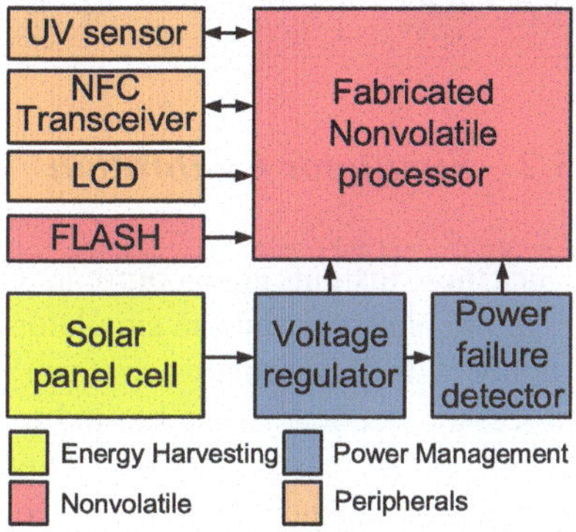

Fig. 26 Block diagram

Table 2 Measured parameters

Parameter	Result	Parameter	Result
Max. clock	25 MHz	Total power	160 μW@1 MHz
Process technology	0.13 μm	Backup energy	23.1 nJ
V_{DD} for core	0.9–1.5 V	Recovery energy	8.1 nJ
Total area	1.015 mm^2	Backup time	7 μs
Energy/Inst	347 pJ	Recovery time	3 μs

design space explored as compared to the MIPS pipeline in our simulation studies. However, our simulation effort encompassed the 8051 processor as well to gain insight to potential sources of differences between the simulation and real system.

The nonvolatile processor based system is interfaced to a solar power panel and a UV sensor as shown in Fig. 25 and 26. The processor is based on a 0.13 μm ROHM CMOS-ferroelectric hybrid process. The PC and all RegFiles are FeRAM-based flip-flops. The flip-flops are realized using an additional backup ferroelectric capacitor (FeCap) for each D flip-flop (DFF) used in the design. When a power failure is detected, the NV control logic backs up the DFFs to the FeCaps. When power is resumed, data is restored from FeCaps to DFFs. All FeCaps are distributed and connected close to their own DFFs; thus, the data backup and recovery can proceed in parallel to reduce the operation time. Table 2 shows the chip specifications. The total power decides the power threshold, and the backup energy decides the energy storage capacitor volume. The capacitor used in the system is 470 nF.

The design process revealed insight to modeling to key aspects in the simulation environment. The clocking network is switched to a lower frequency to transition clock generation from an external oscillator to an internal RC circuit. The external oscillator could become unstable or may not have sufficient power to operate. Further, a lower clock frequency increases the reliability of the FeRAM writes and also reduces peak power consumption. The slower clock impacts the overall backup time as compared to using estimates based on faster operational clock. Similarly, the recovery time should include not only the time required to restore architectural state but the time for the clock generators and power supply grid to be stable.

5.2 Simulator Calibration

Once these insights are incorporated in the simulator, a few kernels were executed on both the platform and the simulator (see Table 3). To model an intermittent power supply, a 1 KHz square waveform power input was fed to the processor and the processor frequency was limited to 3 MHz (the maximum frequency at which it could operate based on power supply when connected to the solar panel). Each kernel was executed 1000 times to obtain overall completion time shown in Table 3. For the stable power case, the simulator and platform mismatch is negligible. For unstable power, the simulator and the platform measurements differ less than 5%. The differences accrue since the simulator averages energy consumed by an instruction to estimate remaining energy for triggers. However, the actual instruction execution exhibits non-uniform activity. Further, the energy storage capacitance models used in the simulation add and decrease in discrete portions unlike the actual design. This validation process for the simulator based on a real design indicates that simulation-based models are fair representation of actual systems.

Table 3 Execution time on simulator and actual platform when using an interrupted power supply generated as a square waveform

	Stable/ms	Interrupted/ms		
Testbench	Measured	Measured	Model	Error
FIR-11	0.626	1.260	1.209	−1.59%
Sqrt	2.620	5.280	5.190	0.81%
KMP	3.573	7.184	7.059	0.77%
FFT-8	4.207	8.460	8.238	−0.13%
Matrix	5.826	11.740	12.021	2.39%
Bubble sort	27.23	54.705	57.236	4.63%

6 Design Guidelines

The complexity of the nonvolatile architecture chosen for a particular application scenario depends on a variety of factors. The input power and the stability of the power supply are two key factors that impact the choice. In addition, the computational complexity of the application and its performance requirements is also important.

6.1 Dependence on Input Power Characteristics

The input signal characteristics play a major role in determining the optimal design, as is evident from our experiments with WiFi power trails under different environment conditions. Figures 27 and 28 demonstrate the performance of the various backup schemes when home and office WiFi sources are used for harvesting energy. For the home environment, a non-pipelined ODSB architecture is the best performing. On the other hand, in the office environment, the more complex OoO processor with the minimum performance loss scheme is desirable. The reason for this behavior is that, the home WiFi signal comprises of a single router, while the office environment consists of several routers of similar signal strengths. A disturbance in the signal would result in input power going to almost zero in the home environment, hence the simplest design with the lowest power threshold is preferred. In contrast, in the office environment, the additional routers continue to supply input power at a relatively similar strength in an uninterrupted fashion, thus allowing for more complex architectures.

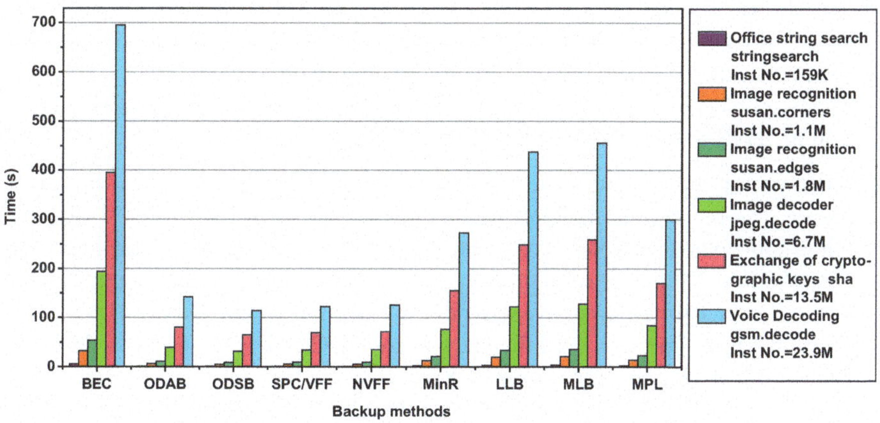

Fig. 27 Execution time with energy scavenged from WiFi home environment

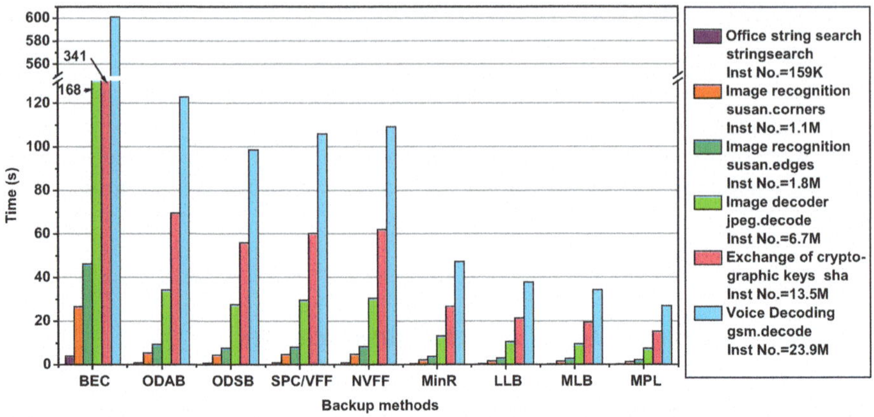

Fig. 28 Executing time with energy scavenged from WiFi office environment

6.2 *Quality-of-Service*

Most of the applications that are expected to run on an energy harvesting platform require an output to within a fixed time period. Quality-of-service (QoS) can be standard to qualify the possibility to achieve that goal. When these systems run on harvesting ambient energy, the unreliable nature of the input source may prevent the QoS demands in some instances.

Figure 29 shows the percentage of instances that meet the QoS demands specified, for two different applications, measurement of ECG and an edge detection algorithm used in vision sensors.

Figure 29 provides an indication of meeting the QoS. For example, in Fig. 29a, the possibility to achieve real-time ECG processing with NP and RF power is 0.92%. Consequently for RF and thermal sources, real-time processing is not possible. For ECG, most of the solar and piezo sources can support 100% QoS.

The baseline configurations for Fig. 29 are listed in Table 4. Table 4 also indicates several methods to optimize the QoS with efficiency:

- From power input view, there are lots of features that can improve the input power. The baseline in the QoS simulations for RF is 10 km from TV stations, but in some places like New York, the average TV station distance is around 3 km [23]. And the RF power strength is in negative square relationship with distance, so we can 11.1 times larger average power, thus improve the QoS to 100%.
- We can try to improve the efficiency of AC–DC, DC–DC, LDO, reduce the capacitor leakage, etc. to improve QoS.
- From output view, try to shrink the technology [24] (baseline is 130 nm CMOS) to get lower power consumption, for example, applying 22 nm FinFET [25] will achieve 100% QoS for real-time ECG in Manhattan. In addition, various methods could be applied to reduce the power: new devices like Tunnel-FET [6], low-

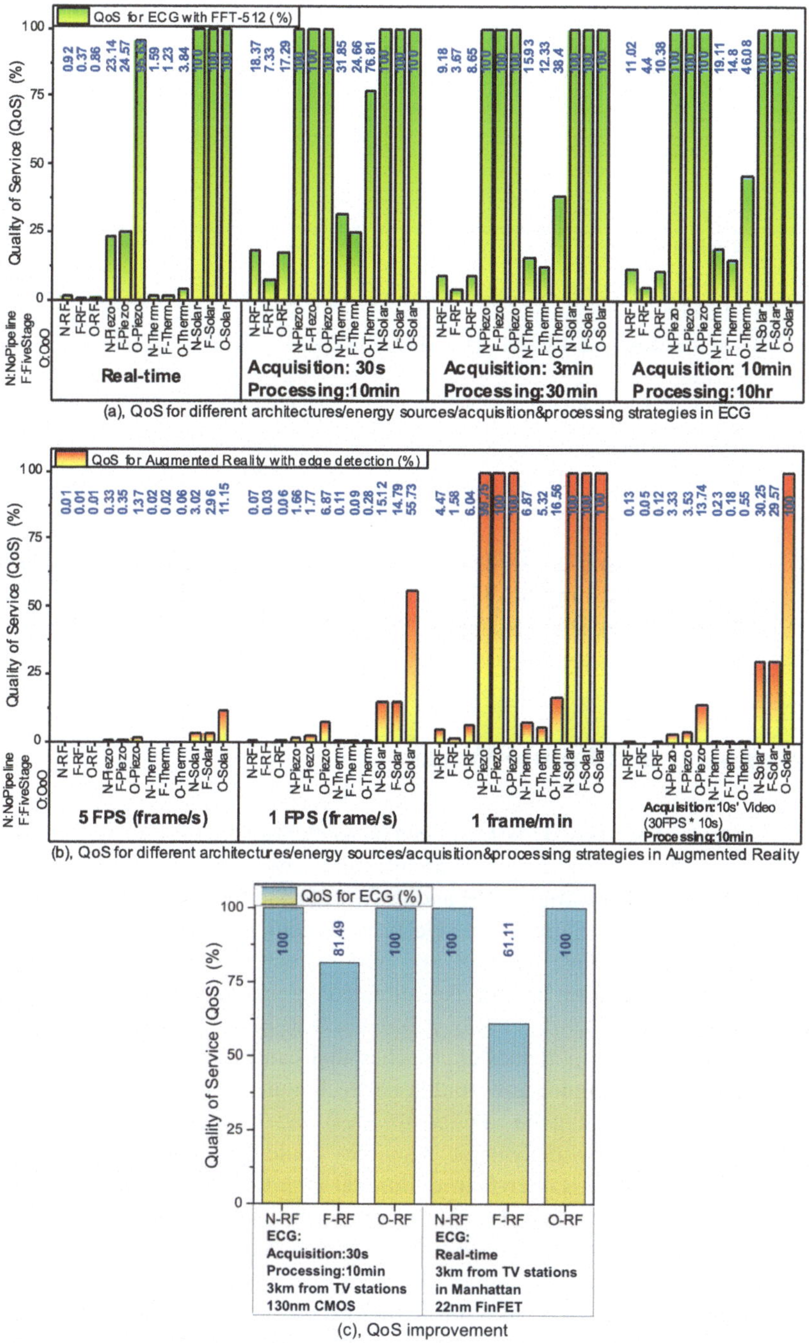

Fig. 29 QoS for ECG/AR, and QoS optimization. (**a**) QoS for different architectures/energy sources/acquisition and processing strategies in ECG. (**b**) QoS for different architectures/energy sources/acquisition and processing strategies in augmented reality. (**c**) QoS improvement

Table 4 Baseline and relationship with QoS improvement

Aspect	Solution	QoS baseline	Efficiency
RF	Antenna gain	6 dBi	α
	Bandwidth	539 M	α
	Distance	10 km	$1/\alpha^2$
Therm	Area	1 cm^2	α
	ΔT	20 °C	α^2
Piezo	Volume	1 cm^3	α
Solar	Area	4 cm^2	α
	Efficiency	28%	α
Circuit	IP matching, AC–DC, DC–DC, LDO, Cap		
Tech.	Shrink Tech.	130 nm	α^2
	FinFET, IG-FinFET, TFET, NC-FET	CMOS	
	DVFS, DATS	Fixed frequency	
	Voltage	0.95 V	$1/\alpha^2$

power circuits like sub-threshold circuits, dark silicon [24], gated clock, dynamic voltage-frequency scaling (DVFS), dynamic-adjusting threshold-voltage scheme (DATS) [26], etc. are some examples.

7 Summary and Future Works

Nonvolatile-processor-based platforms can be an ideal enabler for the IoT and wearable devices. This chapter discusses architectural level designs and optimizations for ambient energy-harvesting NVP and provides design guidelines mapping from power sources to architecture level selection. Various architecture level solutions for non-pipeline, five-stage-pipeline, and out-of-order processor architectures are discussed. The simulation model is calibrated against a fabricated nonvolatile processor.

There are still many potential approaches that can be utilized to optimize the NVP solution, for example, how traditional techniques such as dynamic voltage and frequency scaling can be applied to NVP and how should it be adjusted. A hybrid architecture with dynamic resources could also be useful to adapt to variable power profiles. Rather than traditional architecture methods, a machine-learning-based controller may be able to predict in high quality in control path design. Accelerators for machine learning application level algorithms based on software or hardware implementation can even be merged for both the application and controller. New devices like the tunnel-FET can also be applied to further reduce the power consumption for NVPs. Novel distributed circuits merging both the computation and backup operations can further reduce the backup time and energy.

References

1. Ma, K., Y. Zheng, S. Li, K. Swaminathan, X. Li, Y. Liu, J. Sampson, Y. Xie, and V. Narayanan (2015) "Architecture exploration for ambient energy harvesting nonvolatile processors," *2015 IEEE 21st International Symposium on High Performance Computer Architecture (HPCA)*, pp. 526–537.
2. Ma, K., X. Li, S. Li, Y. Liu, J. J. Sampson, Y. Xie, and V. Narayanan (2015) "Nonvolatile Processor Architecture Exploration for Energy-Harvesting Applications," *IEEE Micro*, **35**(5), pp. 32–40.
3. Ma, K., X. Li, K. Swaminathan, Y. Zheng, S. Li, Y. Liu, Y. Xie, J. J. M. Sampson, and V. Narayanan (2016) "Nonvolatile Processor Architectures: Efficient, Reliable Progress with Unstable Power." *IEEE Micro*, **36**(3), pp. 72–83.
4. Ma, K., X. Li, J. Li, Y. Liu, Y. Xie, J. Sampson, M. T. Kandemir, and V. Narayanan (2017) "Incidental computing on IoT nonvolatile processors," in *Proceedings of the 50th Annual IEEE/ACM International Symposium on Microarchitecture*, ACM, pp. 204–218.
5. Ma, K., X. Li, S. R. Srinivasa, Y. Liu, J. Sampson, Y. Xie, and V. Narayanan (2017) "Spendthrift: Machine learning based resource and frequency scaling for ambient energy harvesting nonvolatile processors," in *Design Automation Conference (ASP-DAC), 2017 22nd Asia and South Pacific*, IEEE, pp. 678–683.
6. Li, X., U. D. Heo, K. Ma, H. Liu, V. Narayanan, and S. Datta (2014) "RF-Powered Systems Using Steep-Slope Devices," *IEEE International New Circuits and Systems Conference.*
7. Shoji, K., Y. Akiyama, M. Suzuki, N. Nakamura, H. Ohno, and K. Morishima (2014) "Diffusion refueling biofuel cell mountable on insect," in *Micro Electro Mechanical Systems (MEMS), 2014 IEEE 27th International Conference on*, IEEE, pp. 163–166.
8. Kim, S., R. Vyas, J. Bito, K. Niotaki, A. Collado, A. Georgiadis, and M. Tentzeris (2014) "Ambient RF Energy-Harvesting Technologies for Self-Sustainable Standalone Wireless Sensor Platforms," *Proceedings of the IEEE*, **102**(11), pp. 1649–1666.
9. Roundy, S., D. Steingart, L. Frechette, P. Wright, and J. Rabaey (2004) "Power sources for wireless sensor networks," in *Wireless sensor networks*, Springer, pp. 1–17.
10. Grezaud, R. and J. Willemin (2013) "A self-starting fully integrated auto-adaptive converter for battery-less thermal energy harvesting," in *New Circuits and Systems Conference (NEWCAS), 2013 IEEE 11th International*, IEEE, pp. 1–4.
11. Leonov, V., T. Torfs, P. Fiorini, and C. Van Hoof (2007) "Thermoelectric converters of human warmth for self-powered wireless sensor nodes," *Sensors Journal, IEEE*, **7**(5), pp. 650–657.
12. Leonov, V., T. Torfs, R. J. Vullers, J. Su, and C. Van Hoof (2010) "Renewable energy microsystems integrated in maintenance-free wearable and textile-based devices: the capabilities and challenges," in *Industrial Technology (ICIT), 2010 IEEE International Conference on*, IEEE, pp. 967–972.
13. Measurement and I. D. C. (MIDC) "http://www.nrel.gov/midc/,".
14. Harpe, P., E. Cantatore, and A. van Roermund (2013) "A 10b/12b 40 kS/s SAR ADC With Data-Driven Noise Reduction Achieving up to 10.1b ENOB at 2.2 fJ/Conversion-Step," *IEEE Journal of Solid-State Circuits*, **48**(12), pp. 3011–3018.
15. Xiaodan, Z., X. Xiaoyuan, Y. Libin, and L. Yong (2009) "A 1-V 450-nW Fully Integrated Programmable Biomedical Sensor Interface Chip," *IEEE Journal of Solid-State Circuits*, **44**(4), pp. 1067–1077.
16. Naderiparizi, S., Z. Kapetanovic, and J. R. Smith (2016) "Battery-free connected machine vision with WISPcam," *GetMobile: Mobile Computing and Communications*, **20**(1), pp. 10–13.
17. series datasheet, C.-X. G. "https://www.cap-xx.com/resource/cap-xx-g-series-datasheets/,".
18. Sheng, X., C. Wang, Y. Liu, H. G. Lee, N. Chang, and H. Yang (2014) "A high-efficiency dual-channel photovoltaic power system for nonvolatile sensor nodes," in *Non-Volatile Memory Systems and Applications Symposium (NVMSA), 2014 IEEE*, IEEE, pp. 1–2.
19. Ma, K., X. Li, H. Liu, X. Sheng, Y. Wang, K. Swaminathan, Y. Liu, Y. Xie, J. Sampson,

and V. Narayanan (2017) "Dynamic Power and Energy Management for Energy Harvesting Nonvolatile Processor Systems," *ACM Trans. Embed. Comput. Syst.*, **16**(4), pp. 107:1–107:23.
20. Natsui, M., D. Suzuki, N. Sakimura, R. Nebashi, Y. Tsuji, A. Morioka, T. Sugibayashi, S. Miura, H. Honjo, K. Kinoshita, S. Ikeda, T. Endoh, H. Ohno, and T. Hanyu (2013) "Nonvolatile logic-in-memory array processor in 90nm MTJ/MOS achieving 75reduction using cycle-based power gating," in *2013 IEEE International Solid-State Circuits Conference Digest of Technical Papers (ISSCC)*, pp. 194–195.
21. Guo, X., E. Ipek, and T. Soyata (2010) "Resistive Computation: Avoiding the Power Wall with Low-Leakage, STT-MRAM Based Computing," *2010 Proceedings of the 37th annual international symposium on computer architecture*.
22. Choudhary, N. K., S. V. Wadhavkar, T. A. Shah, H. Mayukh, J. Gandhi, B. H. Dwiel, S. Navada, H. H. Najaf-abadi, and E. Rotenberg (2011) "Fabscalar: Composing synthesizable RTL designs of arbitrary cores within a canonical superscalar template," in *2011 38th Annual International Symposium on Computer Architecture (ISCA)*, IEEE, pp. 11–22.
23. Fool, T. "http://www.tvfool.com/index.php,".
24. Esmaeilzadeh, H., E. Blem, R. St. Amant, K. Sankaralingam, and D. Burger (2012) "Dark Silicon and the End of Multicore Scaling," *Micro, IEEE*, **32**(3), pp. 122–134.
25. Ma, K., X. Cui, K. Liao, N. Liao, D. Wu, and D. Yu (2014) "Key characterization factors of accurate power modeling for FinFET circuits," *Science China Information Sciences*, pp. 1–13.
26. Xiaoxin, C., M. KaiSheng, L. Kai, L. Nan, W. Di, W. Wei, L. Rui, and Y. Dunshan (2013) "A Dynamic-Adjusting Threshold-Voltage Scheme for FinFETs low power designs," in *Circuits and Systems (ISCAS), 2013 IEEE International Symposium on*, pp. 129–132.

A Fast Joint Application-Architecture Exploration Platform for Heterogeneous Systems

Rafael K. Vivas Maeda, Peng Yang, Haoran Li, Zhongyuan Tian, Zhehui Wang, Zhifei Wang, Xuanqi Chen, Jun Feng, and Jiang Xu

Abstract Computing systems become increasingly heterogeneous by adopting domain-specific accelerators. While heterogeneity provides better trade-offs among performance, energy efficiency, and cost, it adds new dimensions to the already huge design space. Existing design exploration tools rely on time-based analysis. Simulating a heterogeneous computing system using detailed cycle-accurate simulations could last for months, if not years. In this work, we introduce JADE, a joint application-architecture exploration platform for heterogeneous computing systems. JADE targets design space explorations with overall system characteristics such as average performance and energy efficiency instead of detailed cycle-by-cycle behaviors. It uses statistical application models and cycle-accurate architecture models. JADE can speed up design explorations by several orders of magnitude. The statistical application behaviors make it easy to explore heterogeneous systems. JADE can simulate complete systems including processing units, memory hierarchy, interconnect, and peripherals.

1 Introduction

Heterogeneous systems are becoming increasingly popular. They are already pervasive in embedded systems, and are becoming more present in the cloud and high-performance computing with the rise in adoption of GPUs, FPGAs, and other accelerators [5, 13, 33]. Heterogeneity offers better trade-offs for performance, energy efficiency, and cost. However, it adds new dimensions of complexity to the already huge design space. Therefore, we need better methodologies, tools, and models to assist in the design exploration.

The typical evaluation methods rely on time-based techniques such as cycle-accurate simulation [9, 25, 27]. These methods simulate events in a cycle-by-cycle

---

R. K. Vivas Maeda · P. Yang · H. Li · Z. Tian · Z. Wang · Z. Wang · X. Chen · J. Feng · J. Xu (✉)
Hong Kong University of Science and Technology, Kowloon, Hong Kong
e-mail: rkvivasmaeda@ust.hk; jiang.xu@ust.hk

S. S. Bhattacharyya et al. (eds.), *Embedded, Cyber-Physical, and IoT Systems*,
https://doi.org/10.1007/978-3-030-16949-7_9

approach. They can represent with high fidelity the real hardware and are useful when accuracy and precision are required. Some examples include testing, debugging, and accurate benchmarking. However, they do not scale well. These techniques can take months, if not years, to simulate a large-scale system. This is unsuitable for early design explorations when we need to investigate many design points.

In this work, we introduce JADE, a **j**oint **a**pplication-architecture **d**esign **e**nvironment for exploration of heterogeneous systems. JADE targets design exploration considering overall system characteristics such as average performance and energy efficiency instead of detailed cycle-by-cycle behaviors. JADE is composed of two parts, the statistical application models and a cycle-accurate architecture model. The statistical models are obtained from the analysis of the internal structure of the application. We use the statistical models to reproduce the average behavior of the application. This behavior is then used as stimulus to the architecture model instead of the detailed application. The statistical model allows JADE to speed up the simulation by orders of magnitude due to the faster convergence.

Another benefit of the statistical application model is its abstract representation. It allows us to represent the application behavior in an architecture independent way. This abstraction is helpful for heterogeneous systems, in which might co-exist processing units with different instruction sets, and even units with no instructions, such as FPGA. This is especially important in early design stages when compilers and OS might not be available.

JADE can model complete systems including processing units, interconnects, memory hierarchy, and peripherals. It can simulate systems in the scale of tens of thousands of cores. In the current version of JADE, we provide three types of CPU architectures x86_64, ARM-v8, and RISC-V. JADE includes two types of networks, electrical and optical. They can be used to simulate on-chip as well as off-chip networks. We provide several examples of cache-coherence protocols and are relatively easy to add new ones. JADE has built-in power models allowing us to estimate energy consumption. It also includes power management policies such as DVFS. We built the architecture using cycle-accurate models to represent with precision the system performance. Therefore, JADE provides detailed metrics such as application execution time, core utilization, network latency, cache miss, power, energy consumption, and many others.

JADE has been used in several research projects. For instance, Wang et al. studied a novel processor-memory interconnect scheme that combines on-chip and off-chip optical networks [31]. Yang et al. implemented an inter/intra-chip rack-scale network to interconnect processor cores, caches, and memories using silicon photonic network [38]. Li et al. introduced a run-time power delivery management policy for multi-core processors that takes into consideration the workload behavior [18]. Tian et al. studied a collaborative technique for power management using the knowledge gathered in distributed devices [29]. With the kind help of the authors, many of these ideas are also available in JADE. We release JADE for the public use and it is available online at [3].

This chapter is organized into four main sections. We start with an overview of JADE in Sect. 2. Then we detail the two major parts of JADE, the COSMIC[1] heterogeneous benchmarks in Sect. 3 and the architecture simulator in Sect. 4. Then, in Sect. 5 we present two case studies in which JADE has been used.

2 JADE Overview

JADE was initially introduced in [22]. Figure 1 shows an overview of JADE. JADE has two main parts covering application and architecture. The first part is the COSMIC benchmarks, which contains the statistical application models. The second is the cycle-accurate architecture model, implemented as an event-driven simulator. Majority of the simulator is implemented in C++. JADE also comes with a set of tools and configuration templates that allows quick use of the framework. In this section, we describe the major components of JADE, and in later sections, we detail each part.

COSMIC, shown on the left-hand side of Fig. 1, is a heterogeneous multiprocessor benchmark suite [21, 30]. It stands for **c**ommunication-**o**bservant **s**chedulable **m**emory-**i**nclusive **c**omputation. COSMIC currently includes three types of application models. The main type is the statistical application model (SAM). SAM is a statistical representation of the application behavior. In this model, the application is partitioned into several tasks. For each task, we extract its essential characteristics using a mix of static and dynamic profiling. Some of the characteristics we extract are the task dependency, the amount of data shared between tasks, memory locality, and other micro-architecture independent metrics. During simulation, we reproduce the average behavior of the task using a synthesis technique. A key benefit of statistical models is that they converge faster to the average behavior allowing us to reduce the simulation length.

COSMIC also includes a sample of memory traces collected for one run of the application. We refer to the traces as recorded application model. The traces replicate with fidelity the detailed behavior of the application. They are useful for testing and for debugging components of the system. Another model available in the COSMIC is the synthetic (artificial) benchmark. These models are not based on a realistic application. Instead, their behavior is artificially generated. One example of the use of the artificial benchmarks is for sensitivity analysis. For example, to check the system performance when the application characteristics (e.g., computation, memory locality) change.

The architecture model in JADE, shown on top of Fig. 1, includes all necessary components to simulate a complete modern system. It includes processing units, interconnects, memory hierarchy, and peripherals. JADE has two types of network models, electrical and optical. It can simulate on-chip networks (NoC) as well as

[1]COSMIC means **c**ommunication-**o**bservant **s**chedulable **m**emory-**i**nclusive **c**omputation.

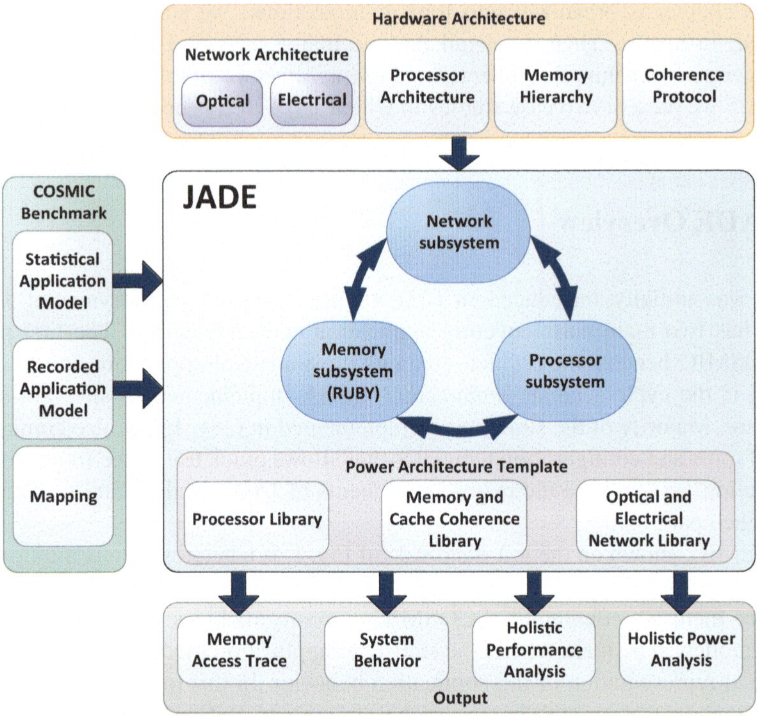

Fig. 1 JADE overview

off-chip networks. JADE can simulate, for example, NoC for multi-core processors as well as network schemes for computer racks. We provide three types of CPU architectures x86_64, ARM-v8, and the open RISC-V instruction set.

The memory hierarchy of JADE is built on top of the Ruby simulator [23]. It has a detailed memory hierarchy including caches, main memories, and coherence protocols. The coherence protocols are specified using a domain-specific language for coherence protocols, SLICC (specification language for implementing cache coherence). In JADE, we provide new coherence protocols in addition to those released in Ruby. One example of protocol provided is a three-level cache using hierarchical coherence protocol. This protocol uses MSI (modified-shared-invalid) for the private caches and MOSI (modified-owned-shared-invalid) for the shared last-level cache. Additionally, it is relatively easy to include new protocols due to the convenient interface provided by SLICC.

JADE has built-in power models with parameters obtained from predictive process design kits. Currently, JADE includes technology models ranging from 7-nm to 90-nm. The power models allow us to obtain power consumption of the system during run-time. JADE uses this information to implement a power management unit (PMU). The PMU includes dynamic voltage and frequency scaling (DVFS) policies, among other features.

Most of the inputs for JADE are simple descriptive text files. We provide several templates that allow users to evaluate a wide range of architectures. Many of the system parameters are configurable through these files and command line options. JADE also includes several tools to assist in the evaluation of the system. For instance, we provide tools to generate network architectures using a simple command line interface, mapping and scheduling tools, and a tool to visualize the coherence protocols in the form of finite state machine (FSM).

JADE reports many system performance metrics. Some examples are the total execution time of the application, per core busy/idle time, cache statistics such as miss rate, power, energy consumption, and network statistics such as throughput and latency. Extra performance metrics can be easily added to the source code.

3 COSMIC Benchmark Suite

COSMIC (communication-observant schedulable memory-inclusive computation) is a benchmark suite for heterogeneous multiprocessors. The three main goals of the COSMIC benchmarks are to: first, provide a convenient model for applications targeting heterogeneous applications, second, increase the speed of the simulation by using statistical simulation, and third, maintain statistical average-level accuracy. In this section, we discuss how the COSMIC benchmarks are obtained.

3.1 Overview

The COSMIC benchmark suite is part of the JADE but we also release them as stand-alone software. It is available online at [2]. Figure 2 shows an overview of COSMIC. The current release of COSMIC includes the application models, some tools commonly used together with the benchmarks, and a C++ API library to allow integration with other software.

The current release of COSMIC has about 20 applications including realistic and artificial applications. Table 1 shows the list of applications provided in COSMIC. Among the realistic applications, we have three classes of benchmarks. The first class we refer as kernels, it includes some common scientific applications and kernel algorithms. The second class is the machine learning applications, including training and testing of fully connected neural networks and deep convolutional neural networks. The third class is the high-performance applications, the APEX benchmarks [1]. For each application, we release the high-level C/C++ implementation, the statistical application models, the recorded traces, and a sample input dataset.

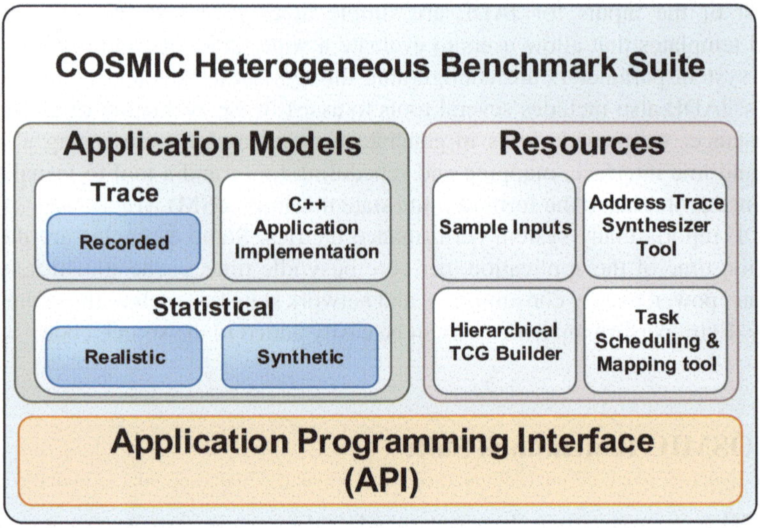

Fig. 2 COSMIC heterogeneous benchmark suite

3.2 Statistical Simulation

Statistical simulation can be decomposed into two parts: profiling and synthesis. In this subsection, we introduce the general idea of statistical simulation. In Sect. 3.4, we give details about the profiling methodology, and in Sect. 3.5, we explain the synthesis procedure.

Figure 3 shows a simplified flow of statistical simulations. The first step is the profiling, where we analyze the algorithm and important application characteristics. The output of the profiling is what we refer to as statistical application model or statistical model. Then the statistical models are used to reproduce the average behavior of the original application. The process of reproducing the behavior is named as synthesis. Finally, the synthesized behavior is used as stimulus for the architecture simulator. Here it could be a trace-simulator, a detailed cycle-accurate simulator (as in JADE), or even others. In the case of a timing simulator, it needs a feedback from the simulator to the synthesis algorithm. This feedback allows the simulator to control the execution speed of the application behavior. In the figure, we also show a logical separation between the synthesis and the simulator. However, they could be implemented as a single monolithic software.

In statistical simulation it is not our intent to clone the original application. In other words, we do not want to obtain an exact copy of the original application after synthesis. Instead, we focus on average behavior. By average behavior, we mean any observable metric of interest. Some examples are average execution time, average memory locality, average task dependencies, and others.

Table 1 List of applications provided in COSMIC

Class	Benchmark	Description
Kernels	FFT	Fast Fourier transform with 1024 complex number inputs
	MD	Computer simulation of physical movements of atoms
	LDPC	Low-density parity-check code encoder with 270 input words
	TURBO	Turbo code decoder with 40-bit data block and 1/3 coding rate
	RSe	Reed–Solomon code encoder
	RSd	Reed–Solomon code decoder
	US	Ultrasound, medical diagnostic algorithm using 2D/3D ultrasound imaging
	RT	Ray-tracing, 3D scene rendering algorithm
Machine learning	ML-FMP	Machine learning for financial market prediction (ML-FMP) using fully connected neural networks
	ML-ALIP	Machine learning for automatic linguistic indexing of pictures (ML-ALIP) using fully connected neural networks
	Cifar-train	Training phase of a deep convolutional neural network for the Cifar-10 dataset
	Facerec-train	Training phase of a deep convolutional neural network for face recognition
	AlexNet-inf	Inference phase of the DCNN with five convolutional layers
Apex	HPCG	Conjugate gradient algorithm Generates a linear system of a three-dimensional heat diffusion problem
	HPCG-prec	HPCG including computation of the preconditioned conjugate gradient
	Pennant	Pennant mini-app. Lagrangian staggered-grid hydrodynamics algorithm on 2-D unstructured finite-volume mesh
	Snap	Performance modelling of a modern discrete ordinates neutral particle transport application
	Stream	Synthetic benchmark measuring the memory bandwidth and a corresponding computation rate for four simple vector kernels

The choice of the metric to observe is important and it depends on the context. In the context of design exploration, we want to evaluate several different architectures. Therefore, it is desirable metrics that are independent of the architecture. For instance, the cache miss rate is one example of metric that depends on the architecture. On the other hand, reuse distance, a metric of memory locality, is architecture independent. It is not always possible to obtain architecture independent metrics. When this happens it is desirable that they are at least weakly dependent.

There are several benefits of using statistical simulation. One of the benefits is about confidentiality. Statistical simulation allows developers to deliver a performance signature of their application while hiding its implementation. This is useful when developers do not want to disclose the source code but they need to know how well the application performs in a specific system.

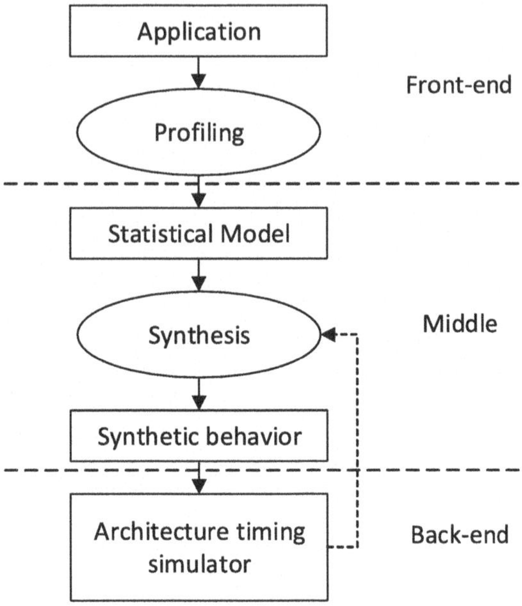

Fig. 3 Statistical simulation flow overview. We can divide statistical simulation into three layers: front-end, middle, and back-end

Another benefit of using statistical is that it is more compact regarding size. This is especially important for large-scale applications and systems. The alternative approach, recording the memory traces, requires a tremendous volume of storage. Therefore, statistical application models can be regarded as a trace compression technique. In fact, many trace compression techniques use some of the ideas used in statistical simulation [16, 20].

Another important benefit is that statistical models converge quickly to the average behavior. Therefore, we can shorten the trace length which reduces the time spent in one simulation run. Later we show that properly choosing the statistical technique allows us to speed up the simulation by three orders of magnitude.

Building statistical simulation comes with a cost. The most obvious cost is the profiling stage. Profiling requires time and storage. Fortunately, it has to be realized only once for each application and can be reused for many simulation runs. Thus, the overhead caused by profiling is small if considered the overall design exploration. Another challenge in statistical simulation is deciding the best model for your purposes. There is no universal model that will be suitable for all cases. For instance, if you want to study memories, you will likely need a locality model. And if you cannot afford to use locality, the model might not accurately represent the real application behavior. In addition, sometimes obtaining a metric that is completely architecture independent is difficult. This might require you to profile a few times, each of them for a particular class of architectures.

3.3 Task Partitioning

The first step of profiling in COSMIC is the task partitioning. In this stage, we want to identify the basic tasks the application has to perform and the dependencies among them. This step is essential as it puts in evidence the parallelism in the application. In the current version of COSMIC, we realize these two steps manually. However, there exist tools that can help identify dependencies automatically. During task partitioning, we take into consideration the algorithm present in the application and one sample implementation in a high-level language (e.g., C/C++). Understanding the algorithm allows us to make better decisions for task partitioning, while the high-level implementation helps us identify the dependencies. There are some general rules to decide the granularity or size of a task. Using a coarse granularity, i.e., large tasks, it might be difficult to depict parallelism in the application.

There are two types of dependencies we are interested in. One is the data dependency, which happens when one task produces data used by others. We capture this dependency by analyzing tasks that have common data in the source code. Some of the dependencies are obvious, for example, when two tasks access the same variable. Others are more difficult to depict such as in pointers. We use static and dynamic profiling to capture the amount of data exchanged between tasks. We use average data-sharing since the sharing between tasks might vary. Data-sharing between tasks is a form of communication between them. Therefore, we use the terms data-sharing and communication interchangeably.

The second type of dependency we are interested in is the control flow dependency. Control flow refers to the sequence in which each of the tasks should be executed. The dependency between them dictates which task precedes another one. For some applications, the control flow might depend on the input dataset. This happens because some tasks might not need to execute depending on the input. We profile the application for several input datasets to ensure that all tasks are visited at least once.

We represent the tasks and their dependencies using a task communication graph (TCG). The TCG is a directed acyclic graph (DAG) where nodes represent tasks and edges represent the dependency between them. For each task in a TCG, we record all the information obtained during profiling. And for each data dependency, we also record the amount of data shared. Figure 4 shows an example of TCG with six tasks. The data dependency is shown in arrows ($\rightarrow$). And the numbers close to the arrows are examples of amount of data-sharing. As can be noted, each dependency dictates which task precedes which task and the amount of communication.

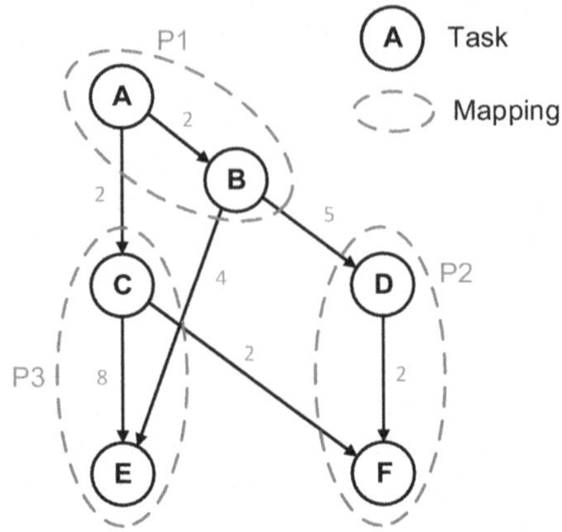

Fig. 4 Example of a TCG with six tasks mapped to three processing units. Each node represents a task and the edges represent the dependencies between them. The numbers close to the edges are examples of amount of data shared between tasks

3.4 Locality Profiling

Memory locality is an abstract concept to refer to the pattern of accessing the memory. There are two types of localities, spatial and temporal locality. Application with high temporal locality accesses recently visited memory blocks with higher frequency. On the other hand, spatial locality refers to the accesses to close memory locations previously visited. There are many metrics to measure locality. Some examples are the footprint (FP), reuse distance (RD), reuse time (RT), miss rate (MR), and others. In fact, these metrics have some higher-order relationship [36]. To completely represent the locality of an application it might need a combination of several of these metrics.

Not all locality metrics are architecture independent. Reuse distance (RD) is one of the metrics that is architecture independent. By construction, it is directly related to the miss rate for least recently used (LRU) caches. RD is one of the dominant metrics in locality analysis. It has many applications such as program characterization, compiler optimization, and analytical cache miss estimation [7, 8, 10, 39, 40]. Another application of the RD is to use it for memory trace synthesis. This technique is used to implement statistical simulation, overcome proprietary code benchmarking, and allow the creation of synthetic benchmarks [4, 6].

Reuse distance has to be computed for a given block size. The common choice of the block size is the cache-line size (i.e., 32B or 64B). In that case, we say that RD is computing the cache-line locality. One of the limitations of the RD is that it can capture only a single locality, such as line locality. For instance, when RD is computed for a 64B-block granularity, it will lose the information of the reuse of pages, typically 4 KB [21]. This is an important drawback since multi-core systems depend on multiple localities. In addition, many modern cache architectures also depend on multiple localities. Some examples are sectored caches [11, 28], footprint caches [15], and unison caches [14]. They are becoming increasingly popular as an effective way to implement large caches. Furthermore, RD is accurate to evaluate

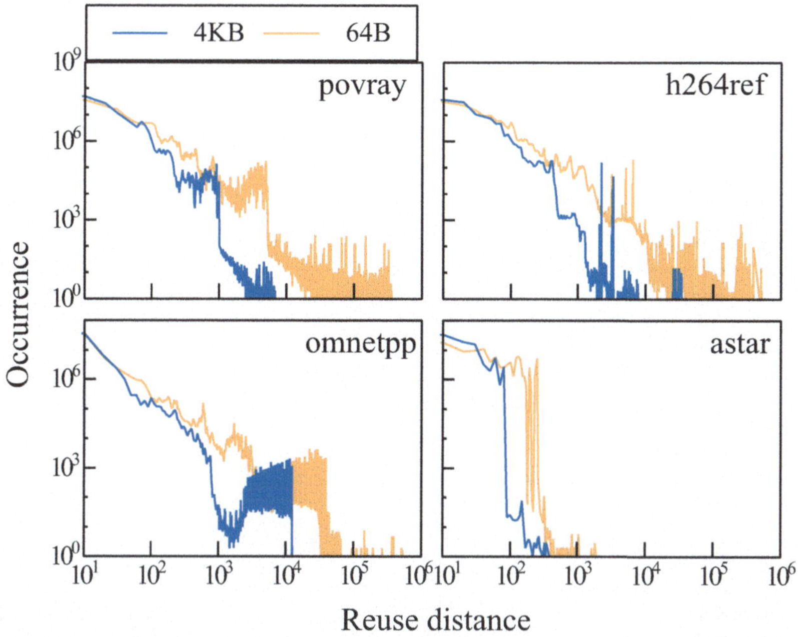

Fig. 5 Locality analysis for two layers: cache lines (64B) and pages (4 KB)

single-level caches, but it performs poorly for multi-level caches due to the week locality representation.

Realistic applications have a considerable difference between the cache-line locality (64B blocks) and page locality (4 KB blocks). Figure 5 shows this behavior for the SPEC CPU2006 benchmarks [12]. We show the RD histogram computed for 64B and 4 KB. As we can note, the page RD differs from the cache line by one order of magnitude. Therefore, due to its limitation, the flat RD is not suitable to model functional units that depend on locality at various granularities. For instance, an RD profiled at 64B will lead to accurate L1 cache miss rate estimations. However, it will not be inaccurate for units that depend on a locality seen at 4 KB, such as translation lookaside buffers (TLB).

To overcome this issue, in COSMIC we use the hierarchical reuse distance (HRD). HRD is a generalization of the RD. It captures locality at multiple granularities using single-profiling single-simulation run. The flat RD model is one particular case of the HRD with a single layer. HRD has higher miss rate accuracy on caches with hybrid line size as well as on conventional designs compared to the flat RD. In addition, HRD has faster statistical convergence and can reduce the simulation length by three orders of magnitude.

Algorithm 1 shows how we can obtain the HRD profile of a task. It receives as input the memory traces T and returns the HRD distribution. The HRD distribution is composed of several histograms, one for each locality captured. In the algorithm,

Algorithm 1 Hierarchical reuse distance profiling. It receives as input the memory traces T, the number of localities to be captured N_l, and the block size for each of the localities W. The output is the HRD distribution which is represented in the histogram $hist$

```
function PROFILE_HRD(T, N_l, W)
    time ← 0
    hist ← 0
    for each t in T do
        dist ← ∞
        l ← 0
        while dist == ∞ and l < N_l do
            block ← ⌊t / W[l]⌋
            dist ← REUSEDISTANCE(Info[l], block, time)
            hist[l][dist] ← hist[l][dist] + 1
            l ← l + 1
        end while
        for j from 0 to N_l do
            block ← ⌊t / W[j]⌋
            UPDATE(Info[j], block, time)
        end for
        time ← time + 1
    end for
    return hist
end function
```

we represent the histograms in the vector $hist$. The algorithm also receives the number of localities (N_l) to capture and their block size (W). In COSMIC, we capture cache-line locality and page locality. In this case, $N_l = 2$ and $W = \{64,4096\}$. The profiler starts the analysis at the layer with smaller blocks of granularity, recording the block-RD for each memory reference. Whenever an infinity RD is found, we record this infinity in the distribution of that layer and also compute the RD for the next one. We stop computing the RD for a memory reference in two cases: when the RD in a layer is not an infinity, or when an infinity is in the last layer. This procedure repeats for all references. Note that only references that cause an infinity RD in one layer can affect the counters in the next. On every infinity RD obtained in one layer, the next layer will capture extra information. This extra information in subsequent layers is lost in the traditional flat RD.

3.5 Synthesis

Synthesis is the process of reproducing the behavior of the application. We can break down the synthesis of the application in two parts. We will start the discussion about the first part: how to generate the behavior of individual tasks. Later we explain the second part which is how to generate the interaction between tasks.

The behavior for each task in COSMIC is linked to the memory locality. In other words, to reproduce the task behavior we need to reproduce its memory locality. All the system performance metrics (miss rate, execution time, network traffic, etc.) are consequences of how the application accesses the memory and the architecture details. We refer to the process of reproducing locality as trace synthesis since we are generating accesses to the memory.

In the previous section, we discussed how to capture the HRD profile of each task. In this section, we discuss the reverse process, the generation. The trace synthesis using the HRD follows a similar procedure as profiling. The high-level algorithm is shown in Algorithm 2. The goal of the algorithm is to generate memory accesses to be issued by the CPU. In the algorithm, the generated addresses are stored in the vector variable $addr$. The algorithm receives the HRD distribution which is composed of several histograms. We represent the histograms with the variable $hist$. The algorithm also needs the number of layers N_l, the block size of each layer W, and the total trace length N.

The general idea is explained as follows. The synthesis starts looking at the layers with finer granularities. First, we generate a random sequence of RD for the first layer following the profile obtained from the original trace. For non-infinity RD

Algorithm 2 Address trace synthesis using *HRD*. It returns the memory traces to be issued by the CPU, represented by $addr$. The algorithm receives the HRD distribution which is composed of several histograms. The histograms are noted with the variable $hist$. The algorithm also needs the number of layers N_l, the block size of each layer W, and the total trace length N

```
function SYNTHESIS_HRD(N_l, W, hist, L)
    for each t in L do
        dist ← ∞
        l ← 0
        RND ← U(0, 1GB)
        while dist == ∞ and l < N_l do
            dist ← RANDOMREUSE(hist[l])
            l ← l + 1
        end while
        if dist == ∞ then
            addr[t] ← RND
        else
            addr[t] ← READHISTORY(Info[l − 1], dist)
            addr[t] ← addr[t] + RND mod W[l − 1]
        end if
        for j from 0 to N_l do
            block ← ⌊addr[t]/W[j]⌋
            UPDATEHISTORY(Info[j], block)
        end for
    end for
    return addr
end function
```

value, the generation is identical to the flat RD. In that case, we reuse the Rth last distinct block address previously used. For an infinite RD, we visit deeper layers until we find a definite RD or there are no more layers. For instance, a common case is when there is a cache-line miss but there is a page hit. In that case, the line RD will be infinity while the page RD will be definite. As soon as we find a definite RD in a layer, we reuse the Rth recently used block and generate a new smaller block within it. In the previous cache-line and page example, we would reuse a page, and generate a random cache-line inside this page. As soon as we encounter a definite RD, we terminate the current generation, and we can start generating a new address. If no definite RD is found, means that we have to generate a completely new random address.

Once the behavior of the tasks has been generated, we need to reproduce the interaction between them. This is accomplished by looking at the task dependencies and the amount of data they share. In COSMIC, the task dependency represents both data and control flow dependency. That means that one task cannot start until all tasks that it depends on finish its execution. Once the task begins running, it will access private data as well as data that are shared with other tasks. We determine the total data-sharing using the respective amount obtained during profiling.

3.6 Convergence Speed

An essential characteristic of statistical simulation is that it quickly converges to the average performance behavior of the original application. This allows reduction of the simulation length, enabling quick design space exploration. In this section, we demonstrate the convergence speed of a single task benchmark. The experiments were performed with the SPEC CPU2006 benchmarks. We evaluate the convergence of the models with respect to the miss rate in a multi-level cache system. We simulate the memory hierarchy using a fraction of the synthesized trace and compare the final cache miss rate with the original full trace length.

We compare three techniques such as the traditional reuse distance (referred to as FLAT), the hierarchical reuse distance (HRD), and the bit Markov chain (BMC). BMC is another type of locality synthesis method similar to reuse distance. The only difference being that in BMC random addresses are generated with the help of a Markov chain model that captures the address distribution [21].

Figure 6 shows the correlation factor and average error for each level cache. The results demonstrate that, for any of the models, the L1 miss rate achieves strong correlation even with a tiny fraction of the synthetic trace length, such as 10^3 memory references. With this simulation length, the average miss rate error is already low, about 7% for FLAT implementations and 5% for hierarchical. As the simulation length increases, the average absolute error can be reduced to less than 2%. For L1 cache, all of the models behave similarly. However, FLAT requires simulation lengths of more than 10^6 and 10^7 to reach strong correlation for L2 and

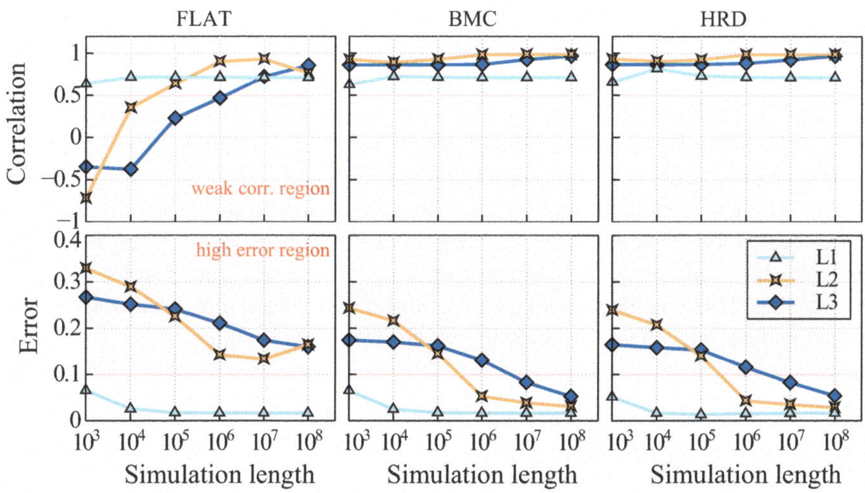

Fig. 6 Convergence for each cache level. Comparing FLAT, BMC, and HRD

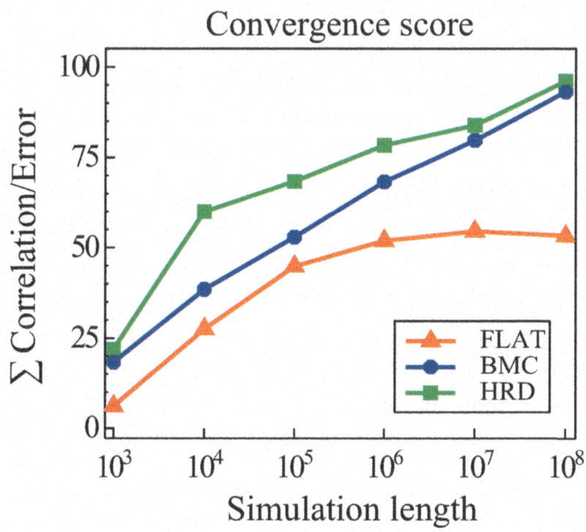

Fig. 7 Convergence score considering all cache levels together

L3, respectively. Furthermore, FLAT does not reach error below 10% for higher level caches.

To compare each of the models, we define the convergence score as the sum $\sum$ correlation/error for all caches. This is justified by the fact that while high correlation desired, a low average error is also important during design exploration. The higher the score, the better is the combination of correlation and accuracy. We plot the score for various simulation lengths in Fig. 7.

The traditional reuse distance is labeled as FLAT, because it has a single layer of locality. The method used in the COSMIC is the HRD. As we can note, FLAT obtains the worst convergence score among all models, due to its high error in L2 and L3. The peak convergence score for FLAT is about 54 for simulation length of 10^7. HRD outperforms this value with much a shorter simulation, 10^4, indicating more than three orders of magnitude faster convergence.

The L1 cache filters out many of the requests issued by the core. This removes some locality from the requests reaching the L2 and L3 caches. Since FLAT has only one layer of locality information it leads to less accurate results in higher level caches. HRD includes extra locality information which minimizes the filtering impact and becomes more accurate to explore multi-level caches.

3.7 *Other Profiling Statistics*

There are other statistics we collect during profiling. For instance, we also record the request type of the traces, i.e., whether the memory access is a read or write (RW) type. We model the request type with a micro-architecture independent model that leads to better accuracy for the eviction traffic in the last-level cache (LLC). The RW is based on a Markov chain (MC) combined with the HRD computation and has traffic error by around 100 times smaller than other naive approaches. Collecting the MC model for the RW comes virtually for free, introducing little overhead during profiling and trace generation [21].

The method is inspired by the way programs access data. We assume that a memory location can be in three different states: new when it has never been touched; clean when it has been read but never written; and dirty when the memory location has been modified. In the beginning, all states are set to new. When a read (or write) operation is performed in a new memory location, its state is set to clean (or dirty). Writing to a clean block causes a state transition to dirty. Once the state reaches dirty, it will never come back to clean again. Thus, writing or reading to a dirty block will remain dirty.

During profiling, we count the number of reads (or write), according to the current state of the memory location, and normalize it with respect to the total trace length. This record gives the respective conditional probability $P(RW|s)$, where s is the state, $s \in \{new, clean, dirty\}$ and RW is the type of operation, $RW \in \{read, write\}$. During synthesis, we select the request type according to the state of the generated location, s. We choose read with probability $P(read|s)$, or write otherwise. Including the proposed RW model is virtually free for both profiling and generation. The only modifications required in Algorithms 1 and 2 are (1) to include a state variable along with the variable holding the unique address, and (2) to update the counters and state as explained. Since the cost of computing the RD is dominated by hashing and searching [26], applying these changes will only have minimal performance degradation.

4 Architecture Simulator

JADE has a timing architecture simulator. The architecture simulator can model complete modern systems including processing units, interconnects, memory hierarchy, and peripherals. It can simulate systems with the scale of tens of thousands of cores. In this section, we describe the main components currently implemented in JADE.

4.1 Processor Models

The current CPU models in JADE have a simplified micro-architecture. Users can change a few parameters such as the number of pipeline stages and instruction issue width. We assume that the number of available functional units is such that it allows all issued instructions to execute in parallel. Also, except for instructions accessing memory, all the other instructions take a constant number of cycles. This simplified CPU is named constant-IPC CPU.

There are a few reasons we only release such a simplified micro-architecture model. One reason is that implementing a detailed CPU model incurs in significant overhead for the simulation. This would limit the speed and also the scalability of the simulator. A second reason is that in practice it is difficult to obtain the details of realistic modern CPUs. Therefore, even if we provided detailed models, users would still need to adjust the source code to model the specific architecture. The third reason is that the CPU micro-architecture is already mature. JADE targets the study of the whole system rather than focused on the CPU alone. Having said that, nothing prevents users from adding new CPU models to the JADE source code.

The processors in JADE contain a few other features. For instance, it includes a memory management unit (MMU), a network interface for off-chip networks, per-core dynamic voltage and frequency scaling (DVFS) support, and others. In JADE, the CPU models itself are ISA free. However, the current benchmarks in JADE have been profiled for ARM-v8, RISC-V, and x86_64. All of which are 64-bit ISAs.

4.2 Memory Hierarchy

JADE models a detailed memory hierarchy including caches, main memories, and coherence protocols. The memory hierarchy models of JADE are built on top of the Ruby simulator [23].

The coherence protocols are specified using a domain-specific language for coherence protocols, the SLICC. In SLICC, the transitions of the coherence protocols are easily expressed based on the current state, incoming event, and the output state. The language has a C-like syntax with a convenient interface to create

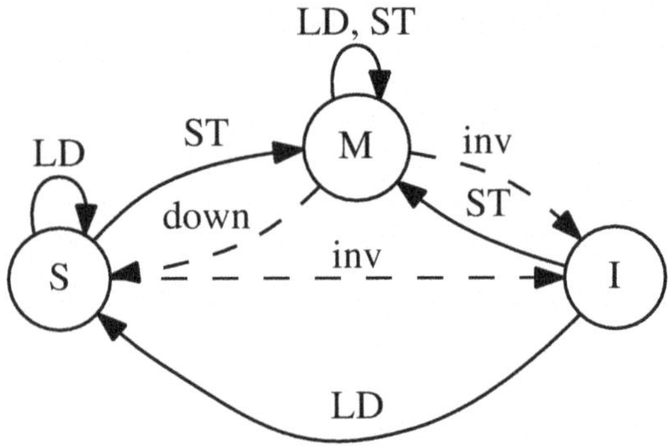

Fig. 8 Example of an MSI cache-coherence protocol for L1 cache. In this diagram, we omit intermediate states and some transitions for the sake of clarity

input and output ports. It is relatively easy to read or develop other coherence protocols.

In JADE, we release new coherence protocols apart from those initially published in the Ruby simulator. One example of protocol included in JADE is a three-level cache hierarchy using the hierarchical coherence protocol. In this protocol, MSI (modified-shared-invalid) is used for the private caches and MOSI (modified-owned-shared-invalid) for the shared last-level cache. This particular protocol has a private L1, private L2, and shared L3 cache. Figure 8 shows a simplified state transition for the MSI protocol. In this figure, the intermediate states as well as some transitions are omitted for the sake of readability. The CPU requests are labeled with capital letters (LD, ST), and the coherence messages are marked with lower case, e.g., downgrade and invalidate messages.

Another example of a possible configuration of the JADE memory hierarchy was initially presented in [31]. It uses a clustered architecture in which a few cores, e.g., four cores, share one L2 cache. Additionally, each core has its own private separate I/D L1 cache, and all cores share the last-level cache. This configuration is currently being integrated into the trunk development of JADE and will soon be available publicly.

4.3 Interconnect

JADE has a comprehensive interconnect model. It can simulate on-chip networks (NoC) as well as off-chip networks. As an example, JADE has been used to study NoC for multi-core processors [37] and network schemes for computer racks [38]. JADE models two types of networks, optical and electrical. On one hand, the electrical network technology is more mature and widely adopted. On the other hand, optical interconnect has been demonstrated to be a promising alternative to achieve high energy efficiency and high bandwidth [17, 24, 32, 34, 35]. In this section, we briefly discuss some of the constructs of the JADE interconnect models.

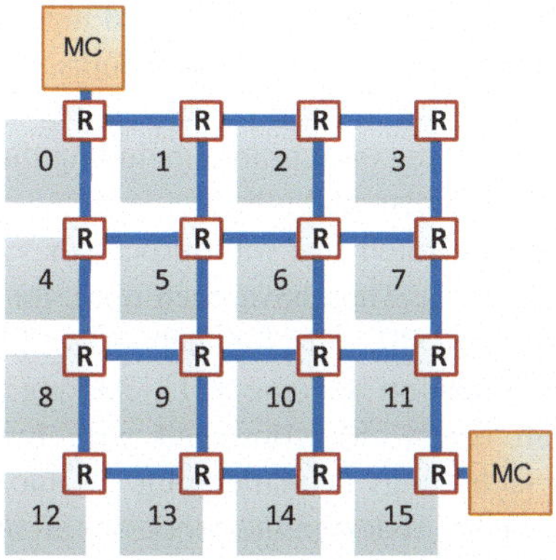

Fig. 9 Example of NoC with a 4 × 4 mesh topology. The external memory controllers are highlighted

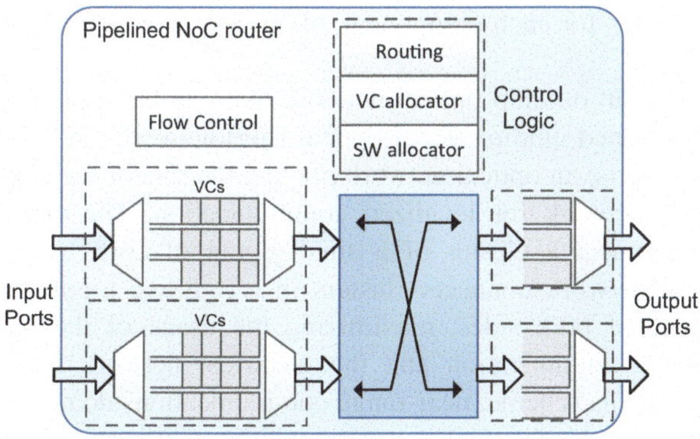

Fig. 10 NoC router block diagram model in JADE

Figure 9 shows an example of an electrical network-on-chip with a 4 × 4 mesh topology. The NoC has three elementary components: links, routers, and network interfaces. Each of these components is detailed at the micro-architecture level. In the models, we account for the precise latency as well as energy consumption. The network topology can be easily configured through intuitive readable text files. The configuration files permit users to set the number as well as the exact position of each component. In JADE, we also include a tool to automatically generate configuration files with a simple command line interface.

Figure 10 shows the NoC router micro-architecture in JADE. It has three pipeline stages: routing computation (RC), virtual channel (VC) allocation, and switch allocation (SA). The router uses a control-flow policy based on credits. In brief words, each output port contains a certain number of credits. Every packet sent through an output port consumes one credit from that port. When the number of credits is zero, the output port is blocked until new credits are received. The output port receives credits from the immediately connected input port, i.e., the respective receiver side. Credits are sent through specialized links, the credit links. For each physical data-link in the NoC, there is a corresponding credit link. Most of the micro-architecture parameters are configurable including the input buffer depth, the number of virtual channels, flit size, and others.

Recent works have demonstrated that multiprocessor systems can benefit considerably from silicon photonics. They have been used, for instance, to implement optical on-chip interconnects [35], off-chip networks [34, 38], and even in a processor-memory communication scheme [31]. Among the benefits of optical interconnects is the high bandwidth combined with low power consumption. However, using this new technology is challenging. To maximize the benefits from optical interconnect, we need to redesign the processor architecture, memories, and peripherals. JADE includes optical network models that permit the study of on-chip and off-chip optical interconnects. Among the optical components we model, we can list the microresonators, conversion from optical-to-electrical (OE) and electrical-to-optical (EO), links, and others. In addition, JADE also accounts for the detailed power losses for each component, allowing accurate estimation of power consumption.

One instance of an on-chip optical network that can be implemented in JADE is the SUOR (sectioned unidirectional optical ring) network [35]. This network is composed of two parts: an optical network playing the role of a global interconnect and an electrical network for localized communication. The electrical network is used to interconnect a cluster of a small count of cores (e.g., four cores), while the optical network connects clusters in a ring-like topology. SUOR tries to take advantage of both sides, maximizing the usage of the optical network for long-distance communication and the electrical network for short-distance communication. In this scheme, near-range communications do not need to pay the price of conversion from electrical-to-optical which consumes power and also adds to the latency. The optical network in SUOR is divided into sections, hence the name, to maximize the total utilization of the waveguide.

JADE can also simulate advanced interconnect schemes such as the I2CON [34]. I2CON is a network architecture targeting multi-chip many-core processors. It is composed of two parts: (1) the network connecting multiple chips, referred as inter-chip network, and (2) the network inside each chip, referred as intra-chip network (or on-chip network). Both the inter and the intra-chip networks use some form of optical interconnect. The intra-chip network is a clustered optical network such as the SUOR. On the other hand, the inter-chip network uses an optical interconnect scheme that connects clusters of different chips and is floorplanned in a way that avoids waveguide crossings. In the inter-chip network, there are N data channels coordinated by one centralized arbiter.

4.4 Support

JADE makes use of the built-in power models to implement a power management unit (PMU). The PMU includes three subsystems, dynamic voltage and frequency scaling (DVFS), energy consumption measurement, and power delivery system (PDS) models.

Currently, JADE can apply DVFS to processor cores, network-on-chip routers, and network interfaces. However, by default, only DVFS of processor cores is enabled. The DVFS uses a run-time policy that takes into consideration the load of the system. It adjusts the voltage along with the frequency to reduce power consumption. The voltage and frequency pairs are obtained from realistic voltage regulators and device parameters. In addition, the operating frequency for each component can be easily modified. In the default configuration, all components have the same frequency of 2 GHz.

4.5 Power Models and Device-Level Models

JADE integrates power model libraries to provide holistic power analysis for various configurations of architecture, and technology nodes. We use McPAT [19] for power estimations for some of the components such as cache and the CPU. The power library includes technology nodes initially released by McPAT and others extracted from device models of advanced predictive process design kits. Currently, JADE supports 7, 22, 32, 45, 65, and 90-nm technology nodes. JADE reports both the static and dynamic energy for all components including processor cores, on-chip and off-chip routers, caches, and others.

JADE also simulates process variations by emulating variations on the conditions for the cores. Some examples of conditions we vary from core to core are the operating frequency, dynamic power, and static power. We obtain the variations based on two device-level parameters: the effective gate-length (L_{eff}) and the threshold voltage (V_{th}). They are varied as a normal distribution spatially correlated, and we use a spherical function to generate covariance between cores.

4.6 Implementation

JADE is implemented as an event-driven simulator written in C++. The code is highly configurable. Most of the parameters can be configured either by setting the configuration files or by command line. JADE is in constant development, and we release new features and fixes quite regularly. We also welcome and appreciate the help of other users in reporting bugs and implementing features.

Adding new components or replacing the existing parts is relatively simple. All active objects derive from a consumer class and need to implement a wake-up function. There are two different types of events that trigger the wake-up function of objects. The first one is the explicit object scheduling. In this case, either the component itself or other components schedule the wake-up function for the future. This approach is commonly used to model state transitions inside the component. The second type of event is more convenient to model communication between components. This is only applicable when the component has input ports. In such case, when one component sends data to another, the input port will automatically schedule the wake-up function of the receiver.

5 Case Studies

In this section, we show a few examples of research that has been carried using JADE. Each subsection discusses a different project. It starts with the title, authorship, and publication details. We summarize the main topics and achievements of each research.

5.1 *MOCA: An Inter/Intra-Hip Optical Network for Memory*

For the complete details of this study, please refer to Wang et al. "MOCA: an inter/intra-hip optical network for memory," DAC 2017 [31].

As mentioned earlier, optical interconnects are a promising alternative to obtain high bandwidth and low power. In this work, the authors propose a new optical network to interconnect processors and memories. There are three major contributions. First, they introduce a new memory organization optimized for optical interconnect. Second, they integrate multiple processor cores and the memories with a unified on-chip/off-chip optical network. Third, they use wavelength division multiplexing and time division multiplexing (WDM/TDM) to increase utilization of the link and reduce the power.

In MOCA, the interconnect between processors chips and memory chips is implemented with optical links. To extract the maximum benefit of this optical interconnect we need to redesign the whole system architecture. More especially, it is required to co-design the processor architecture and the memory architecture. On the processor side, the authors tailor the on-chip network as well as the memory controllers. On the memory side, the authors customize the memory organization to extract the maximum benefit from the optical interconnects.

Typical memory controllers consist of two parts the front-end engine (FE) and transaction engine (TE). The front-end engine performs rescheduling while the transaction engine translates commands to memory signals, in other words, the transaction engine is responsible for generating the control signals, address signals, and data signals with the proper timing. In traditional systems, the front-end engine and the transaction engine are combined into a single unit, and it is located inside the processor chip. In MOCA, the FE and TE are physically separated. FE is located in the processor chip while the TE is located in the memory chip.

The processor in MOCA is organized with multiple clusters of a small count of processing units, e.g., four. Each processing unit has a private L1 I/D cache, and the L2 cache is shared among the cores in the cluster. MOCA has the best performance when the optical off-chip interconnect is used in combination with the optical on-chip network (NoC). The organization of the optical NoC is similar to SUOR. Requests issued by the destination outside a cluster are routed through the optical NoC in a ring-like topology. In the case of memory requests, the message is sent to the front-end engine also using the optical NoC. The front-end engine redirects the request to the proper memory chip using the optical off-chip links.

Figure 11 shows the logical representation of the processor chip in MOCA. The way the components are laid physically will depend on the technology used, for instance, 2.5D integration or 3D integration. The differences and trade-offs between the two are discussed in their work [31]. This figure shows an example of four clusters of processing units, and four front-end engines (equivalent to the memory controllers). In the figure, we only show one link for each front-end engine, but in fact, there will be multiple links. The same happens for the optical on-chip network, in which there will be several links to complete the network. For the sake of simplicity, we also omit EO and OE converters and serializers/deserializers.

Figure 12 shows an overview of the memory chip organization in MOCA. It has an optical interface composed of several optical links. The optical transceiver performs conversions between the optical and electrical domains and uses WDM and TDM to reduce power and increase the utilization of each link. The requests

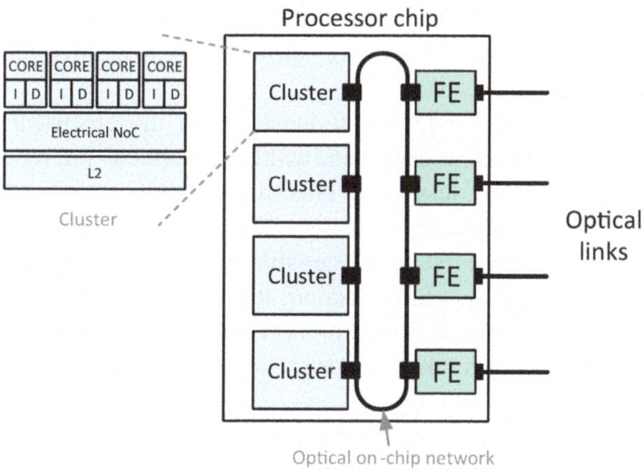

Fig. 11 Example of a processor chip in MOCA. Front-end engine (FE) is responsible for forwarding requests to and from the memory chips. The processor is organized in clusters of a small count of cores. The clusters are connected through a sectioned optical ring-like network

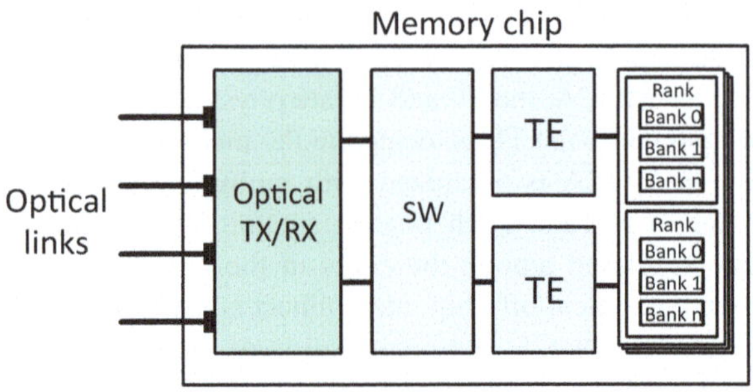

Fig. 12 Example of memory chip organization in MOCA. The optical transceivers (TX/RX) perform conversion between optical and electrical domains. There are two transaction engines (TE). The memory is organized in a hierarchy manner with ranks and banks

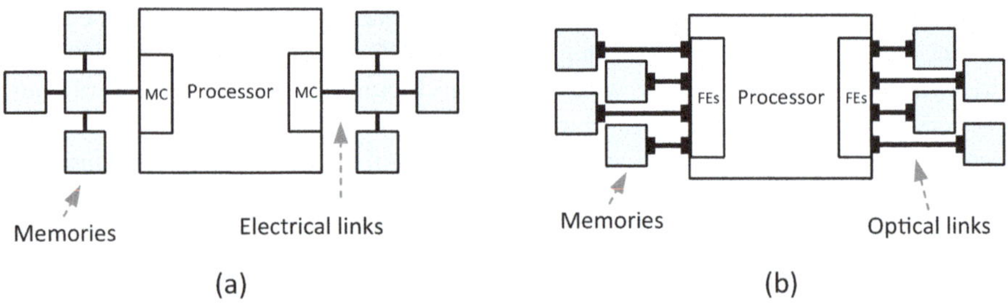

Fig. 13 Typical processor-memory interconnect using (**a**) HMC and (**b**) MOCA

from the processor are received through the optical transceivers and arrive at the corresponding transaction engine passing the electrical switch. Then the TE performs the traditional DRAM signaling to access the memory and issue the response if necessary. In the figure, we show the typical hierarchical memory organization with ranks and banks.

Optical links have higher bandwidth-density than the electrical counterparts. Therefore, to achieve the same total bandwidth, the optical interconnect requires fewer pins and consequently less area. This allows MOCA to use more memory channels per processor chip, each of them connected to multiple memory chips. This has a twofold benefit. First, it will enable accessing more memory chips in parallel achieving higher bandwidth. Second, it reduces the latency because of the smaller number of hops to reach the memory. Figure 13 shows how (a) the typical memory interconnect using hybrid memory cube (HMC) compares to (b) the one with optical interconnects in MOCA.

The authors used JADE to evaluate their proposed architecture. They compared MOCA with the equivalent HMC implementation. For a 256-core processor, MOCA achieves more than 160% higher memory bandwidth than HMC due to the ability to access more memory chips in parallel. The authors also evaluate the total latency to access the memory. In their evaluation, they account for the latency in the

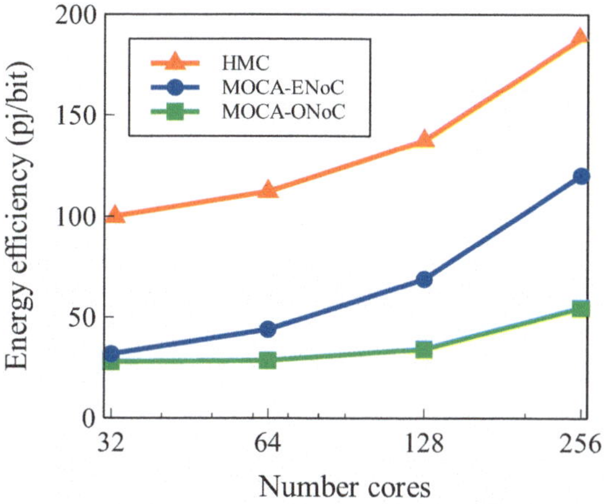

Fig. 14 Energy efficiency comparison of HMC and MOCA

network-on-chip, the off-chip communication, and the memory access latency. The experiment showed that MOCA not only reduces the average latency by 75% but also reduces the variation. The combined effect of higher bandwidth and lower latency reduces the application execution time by 2.6 times on average.

Another important aspect is the energy efficiency. The optical interconnect requires less energy per bit transmitted. Figure 14 compares the energy efficiency of HMC and MOCA. It shows the total energy spent per bit on systems with 32, 64, 128, and 256 cores. In this figure, we show two implementations of the MOCA, one using optical network-on-chip (MOCA-ONoC) for the processor, and another using electrical network-on-chip (MOCA-ENoC). The use of optical off-chip interconnects in MOCA makes it more energy efficient than HMC. This happens for either optical or electrical NoC. The difference between MOCA-ENoC and MOCA-ONoC is more evident for an increased number of cores. In the experiment realized, MOCA-ENoC is 1.6× more energy efficient than HMC. The best choice of architecture is when the off-chip optical network is combined with optical network-on-chip (MOCA-ONoC). In that situation, it is possible to achieve 3.6× more energy efficiency.

5.2 RSON: An Inter/Intra-Chip Silicon Photonic Network for Rack-Scale Computing Systems

For the complete details of this study, please refer to Yang et al. "RSON: an inter/intra-chip silicon photonic network for rack-scale computing systems," DATE 2018 [38].

RSON is a network architecture for rack servers composed of tens of server nodes, where each server node is assumed to have a processor chip, local memories, and other peripherals like storage units. RSON uses optical interconnect to imple-

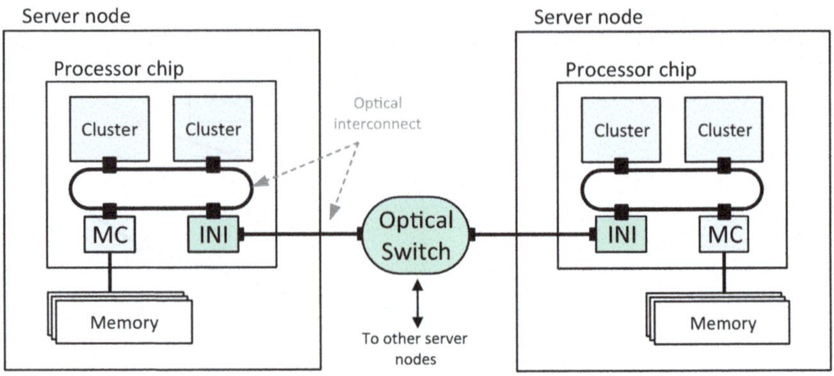

Fig. 15 Typical RSON interconnect. Optical interconnect is used inside the processor chip and in the inter-node network. The optical switch connects tens of server nodes. Each processor chip has an inter-node interface (INI) to connect to the network

ment both the inter-node network and the on-chip network inside each processor. In the literature, they refer to the inter-node network as inter-chip network while the processor on-chip network is referred as intra-chip network. RSON takes advantage of silicon photonics to provide an efficient communication scheme among processor cores, caches, local memories, and remote memories.

Figure 15 shows the logical overview of the RSON architecture. In this figure, we show two nodes interconnected through a multi-port optical switch. In RSON, each processor chip has an inter-node interface that allows a direct connection between the processor chip and the optical switch. This was specially designed to allow an efficient integration between the inter-node network and the processor on-chip network. One of the operations that benefit from this scheme is the transfer between memories located in different nodes, usually performed by the remote direct memory access (RDMA) unit.

RSON uses circuit-switching because it is more convenient for optical interconnects. Therefore, to perform remote memory transfers, the first step needed is to set up a path connecting the local memory and the remote memory. This path reservation will involve the local processor chip, the inter-node network, and the remote processor chip. The processors will need to reserve the on-chip path connecting the memory controller to the respective inter-node interface, which might include the electrical NoC and the optical NoC. On the other hand, the optical switch will be responsible for reserving a path in the inter-node network. In their work, the authors present a novel algorithm to hide the latency due to path reservation for RDMA operations. Once the circuit is set, the transfers can be performed making use of the high bandwidth offered by the optical interconnect.

The authors of RSON also used JADE to evaluate their proposed architecture. They discuss the energy efficiency, total bandwidth, latency, and total execution time. One of the findings is that RSON has about seven times higher performance-per-energy (PPE) score than traditional network architectures using InfiniBand.

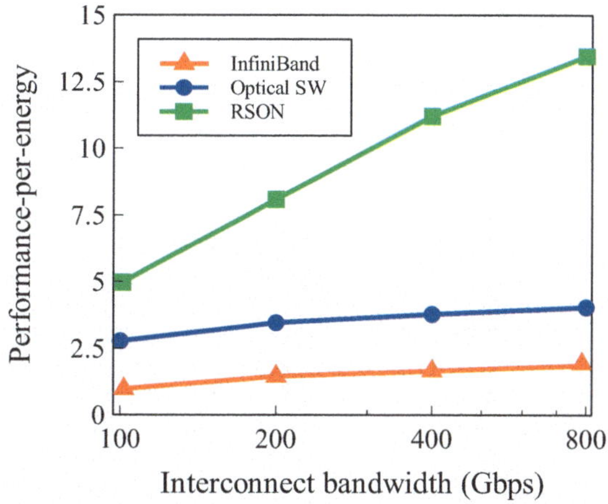

Fig. 16 Performance-per-energy comparison between InfiniBand, RSON, and the traditional network using optical switch (SW) only

Also, the authors discovered that RSON has about 85% lower energy per bit. This shows that RSON is a promising solution to obtain high energy efficiency in rack-scale computing. The PPE score is summarized in Fig. 16, where it shows how the PPE varies by increasing the total bandwidth in the interconnect. The figure compares three network technologies (1) InfiniBand, (2) conventional network using optical switch only, and (3) the RSON.

6 Conclusion

In this work, we introduced JADE a joint architecture/application design environment. JADE targets design explorations considering overall system characteristics such as average performance and energy efficiency instead of detailed cycle-by-cycle behaviors. JADE distinguishes from other simulators due to the statistical application models and the comprehensive architecture models. Among the benefits of statistical simulation are confidentiality, compact representation, and simulation speed. JADE is composed of two parts, the COSMIC benchmarks and the architecture simulator enabling a holistic study of application and architecture. In this chapter, we presented two examples of studies carried out in the JADE simulator. The first is a novel processor-memory interconnect architecture, and the second is a rack-scale network. These studies combine electrical and optical interconnects to maximize system performance and efficiency.

References

1. APEX benchmarks. http://www.nersc.gov/research-and-development/apex/apex-benchmarks. Accessed on 2016-05-30
2. COSMIC benchmarks (BDSL). https://www.ece.ust.hk/~eexu/COSMIC.html. Accessed: 2018-07-26
3. JADE: Joint Application-Architecture Design Environment (BDSL). https://www.ece.ust.hk/~eexu/JADE.html. Accessed: 2018-07-26
4. Awad, A., Solihin, Y.: STM: Cloning the spatial and temporal memory access behavior. In: Proceedings - International Symposium on High-Performance Computer Architecture, pp. 237–247 (2014). https://doi.org/10.1109/HPCA.2014.6835935
5. Bailey, D.G.: Design for embedded image processing on FPGAs. John Wiley & Sons (2011)
6. Balakrishnan, G., Solihin, Y.: WEST: Cloning data cache behavior using stochastic traces. Proceedings - International Symposium on High-Performance Computer Architecture, pp. 387–398 (2012). https://doi.org/10.1109/HPCA.2012.6169042
7. Berg, E., Hagersten, E.: Statcache: a probabilistic approach to efficient and accurate data locality analysis. In: Performance Analysis of Systems and Software, 2004 IEEE International Symposium on-ISPASS, pp. 20–27. IEEE (2004)
8. Beyls, K., D'Hollander, E.: Reuse distance as a metric for cache behavior. In: Proceedings of the IASTED Conference on Parallel and Distributed Computing and systems, vol. 14, pp. 350–360 (2001)
9. Binkert, N., Beckmann, B., Black, G., Reinhardt, S.K., Saidi, A., Basu, A., Hestness, J., Hower, D.R., Krishna, T., Sardashti, S., et al.: The gem5 simulator. ACM SIGARCH Computer Architecture News **39**(2), 1–7 (2011)
10. CaBcaval, C., Padua, D.A.: Estimating cache misses and locality using stack distances. In: Proceedings of the 17th annual international conference on Supercomputing, pp. 150–159. ACM (2003)
11. Cuesta, B., Cai, Q., Hyuseinova, N., Ozdemir, S., Nicolaides, M., Zyulkyarov, F.: Sectored cache with hybrid line granularity (2014). US Patent App. 13/729,523
12. Henning, J.L.: Spec cpu2006 benchmark descriptions. SIGARCH Comput. Archit. News **34**(4), 1–17 (2006). https://doi.org/10.1145/1186736.1186737. http://doi.acm.org/10.1145/1186736.1186737
13. Herbordt, M.C., VanCourt, T., Gu, Y., Sukhwani, B., Conti, A., Model, J., DiSabello, D.: Achieving high performance with FPGA-based computing. Computer **40**(3), 50–57 (2007). https://doi.org/10.1109/MC.2007.79
14. Jevdjic, D., Loh, G.H., Kaynak, C., Falsafi, B.: Unison cache: A scalable and effective die-stacked dram cache. 2014 47th Annual IEEE/ACM International Symposium on Microarchitecture (Micro), 25–37 (2014). https://doi.org/10.1109/MICRO.2014.51
15. Jevdjic, D., Volos, S., Falsafi, B.: Die-stacked DRAM caches for servers. ACM SIGARCH Computer Architecture News **41**(3), 404 (2013). https://doi.org/10.1145/2508148.2485957
16. Johnson, E.E., Ha, J., Zaidi, M.B.: Lossless trace compression. IEEE Transactions on Computers (2), 158–173 (2001)
17. Krishnamoorthy, A.V., Ho, R., Zheng, X., Schwetman, H., Lexau, J., Koka, P., Li, G., Shubin, I., Cunningham, J.E.: Computer systems based on silicon photonic interconnects. Proceedings of the IEEE **97**(7), 1337–1361 (2009). https://doi.org/10.1109/JPROC.2009.2020712
18. Li, H., Xu, J., Wang, Z., Maeda, R.K.V., Yang, P., Tian, Z.: Workload-aware adaptive power delivery system management for many-core processors. IEEE Transactions on Computer-Aided Design of Integrated Circuits and Systems, pp. 1–1 (2017). https://doi.org/10.1109/TCAD.2017.2778080
19. Li, S., Ahn, J.H., Strong, R.D., Brockman, J.B., Tullsen, D.M., Jouppi, N.P.: McPAT 1.0: An integrated power, area, and timing modeling framework for multicore architectures. 2009. Micro-42. (c), 1–40 (2009). https://doi.org/10.1145/1669112.1669172

20. Luo, Y., John, L.K.: Locality-based online trace compression. IEEE Transactions on Computers (6), 723–731 (2004)
21. Maeda, R.K.V., Cai, Q., Xu, J., Wang, Z., Tian, Z.: Fast and accurate exploration of multi-level caches using hierarchical reuse distance. In: 2017 IEEE International Symposium on High Performance Computer Architecture (HPCA), pp. 145–156 (2017). https://doi.org/10.1109/HPCA.2017.11
22. Maeda, R.K.V., Yang, P., Wu, X., Wang, Z., Xu, J., Wang, Z., Li, H., Duong, L.H.K., Wang, Z.: JADE: A heterogeneous multiprocessor system simulation platform using recorded and statistical application models. In: Proceedings of the 1st International Workshop on Advanced Interconnect Solutions and Technologies for Emerging Computing Systems, AISTECS '16, pp. 8:1–8:6. ACM, New York, NY, USA (2016). https://doi.org/10.1145/2857058.2857066
23. Martin, M.M.K., Sorin, D.J., Beckmann, B.M., Marty, M.R., Xu, M., Alameldeen, A.R., Moore, K.E., Hill, M.D., Wood, D.A.: Multifacet's general execution-driven multiprocessor simulator (GEMS) toolset. ACM SIGARCH Computer Architecture News **33**(4), 92–99 (2005)
24. Miller, D.A.: Optical interconnects to electronic chips. Applied optics **49**(25), F59–F70 (2010)
25. Miller, J.E., Kasture, H., Kurian, G., Gruenwald, C., Beckmann, N., Celio, C., Eastep, J., Agarwal, A.: Graphite: A distributed parallel simulator for multicores. In: High Performance Computer Architecture (HPCA), 2010 IEEE 16th International Symposium on, pp. 1–12. IEEE (2010)
26. Niu, Q., Dinan, J., Lu, Q., Sadayappan, P.: PARDA: A fast parallel reuse distance analysis algorithm. In: Proceedings of the 2012 IEEE 26th International Parallel and Distributed Processing Symposium, IPDPS 2012, pp. 1284–1294 (2012). https://doi.org/10.1109/IPDPS.2012.117
27. Sanchez, D., Kozyrakis, C.: Zsim: Fast and accurate microarchitectural simulation of thousand-core systems. In: ACM SIGARCH Computer architecture news, vol. 41, pp. 475–486. ACM (2013)
28. Seznec, A.: Decoupled sectored caches: conciliating low tag implementation cost and low miss ratio. In: Computer Architecture, 1994. Proceedings the 21st Annual International Symposium on, pp. 384–393 (1994). https://doi.org/10.1109/ISCA.1994.288133
29. Tian, Z., Wang, Z., Xu, J., Li, H., Yang, P., Maeda, R.K.V.: Collaborative power management through knowledge sharing among multiple devices. IEEE Transactions on Computer-Aided Design of Integrated Circuits and Systems, pp. 1–1 (2018). https://doi.org/10.1109/TCAD.2018.2837131
30. Wang, Z., Liu, W., Xu, J., Li, B., Iyer, R., Illikkal, R., Wu, X., Mow, W.H., Ye, W.: A case study on the communication and computation behaviors of real applications in NoC-based MPSoCs. In: 2014 IEEE Computer Society Annual Symposium on VLSI, pp. 480–485 (2014). https://doi.org/10.1109/ISVLSI.2014.36
31. Wang, Z., Pang, Z., Yang, P., Xu, J., Chen, X., Maeda, R.K.V., Wang, Z., Duong, L.H.K., Li, H., Wang, Z.: MOCA: An inter/intra-chip optical network for memory. In: 2017 54th ACM/EDAC/IEEE Design Automation Conference (DAC), pp. 1–6 (2017). https://doi.org/10.1145/3061639.3062286
32. Wang, Z., Xu, J., Yang, P., Wang, Z., Duong, L.H.K., Chen, X.: High-radix nonblocking integrated optical switching fabric for data center. Journal of Lightwave Technology **35**(19), 4268–4281 (2017)
33. Wolf, M.: Computers as components: principles of embedded computing system design. Elsevier (2012)
34. Wu, X., Xu, J., Ye, Y., Wang, X., Nikdast, M., Wang, Z., Wang, Z.: An inter/intra-chip optical network for manycore processors. IEEE Transactions on Very Large Scale Integration (VLSI) Systems **23**(4), 678–691 (2015)
35. Wu, X., Xu, J., Ye, Y., Wang, Z., Nikdast, M., Wang, X.: Suor: Sectioned undirectional optical ring for chip multiprocessor. ACM Journal on Emerging Technologies in Computing Systems (JETC) **10**(4), 29 (2014)
36. Xiang, X., Ding, C., Luo, H., Bao, B.: HOTL: A higher order theory of locality. SIGARCH Comput. Archit. News **41**(1), 343–356 (2013). https://doi.org/10.1145/2490301.2451153

37. Yan, G., Wu, N., Zhou, Z.: A novel non-cluster based architecture of hybrid electro-optical network-on-chip. IAENG International Journal of Computer Science **44**(3) (2017)
38. Yang, P., Pang, Z., Wang, Z., Wang, Z., Xie, M., Chen, X., Duong, L.H.K., Xu, J.: RSON: An inter/intra-chip silicon photonic network for rack-scale computing systems. In: 2018 Design, Automation Test in Europe Conference Exhibition (DATE), pp. 1369–1374 (2018). https://doi.org/10.23919/DATE.2018.8342226
39. Zhong, Y., Orlovich, M., Shen, X., Ding, C.: Array regrouping and structure splitting using whole-program reference affinity. ACM SIGPLAN Notices **39**(6), 255–266 (2004)
40. Zhong, Y., Shen, X., Ding, C.: Program locality analysis using reuse distance. ACM Trans. Program. Lang. Syst. **31**(6), 20:1–20:39 (2009). https://doi.org/10.1145/1552309.1552310

The Journey of a Project Through the Eyes of a Smart Camera

Burak Ozer

I met Marilyn Wolf in 1998 when I was a Ph.D. student in the New Jersey Center for Multimedia Research. I worked with her first as a student for my Ph.D. thesis and then as a researcher in her group at Princeton University. Since then, we have been business partners on several projects and most importantly friends throughout all these years. Her guidance at each step of my early research, her visionary and innovative ideas, and her ability to foresee next technological breakthroughs led us to develop the first "smart camera" system with extended features that can be applied to different applications. In this essay, I will focus on the critical stages of our project journey from research and development at Princeton University to commercialization for train stations and other application areas. Marilyn has been working on many different topics related to computers with other festschrift participants. I would like to tell you the story about how we converted "dummy cameras" to "smart cameras," what we achieved and learned, and what can still be done in this area. The article will discuss different platforms and algorithms that we developed throughout the years for different application areas, real-world problems that we faced during commercialization, and current developments that cite our work.

1 Smart Cameras

In "smart cameras" [1], it is stated that "smart cameras perform signal and image processing where the signal is captured and where signal quality is best," which is basically the most important advantage of the smart cameras. A stand-alone smart

B. Ozer (✉)
CTO, Pekosoft LLC, Ambler, PA, USA
e-mail: bozer@pekosoft.com

S. S. Bhattacharyya et al. (eds.), *Embedded, Cyber-Physical, and IoT Systems*,
https://doi.org/10.1007/978-3-030-16949-7_10

camera that processes, extracts, and sends only the necessary information to the end user will remove the need for manual work and greatly reduce bandwidth and data amount that needs to be processed by the end user. This advantage also reduces data loss in case of emergencies, e.g., network interruption during a storm. In the same book, smart cameras are classified as single-chip, embedded (e.g., phone camera), stand-alone, compact system, and distributed smart cameras. Our first smart camera testbed that I will discuss here is a compact system smart camera.

Our journey started in January 2001 at Princeton University. Marilyn's idea on how to embed smart cameras was a breakthrough in computer vision that combines vision algorithms, real-time video processing, and embedded systems. While dummy cameras capture images, smart cameras capture high-level descriptions of the scene and analyze what they see. Although there were a wide variety of applications including human and animal detection, surveillance, motion analysis, facial identification, etc., we decided to use gesture recognition as our first application. Smart cameras leverage very large-scale integration to meet real-time performance conditions in a low-cost, low-power system with substantial memory. Moving well beyond pixel processing and compression, smart camera systems can run a wide range of algorithms to extract meaning from streaming video.

As the first step towards smart cameras, we decided to use fast and reliable video capture boards for video processing. As our testbed, we used a PC as host, two TriMedia video capture boards with 100-MHz Philips TriMedia TM-1300 processors, TriMedia SDK, and analog camcorders. First application we developed was to detect skin areas in real-time. Modeling human body, finding body parts, and as our ultimate goal, recognizing human gesture in real-time by the camera without any human interaction were the major cornerstones of our smart camera project. The initial gesture recognition algorithm was able to find body parts of a person, model each body part with ellipses, and identify his/her simple gestures in real-time. Our paper published in computer magazine [2] summarizes the basic algorithmic steps and our optimization techniques to increase frame rate, reduce latency, and use efficient memory. This board-level system was a critical first step in the design of a highly integrated smart camera. We published many architecture and algorithm related papers by using this initial testbed. We looked into parallelism, pipelining, different optimization techniques, as well as 3D modeling of the human body parts [3–6].

2 Start-up and Commercialization

After 9/11 events, the security and surveillance markets were flooded with companies introducing new ideas. Several companies started working on gesture recognition related applications by using cameras. Especially, crowded area monitoring became one of the hottest topics. We were several steps ahead of these recently formed companies due to our expertise in smart camera algorithms and embedded system knowledge. After discussing the IP issues with the university, we decided to commercialize our technology as a start-up company from Princeton University. The second part of our journey started after we formed Verificon Corporation. We started working with Yokogawa Electric, a Japanese company, to install the gesture recognition system at train stations around Tokyo metro area.

During commercialization of the software, we designed the algorithms for RISC processors for Windows and Linux environments as well as for FPGA acceleration. The latest system was running on DaVinci evaluation boards. One of the major problems was efficient system design for real-time processing. Processing enormous amount of data obtained from a very crowded scene in real-time is a hardware challenge. Some other challenges we faced during deployment are changing camera characteristics that vary from unit to unit, changing lighting (low/high light levels, high contrast, flicker effect due to fluorescent light, shadow, automatic exposure correction), changing color balance, camera motion (due to pan-tilt zoom, wind, shaking due to another object, e.g., train), perspective change, camera calibration, reflection, abrupt lighting changes, and occlusion.

Train station lighting creates unique lighting challenges. One of the basic problems is the flickering effect due to fluorescent lighting. The flicker effect causes a constant lighting change especially in the train stations on the reflective surfaces. Another challenge is the shadow elimination. Shadows, especially on low saturated areas, are very hard to eliminate. The lighting as well as camera parameter changes affects the gesture detection rate directly, e.g., when a train arrives at a platform the platform camera parameters are changing rapidly which causes poor detection rates. Mixed indoor–outdoor lighting creates many detection challenges. The background extraction may change for several reasons depending on the objectives. For example, in one application, the objective may be to classify standing passengers as foreground objects to gather information such as how long a passenger spends at a certain location while in other application passengers standing at another location may be considered as background objects. Calibration of each camera is another challenge in a crowded train station. There are several calibration algorithms in the literature; however, a detailed calibration algorithm that takes all the camera parameters into account is hard to accomplish. The camera position and the topology of the train station: escalators, stairs, and ramps make the calibration a more challenging problem. Tracking is affected by occlusion. Especially in a crowded train station, tracking multiple persons for a period of time is a very difficult task. The algorithm should be able to find the gesture of the person(s) during this tracking period and identify different gestures of the same person. Our solutions to the above-mentioned problems can be found in our paper [7]. In addition to algorithm design challenges, implementation issues such as converting floating point calculations to fixed point calculations were another major challenge at this stage. Second generation software was running on Linux platforms in real-time. The third generation software was adapted to run on DaVinci video processing boards with TMS320C64 fixed point DSP. The fourth generation software was planned to run on custom-designed hardware designed by the railway company itself. We also carried parts of the algorithm on FPGA platform and the reader can find details of this effort in our paper [8]. The output (e.g., ID of the individual, his/her gesture, position in 3D, speed of movement (in meters/frames), unattended object, person's height and width (in meters), number of people, flow direction of people, etc.) can be sent to a remote location with an IP address. The system is used for indoor and outdoor train platforms as well as inside the stations. A platform, e.g., an area of the size 200 m by 50 m is usually covered by one camera. Each camera is connected to a video processing board. Oblique views allow a single camera to cover a larger area, reducing the number of cameras required. The deployed system passed 9 phases of evaluation. Each phase has been evaluated by Yokogawa and railway company, and Verificon adjusted its algorithm and developed new algorithms to achieve the pre-set specifications by Yokogawa. Parallel to the development and research phases of the commercialization of the smart camera system, we published several papers and filed patents related to our research. Marilyn's leadership and her academic background enabled us to publish many conference papers, journals, and patents and in the meantime helped us to protect the company IP.

Recently, tracking algorithms and object detection algorithms are getting more and more complex due to surveillance requirements. There are many tracking, surveillance, self-driving car applications based on NNs and other algorithms with heavy processing loads. Gesture recognition needs more detailed analysis than tracking and will increase the processing load further. Even today, it is still hard for embedded systems to process real-time data for gesture related applications. On the other hand, recent developments in cheap and affordable general-purpose hardware like Raspberry Pi enable the user to use same hardware for different applications in a cost-effective way. Instead of a single task designated smart camera system, the general-purpose hardware can be used for different tasks. Especially government institutions require the use of existing "dummy" cameras for non-critical applications. The fastest, most reliable, and cheapest way to convert these cameras into smart cameras can be achieved by using general-purpose hardware with the existing camera for non-critical applications.

3 Next Step, Retail Stores

Our Verificon journey has ended after Marilyn accepted a position at GATech. I started using my expertise about video processing at different application areas as a consultant and as a partner of several start-up companies. I looked into different markets outside of the USA, and designed algorithms/systems for counting people inside retail stores by using cameras connected to very cheap multi-purpose hardware like Raspberry Pi. Due to customer request, we were able to use several programs on the same hardware without changing the camera setup for different applications. For example, same camera setup can be used for tracking customers inside the store, generating heatmaps, and counting. PI's camera is connected to the GPU directly, which enabled us almost real-time video processing for 320×240 frame sizes. There were several bumps during deployment, one of them was covering a wide area with a single camera. Although off-the-shelf fish-eye lenses are

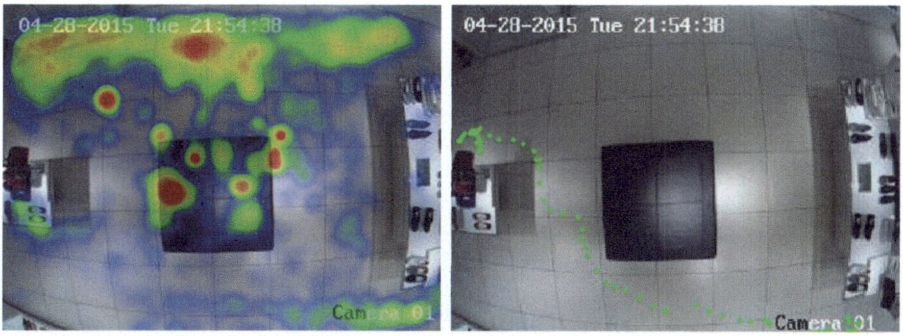

Fig. 1 Left: Heatmap shows store location where the customers spent the most time for a five hour period. Red areas represent the busiest locations. Right: Same camera tracks and displays a single customer trajectory

available, we had to find solution for camera distortion caused by the lens. Heating was another design issue, spotlights inside the retail stores generate excessive heat around ceiling area. We had to design the casing accordingly to allow the hardware to work within operating temperature range. There were several issues on the application side as well, for example, inside a shoe store, mirrors on the walls, and under the seats cause problems for detection/tracking. Number of doors, door width, and door type are also factors that affect algorithms, and finding a general solution that can work for all store types was a challenge. Store setup and display change within the store was another problem that was affecting system performance. Figure 1 (left) shows the heatmap in a shoe store after a five hour detection period. Figure 1 (right) is the tracking result of a single customer from entrance to the back of the store. Figure 2 is the hourly counter result (number of people entering to the store) for two consecutive days. Some stores required to use same cameras for analyzing different customer data, e.g., number of people checking the store display outside of the store versus number of customers entering the store. Another advantage of using a raspberry board is the storage capability of the board. We stored processed frames and output data for several days. In case of network interruption and power loss the data is also stored on the camera side before it is transferred to the central processing unit and database.

4 Water Management

More recently, I started using multi-purpose processing boards with cameras for water level and water velocity measurement in Asia. The need for water monitoring is becoming more and more important around the world which requires cost effective and robust water control systems to limit the loss of human lives, crops, property, and livestock. Depending on NOAA, extreme weather and climate events have increased in recent decades. Although there are multiple techniques for water

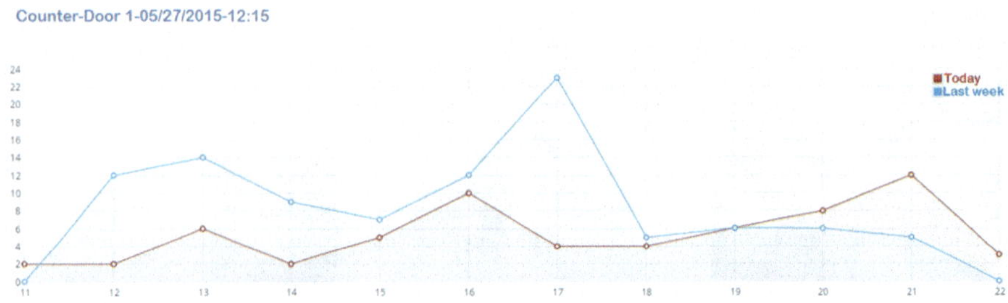

Fig. 2 Hourly customer number for two consecutive days

management, the prior art depends on traditional measuring tools and methods. Most of these are manual sensors and needed to be setup for each measurement. They require extensive maintenance and are only focused on a specific location and on a specific task.

I developed a water level (WL) and surface water velocity (WV) measurement system based on smart cameras for use around rivers, lakes, canals, harbors, dams, and all other water related areas as well as flood risk regions. The principle of this system is to monitor water and water movement by smart video cameras and alert the first responder team by calling attention to the risky areas. Our objective is to have a solution that is modular and adaptable, easy to use, stable, real-time, and with multi-region detection capability. The current software of our proposed platform is being used in Taiwan and Japan to find WL and WV in several flood prone areas. Our approach is to use both water level data and water velocity data to analyze and derive water related characteristics and make predictions. Video based system will trigger alarm(s) based on the user benchmarks and thresholds as well as enable the user to see the actual water body which can't be done by other traditional water sensor systems. Unlike some video surveillance systems, the proposed system does not need any expensive and maintenance dependent gauges and boards. As the end-product, I built a "smart" camera system based on our previous expertise with:

- Camera module in a hardened enclosure.
- Communication/networking between smart camera output and the central processing unit/database.
- Solar panels and battery for the camera module and IR light if the system needs to work as a stand-alone system depending on the application.
- Database for storing frames/date/time/water level/water velocity, etc.
- Software for connecting low level data to high-level semantics and data interpretation.

Selecting the right camera type, camera filter, and processing module are major factors which can affect the output of the system. Cost-effective and multi-purpose video processing boards are used depending on the location with the appropriate lens. It is also crucial to choose the right filters, e.g., IR filters, wide or directional lens, etc. for the application.

The water gauge area is usually not well defined in the real-world environment. The traditional striped water gauges are hard to see by the camera since the camera is located far from the gauge or the paint on the gauge may get worn off due to weather conditions. Some gauges are corroded. Some spots even don't have gauges. In some cases, a close-up detection of the region of interest (ROI) area is preferred. Cropping the ROI area and zooming in to the cropped area is desirable under these circumstances. Another important aspect is removing the noise due to different weather conditions. The system is equipped to check possible camera blockage/heavy fog/heavy smoke to minimize false alarms in a pre-defined interval. Automatic camera calibration is a crucial step in the process. To find water and non-water areas, different algorithms are implemented. Especially man-made canals and water ways have uniform color which is very similar to water color even under different weather conditions. Pre-processing methods such as histogram equalization, color sharpening, and other image enhancement algorithms are used to eliminate several factors, e.g., rain drops, snow, etc. The overall smart camera system can be seen in Fig. 3. There are different systems deployed at different locations, for example, in Taiwan, although we deployed Raspberry boards with Raspberry cameras, the government requires us to use existing CCTV cameras with PC or Raspberry board at several locations. We had to work around this problem in several cases. There is no single solution for this type of deployment. Even at the same location the requirements may change dramatically.

Some challenges we faced during this type of smart camera deployment are:

- Removing the noise due to different weather conditions, e.g., fog/smoke, heavy rain/snow.
- Calibration. Conversion of the camera coordinates into world coordinates without any markers during setup is a challenge.

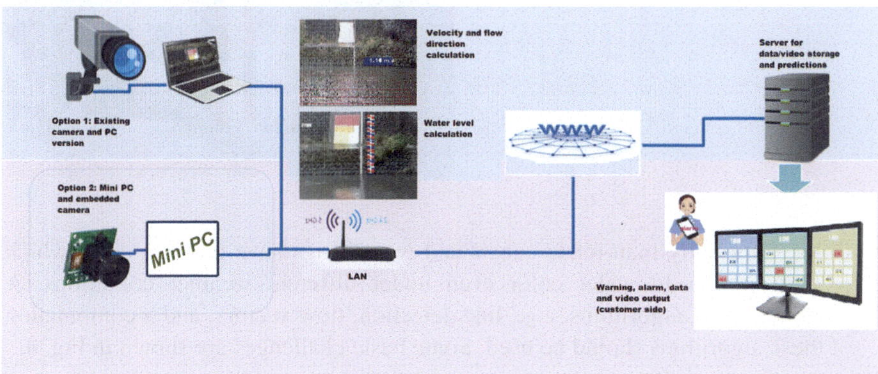

Fig. 3 Overall system: Input video frames are processed in real-time on the processing board or on a PC based on the user needs. Water level and/or water velocity results are sent from the field to a central processing center via network. Data is stored and interpreted at the center to give warning/alarm levels based on the thresholds

Fig. 4 Natural factors can affect the detection performance in different ways. Spider webs, heavy snow, and fallen branches are some of these factors

- Color. Especially man-made canals and water ways have uniform color which is very similar to water color even under different weather conditions. A combination of algorithms, e.g., line detection, flow vectors, and a combination of these algorithms should be used. Some basic challenges are shown in Fig. 4.

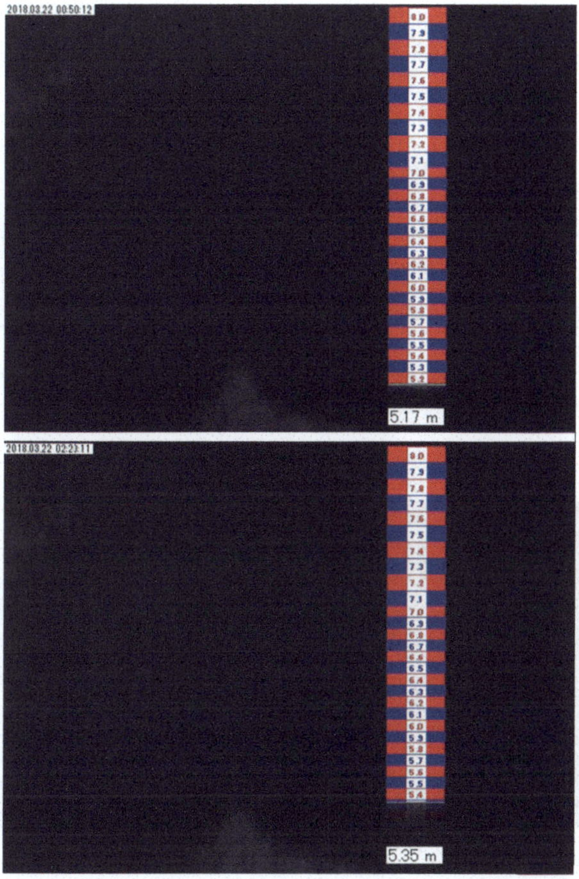

Fig. 5 Japan site—canal area around Tokyo. Night view by using IR light. There is no reflection board on the canal wall. The virtual gauge (GUI) is for visual purposes only. Time stamps are in the upper left corner. The system runs on solar panels/battery as a stand-alone system and shows results every 10 min when it rains. Otherwise, the system is idle

- Sudden changes due to foam, waves, etc. and gaps due to wiper, branches in the water, tree leaves before the camera, etc. need to be considered by the system.
- Power management.
- Network security and video transfer.

Some results from the deployed systems are shown in Figs. 5, 6, and 7.

Figure 8 shows our stand-alone smart camera system.

Fig. 6 Japan—River in Niigata: Left: Benchmarks are installed on the wall before testing the system. Middle and right: Water level at different times and the visual imaginary gauge displayed by GUI

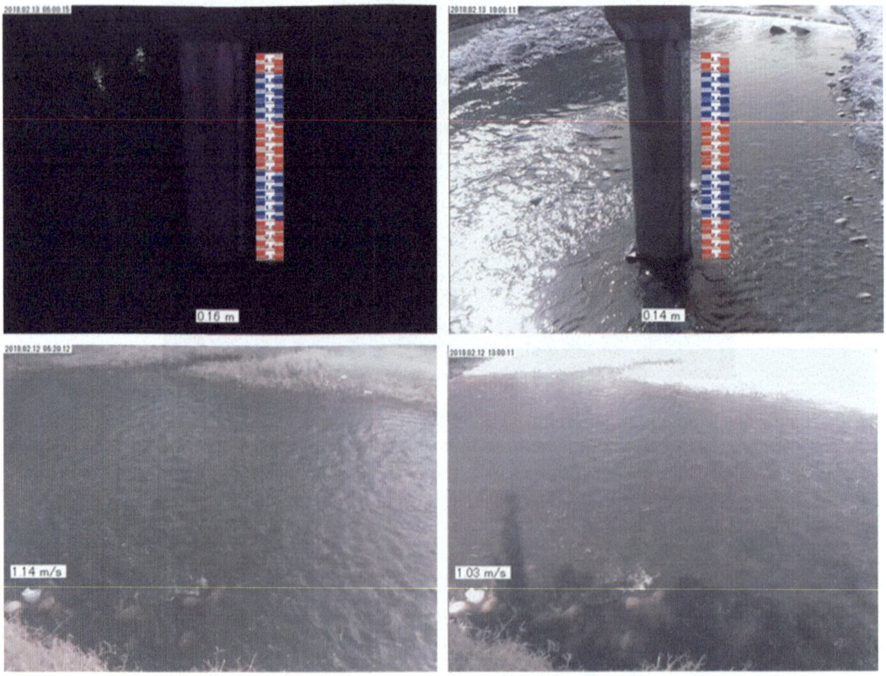

Fig. 7 Taiwan deployment around Taipei. Top: Water level. Bottom: Velocity

Fig. 8 Stand-alone smart camera for water level and water velocity measurement

5 Epilogue

Marilyn Wolf's vision and her innovative thinking light our way as researchers in many aspects. I can't thank her enough suggesting me to work on smart cameras. After almost 20 years, I still feel the excitement as a researcher and as an entrepreneur when I look back. There is still much to do in smart camera area. Especially using cheap and affordable smart cameras for everyday applications should be our goal as researchers. There is an anecdote with which I would like to conclude this schrift. When I was a post-doc, I went to Marilyn's office at Princeton, we were discussing about a smart gun project (a gun connected to a smart camera) that some researchers were proposing to develop. Marilyn's response was to let the researcher run in front of the smart gun and test the system by him/herself! The debate goes on today, should we rely on smart cameras, smart cars, smart machines? There are no easy answers for these questions, but researchers like Marilyn will always push the innovation limits. Thank you, Marilyn!

References

1. Ahmed Nabil Belbachir (Editor), "Smart Cameras", Springer, 2010.
2. W. Wolf, B. Ozer, T. Lv, "Smart Cameras as Embedded Systems", IEEE Computer, Volume: 35, Issue: 9, Sep 2002.
3. T. Lv, B. Ozer, W. Wolf, "Exploiting parallelism in media processing using VLIW processor", Proceedings ICIP, International Conference on Image Processing, October 2003.
4. T. Lv, B. Ozer, W. Wolf, "VLSI architectures for distributed smart cameras", Proceedings ITRE2003, September 2003.
5. T. Lv, B. Ozer, W. Wolf, "Parallel architecture for video processing in a smart camera system", Signal Processing Systems (SIPS '02), November 2002.
6. B. Ozer, T. Lv, W. Wolf, "Real-Time Analysis of Human Body Parts and Gesture-Activity Recognition in 3D", In book: 3D Modeling and Animation, January 2004.
7. B. Ozer, M. Wolf, "A Train Station Surveillance System: Challenges and Solutions", IEEE Conference on Computer Vision and Pattern Recognition Workshops, 2014.
8. Schlessman, J.; Cheng-Yao Chen; Wolf, W.; Ozer, B.; Fujino, K.; Itoh, K., "Hardware/Software Co-Design of an FPGA-based Embedded Tracking System", Computer Vision and Pattern Recognition Workshop, June 2006.

Rotators in Fast Fourier Transforms

Fahad Qureshi, Jarmo Takala, and Shuvra Bhattacharyya

Abstract This chapter discusses architectures for computing the rotations in fast Fourier transforms. There are two principal methods, which can be exploited: general complex multipliers or multiplier-less techniques. We describe different architectures, each with different advantages, indicating that the final selection depends on the requirements of the application at hand.

1 Introduction

A rotation is a multiplication by a complex number whose magnitude is one, i.e., a transformation that describes a circular movement with respect to a point [7, 11, 30, 32]. The rotation of a complex number $(x + iy)$ by an angle α can be defined as:

$$\begin{bmatrix} X \\ Y \end{bmatrix} = \begin{bmatrix} \cos\alpha & -\sin\alpha \\ \sin\alpha & \cos\alpha \end{bmatrix} \begin{bmatrix} x \\ y \end{bmatrix}, \tag{1}$$

where X and Y are the real and imaginary parts of the result, respectively. Thus, the rotation can be written as:

$$X + iY = (x\cos\alpha - y\sin\alpha) + i(y\cos\alpha + x\sin\alpha). \tag{2}$$

F. Qureshi (✉) · J. Takala
Tampere University, Tampere, Finland
e-mail: fahad@tuni.fi; jarmo.takala@tuni.fi

S. Bhattacharyya (✉)
University of Maryland, College Park, MD, USA

Tampere University, Tampere, Finland
e-mail: ssb@umd.edu

S. S. Bhattacharyya et al. (eds.), *Embedded, Cyber-Physical, and IoT Systems*,
https://doi.org/10.1007/978-3-030-16949-7_11

Let C and S be representations of $\cos\alpha$ and $\sin\alpha$ with a finite number of bits. A rotation of complex number, i.e., a rotation in complex plane, can be presented as

$$\begin{bmatrix} X_Q \\ Y_Q \end{bmatrix} = \begin{bmatrix} C & -S \\ S & C \end{bmatrix} \begin{bmatrix} x \\ y \end{bmatrix}, \tag{3}$$

where X_Q and Y_Q are the result of the rotation, which includes the error resulting from quantization of the coefficients. Many digital signal processing algorithms require to carry out rotations of complex numbers by the given angles with respect to the origin. This is the case with *fast Fourier transforms* (FFT) [26], discrete cosine transforms, and lattice filters [23].

2 Rotations in FFT

The fast Fourier transform is one of the most important tools in digital signal processing. It is based on discrete Fourier transform and used to convert time domain signals to frequency domain [3, 26, 28]. The Fourier transform is defined for continuous signals, while the modern signal processing mainly considers digital systems, and the *discrete Fourier transform* (DFT) is used instead. The DFT transforms a finite sequence of equally spaced samples to a corresponding frequency domain representation as follows:

$$X[k] = \sum_{n=0}^{N-1} x[n] W_N^{kn}, \qquad k = 0, 1 \ldots, N-1, \tag{4}$$

where N denotes the length of DFT, i.e., the number of points of the DFT, $x[n]$ and $X[k]$ are the input and output samples, respectively. Note that both the signals are discrete in nature. The complex-valued coefficients W_N are called as *twiddle factors* and are defined as

$$W_N = e^{-j2\pi/N} = \cos(2\pi/N) - j\sin(2\pi/N), \tag{5}$$

where j denotes the imaginary unit.

The original signal $x[n]$ can be recovered from $X[k]$ with the aid of an *inverse discrete Fourier transform* (IDFT):

$$x[n] = \frac{1}{N} \sum_{k=0}^{N-1} X[k] W_N^{kn}, \qquad k = 0, 1, \ldots, N-1. \tag{6}$$

The arithmetic complexity of the DFT in (4) is $O(N^2)$. However, DFT contains redundant computations and several methods have been introduced for avoiding such a redundancy, thus reducing the complexity. Any algorithm computing DFT with less than $O(N^2)$ complexity is called as a fast Fourier transform. The most popular FFT is the Cooley–Tukey algorithm [3], which uses divide-and-conquer paradigm to decompose DFT into a set of smaller DFTs. Especially, the Cooley–Tukey principle says that an N-point DFT ($N = PQ$) can be computed with the aid of a P-point DFT and a Q-point DFT. By exploiting the periodicity of the twiddle factors:

$$W_N^{kQ} = W_P^k;\ W_N^{kP} = W_Q^k;\ W_N^{kPQ} = 1,$$

the radix-Q FFT can be expressed as:

$$\begin{aligned} X(Qk_1 + k_2) &= \sum_{n_1=0}^{P-1} \sum_{n_2=0}^{Q-1} x\,(n_1 + Pn_2)\, W_P^{n_1 k_1} W_N^{n_1 k_2} W_Q^{n_2 k_2} \\ &= \underbrace{\sum_{n_1=0}^{P-1} \left[\underbrace{\left(\sum_{n_2=0}^{Q-1} x(n_1 + Pn_2) W_Q^{n_2 k_2} \right)}_{Q-point\ DFT} \underbrace{W_N^{n_1 k_2}}_{twiddle factor} \right] W_P^{n_1 k_1}}_{P-point\ DFT}. \end{aligned} \tag{7}$$

The two summations indexed by n_1 and n_2 are referred to as inner and outer DFTs. As a result, an N-point DFT is broken down into P-point and Q-point DFTs. The output of the inner DFT is multiplied by $W_N^{n_1 k_2}$, which is called as a twiddle factor multiplication. The scheme of decomposition is shown in Fig. 1, where the left and right sides represent the P-point inner DFT and Q-point outer DFT, respectively. Between those DFTs, a twiddle factor multiplication indicates a rotation by $W_N^{\phi} = e^{-j\frac{2\pi}{N}\phi}$. The arithmetic complexity is reduced from $O(N^2)$ of the DFT in (4) to $O(N \log N)$.

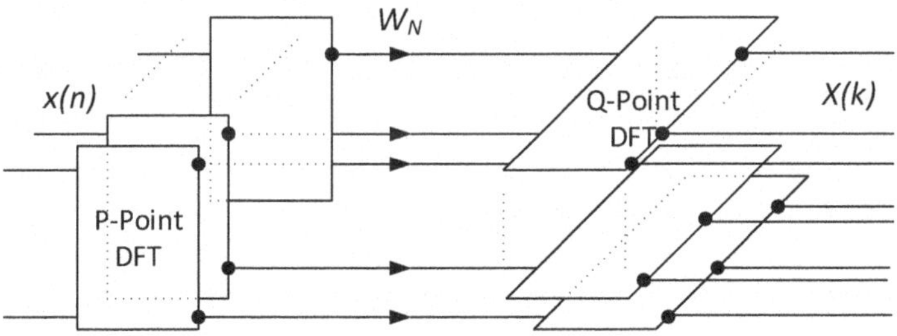

Fig. 1 Decomposition scheme of Cooley–Tukey FFT algorithm

If the length N is not a prime, the Cooley–Tukey principle can be applied iteratively and then the DFT is computed with the aid of several smaller DFTs. In particular, if the DFT length is a power of a prime, i.e., $N = P^q$, then the N-point DFT can be computed with the aid of P-point DFTs constructed in q computing stages. As the resulting fast algorithm contains only P-point DFTs, it is called as a *radix-P FFT*.

The most popular approach is radix-2 FFT algorithm, where the DFT is decomposed recursively until the entire algorithm is computed with the aid of two-point DFTs. The advantage is that that the two-point DFT can be computed with trivial twiddle factors, thus multiplications can be avoided. Another popular algorithm is radix-4 FFT as twiddle factors contain only 1, -1, j, and $-j$, thus again no multiplications are needed.

The recursive application of Cooley–Tukey principle can be done by starting from the time domain sequence, which results in a decimation-in-time (DIT) algorithm as shown in Fig. 2. In a similar fashion, the decomposition process can be started from the frequency domain sequence resulting in a decimation-in-frequency (DIF) algorithm as shown in Fig. 3. The numbers between each stage represent the rotations. Each of these values (ϕ) corresponds to a rotation by the twiddle factor [6]. It should be noted that $W_{16}^2 = W_8^1$; $W_{16}^4 = W_4^1$.

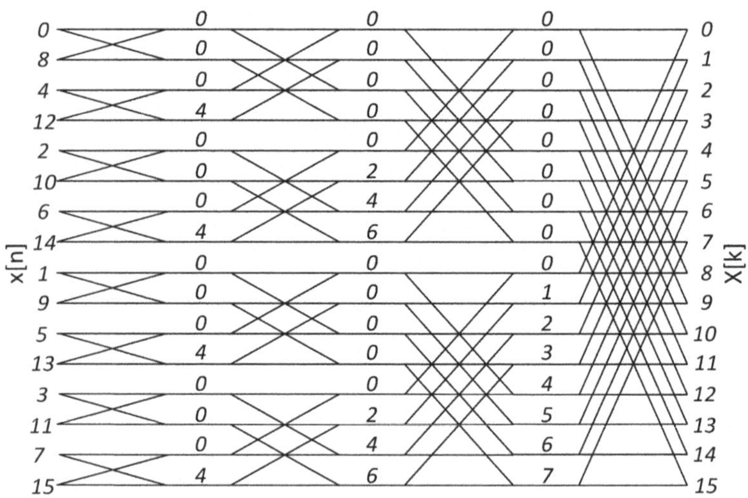

Fig. 2 16-point radix-2 DIT FFT flow graph. The number between the stages indicates rotation, i.e., value ϕ in twiddle factor multiplication W_{16}^{ϕ}

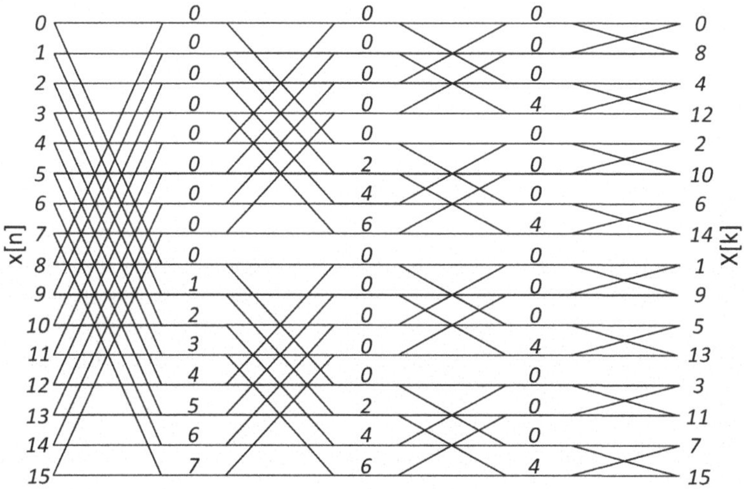

Fig. 3 16-point radix-2 DIF FFT flow graph. The number between the stages indicates rotation, i.e., value ϕ in twiddle factor multiplication W_{16}^{ϕ}

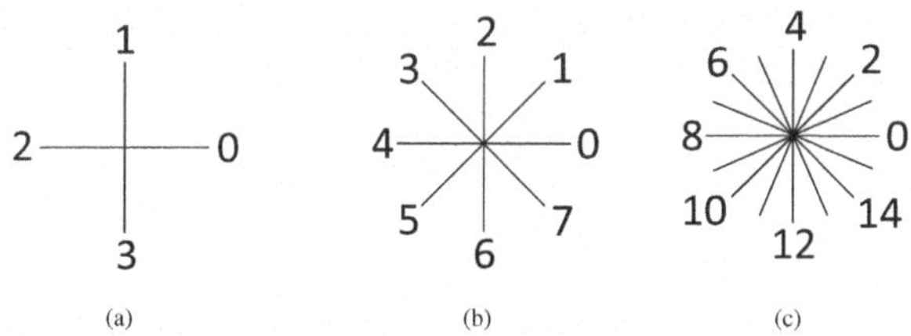

Fig. 4 Twiddle factors: (**a**) $W_4(W_4^0, W_4^1, W_4^2, W_4^3)$, (**b**) $W_8(W_8^0, W_8^1, \ldots, W_8^7)$, and (**c**) $W_{16}(W_{16}^0, W_{16}^1, \ldots, W_{16}^{15})$

2.1 Twiddle Factors

Each input of an FFT stage is rotated by a different angle, which is determined by the twiddle factor $W_L^{\phi} = e^{-j2\pi\phi/L}$. The parameter L is a constant for each stage and determines the number of possible rotations in that stage. These angles are based on the division of the circumference into L equal parts. The exact angle is determined by the parameter ϕ, which is a natural number $\{0, \ldots, L-1\}$ [7, 11, 30, 32].

The complexity of the rotator is determined by the value of L. A small value is desirable, because it results in a simple twiddle factor architecture. An example of twiddle factors W_4, W_8, and W_{16} is shown in Fig. 4.

The simplest rotator is $W_L = W_4$, which includes only trivial rotations $(0, \frac{\pi}{4}, \pi,$ and $\frac{3\pi}{4})$. Trivial rotations are characterized by the fact that they can be calculated by simply exchanging the real and imaginary parts of the input and/or changing their sign [11]. Thus, the criterion used to select the algorithm is to minimize the number of large twiddle factors and maximize the number of trivial rotators (W_4), which are the cheapest ones. There are also trade-offs between the number of large (W_{64} and larger) and small (W_8, W_{16}, and W_{32}) twiddle factors.

3 Rotation in Fixed-Point Arithmetic

Ideally the arithmetic operations in the FFT algorithm should be represented with infinite precision. However, in practice, temporary values are stored in registers with a finite capacity. There exist different ways to approximate real numbers using a finite number of digits. One of the most common ways for embedded systems is two's complement fixed-point (FxP) arithmetic [5, 27].

For angle α, it is always true that: $\cos\alpha \in [-1, 1]$ and $\sin\alpha \in [-1, 1]$. Therefore, it is assumed that C and S are also numbers in the range $[-1, 1]$ with rounding to a certain number of fractional bits, b. According to the general interpretation of the coefficients, the magnitude of the rotation is always one, independently on the number of bits, b, as it happens in the definition of the rotation.

As C and S contain b bits, they can also be considered integers in two's complement representation in the range of $[-2^{b-1}, 2^{b-1}]$. Accordingly, the values of the cosine and sine components of the angle α will be

$$\begin{aligned} C &= R(\cos\alpha + \epsilon_c); \\ S &= R(\sin\alpha + \epsilon_s), \end{aligned} \tag{8}$$

where ϵ_c and ϵ_s are the relative quantization errors of the cosine and sine components, respectively, and R is the scaling factor of the coefficients being the error rotation [9].

4 Rotations in FFT Architectures

Closer investigation of signal flow graphs of FFT algorithms indicates that the rotations in the signal flow graph have some systematic properties, which can be exploited in implementations. The way how these properties can be exploited depends on the mapping from the signal flow graph onto the processing units. The main differences are that how many samples are processed in parallel and how many different twiddle factors each rotator must support. We can identify three principal types:

1. Single Branch with Multiple Rotations. A general scenario is to compute rotations on the data that flow through a single branch, where different pieces of data rotated by different coefficients are shown in Fig. 5a. This scenario is mostly used in single/multiple feedback pipelined and iterative FFT architectures [11].

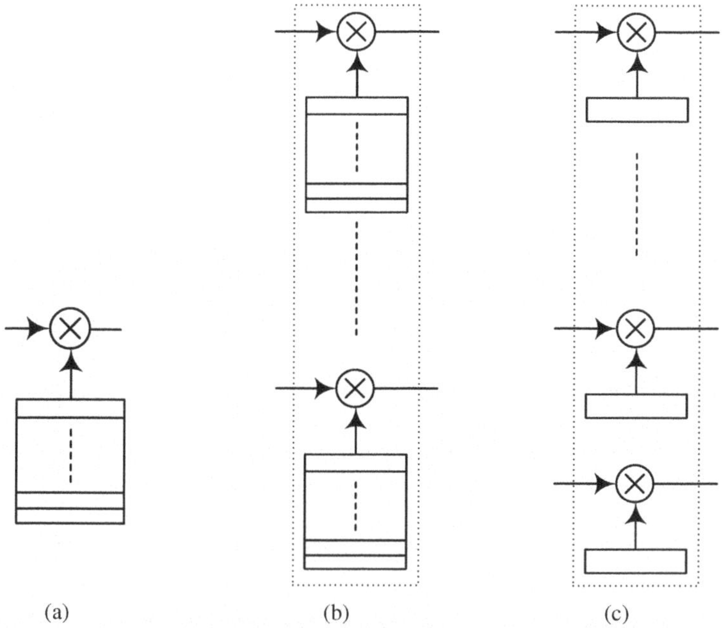

Fig. 5 Rotator scenarios: (**a**) Type 1, (**b**) Type 2, and (**c**) Type 3

2. Multiple Branches, Multiple Rotations for Each Branch. The most typical case consists of several branches with several rotations for each of them, as shown in Fig. 5b. This scenario is part of feed forward pipelined FFT architectures [11].
3. Multiple Branches, Single Rotation for Each Branch. This scenario is applied on fully parallel FFT architecture, where data flow is parallel and each branch may carry out a rotation by a different coefficient [11] as shown in Fig. 5c.

The Type 1 and Type 2 are also referred to as *multiple constant rotations* (MCR) and Type 3 is called as a *single constant rotation* (SCR) [11].

5 Rotator Units

As discussed earlier, the FFT signal flow graphs can be mapped on processing elements with different methods indicating different properties for the units. In this section, we discuss the implications to the rotators due to these mappings and identify three different classes: rotators based on general complex multipliers, constant rotators, and CORDIC rotators.

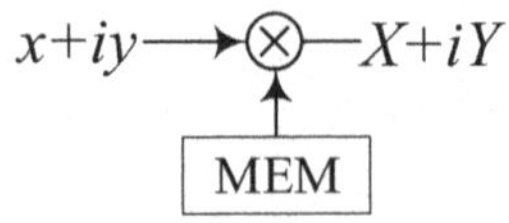

Fig. 6 Rotation architecture based on complex multiplier

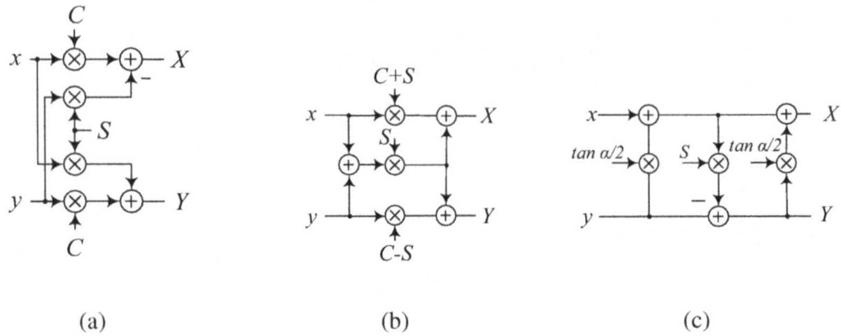

Fig. 7 Rotation based on complex multipliers: (**a**) standard, (**b**) modified I, and (**c**) lifting-based

5.1 *Rotators Based on General Complex Multiplier*

The most popular approach to implement the rotation is to use complex multiplier and lookup table for storing the sine and cosine components of the rotation angle as shown in Fig. 6 [30]. A straightforward method is to implement the complex multiplication with four real multipliers and two adders as depicted in Fig. 7a. The complex multiplication can be realized more efficiently by exploiting the fact that this is a rotation. Such methods reduce the number of real multipliers from four to three [30]. Among them, the most alluring ones are the following:

$$\begin{aligned} X &= x(\cos\alpha + \sin\alpha) - (x + y) \cdot \sin\alpha; \\ Y &= x(\cos\alpha + \sin\alpha) + (y + x) \cdot \sin\alpha \end{aligned} \tag{9}$$

and

$$\begin{aligned} X &= (x + y)\cos\alpha - y \cdot (\cos\alpha + \sin\alpha); \\ Y &= (x + y)\cos\alpha - x \cdot (\cos\alpha - \sin\alpha). \end{aligned} \tag{10}$$

Both these cases consist of a common term in the equations for X and Y that only need to be computed once. Thus, when $(C + S)$ and $(C - S)$ are precomputed, these cases require three real-valued multiplications and five real-valued additions. The architecture for (9) is shown in Fig. 7b.

Another approach can be applied only in the case of rotation, which is called lifting-based rotators [2]. The rotation in (1) can be written in matrix form as follows:

$$\begin{bmatrix} X \\ Y \end{bmatrix} = \begin{bmatrix} \cos\alpha & \sin\alpha \\ -\sin\alpha & \cos\alpha \end{bmatrix} \times \begin{bmatrix} x \\ y \end{bmatrix}. \tag{11}$$

We can apply the lifting approach to the previous equation and the following form is obtained:

$$\begin{bmatrix} X \\ Y \end{bmatrix} = \begin{bmatrix} 1 & 0 \\ 0 & -1 \end{bmatrix} \cdot \begin{bmatrix} 1 & -\tan\frac{\alpha}{2} \\ 0 & 1 \end{bmatrix} \cdot \begin{bmatrix} 1 & 0 \\ \sin\alpha & 1 \end{bmatrix} \cdot \begin{bmatrix} 1 & \tan\frac{\alpha}{2} \\ 0 & -1 \end{bmatrix} \times \begin{bmatrix} x \\ y \end{bmatrix}. \tag{12}$$

By further computing the matrices, the following equations are obtained:

$$\begin{aligned} X &= x - x\sin\alpha\tan\frac{\alpha}{2} + 2y\tan\frac{\alpha}{2} - y\tan^2\frac{\alpha}{2}\sin\alpha; \\ Y &= y - x\sin\alpha - y\sin\alpha\tan\frac{\alpha}{2}. \end{aligned} \tag{13}$$

These equations can be realized with the structure depicted in Fig. 7c. This approach requires also three real-valued multiplications and three real-valued additions. However, there is no need for an additional coefficient as it simply replaces the $\cos\alpha$ with $\tan\frac{\alpha}{2}$ in memory.

This type of rotator can be applied in all types of FFT architecture scenarios mentioned earlier. Figure 5b and c requires more than one rotation, which are used in signal flow graphs and parallel pipelined FFT architectures, respectively. In those scenarios, the complexity of rotation memory can be reduced by applying appropriate techniques. Rotation memory is used to store the coefficients C and S of rotation. There are a number of techniques to store the coefficients in the memory according to the requirement of FFT architectures [31]. The simplest approach is to store all the coefficients of rotations in the memory without considering any optimization technique as shown in Fig. 8. This results in a large rotations memory especially for large FFTs. It should be noted that this scheme possibly stores the same rotations in several locations as the mapping is from the computing stages of FFT algorithms.

A possible simplification is to use an address generator that generates the row address for the corresponding angle. As a result, we only need to store the coefficients once in the memory as shown in Fig. 9. For the case $L = N$, we will need to store many but not all values, still using N possible words even though many

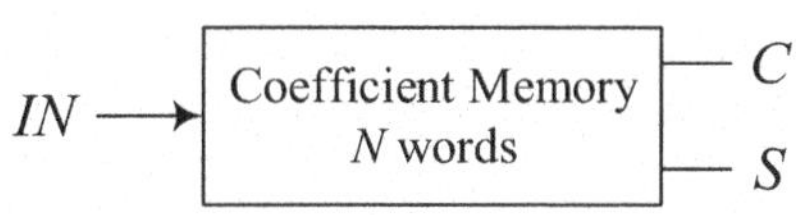

Fig. 8 Coefficient memory for storing all rotations of twiddle factor

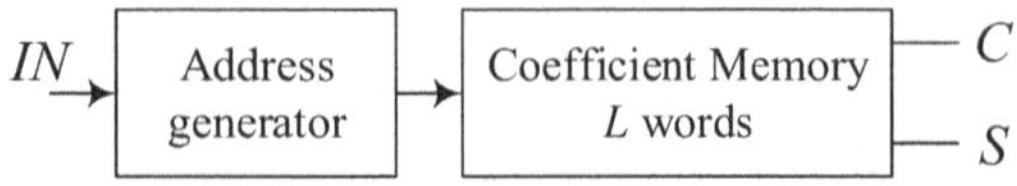

Fig. 9 Coefficient memory with address generation

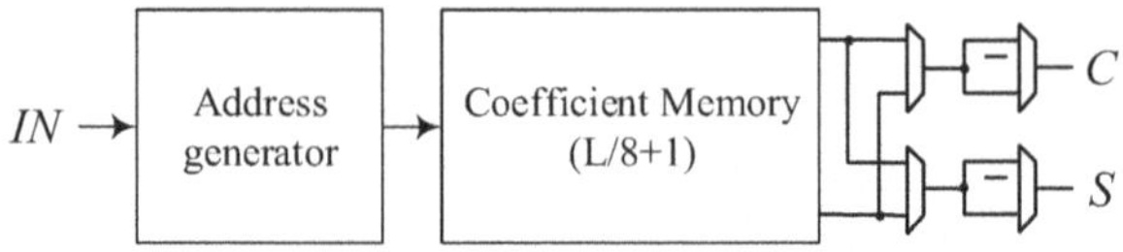

Fig. 10 Coefficient memory with address generation and octave symmetry circuit

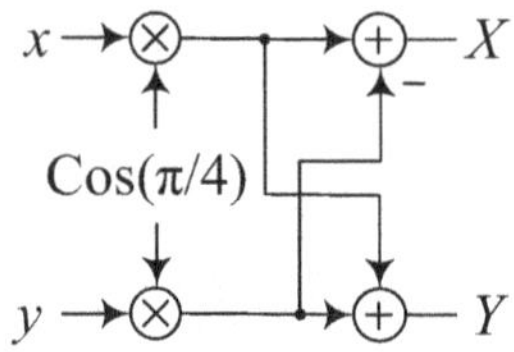

Fig. 11 Single constant rotation by $\cos(\frac{\pi}{4})$

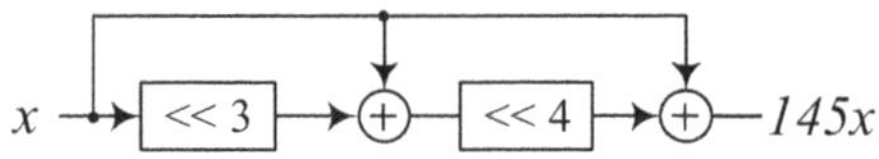

Fig. 12 Simplified constant multiplication

can be set as "don't cares." Due to this fact, one can expect the resources used to the lookup table to be reduced compared to the previous approach, given that the synthesis tool can benefit from it.

Another modification, proposed in [16], is to use the well-known octave symmetry to only store twiddle factors for $0 \leq \phi \leq \frac{\pi}{4}$. The additional cost is an address mapping circuit as discussed in the previous section as well as multiplexers to interchange the real and imaginary parts and possible negations. The main benefit is that only $\frac{L}{8} + 1$ words are required to be stored. The resulting architecture is illustrated in Fig. 10.

5.2 *Constant Rotators*

Typically the real-valued multipliers discussed in the previous section can be replaced by shift-and-add multiplication as shown in [15, 22, 39]. This is especially useful when the rotator needs to support only a single rotation angle.

Let us consider a rotation by $\frac{\pi}{4}$, which requires only one coefficient as $\cos\frac{\pi}{4} = \sin\frac{\pi}{4}$. Hence, each real, x, and imaginary, y, parts of the operand are multiplied separately by $\cos\frac{\pi}{4}$ and the multiplier outputs are added and subtracted as depicted in Fig. 11. When an input signal is multiplied by one constant, an optimal single constant multiplier from [13] can be used for the implementation. The internal architecture of a constant multiplier is shown in Fig. 12, where the input operand x is multiplied with a coefficient 145, the result being $145x$.

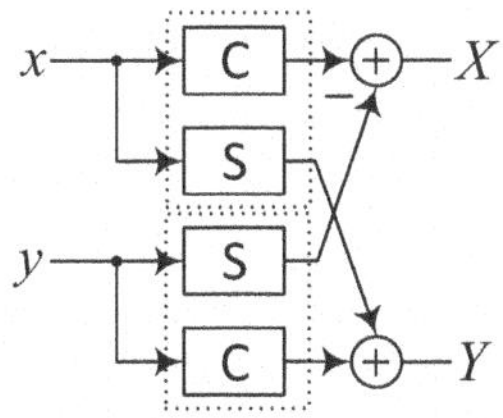

Fig. 13 Rotation based on constant multiplication

In terms of complexity, the shift operations are free as they only reduce the number of adders in the implementation of the constant multiplications. A general block diagram of the rotator is shown in Fig. 13. This consists of two constant multiplication blocks by dashed blocks. Each dashed block consists of two constants, C and S, which are implemented by shift-and-add. Hence, both the multipliers sharing the same input can be simultaneously realized using a *multiple constant multiplication* (MCM) technique [4].

The constant rotators can be applied for all the FFT rotation scenarios. However, this architecture is no more efficient for large values of L; the higher the value of L, the greater number of constant coefficients are needed implying more complex shift-and-add circuits. As a result, this architecture can be used efficiently for twiddle factors W_8, W_{16}, and W_{32} [11, 25, 32]. The maximum number of rotations in the twiddle factor set is $0, \ldots, L-1$. As discussed earlier, thanks to the symmetries of the angles on the complex plane, for L-point twiddle factor only the $L/8+1$ angles in the range $[0, \pi/4]$ need to be considered. This is due to the fact that the rest of rotations of twiddle factor can be formed from these angles by interchanging the real and imaginary parts of the input and output data and/or the signs of the outputs. The design of constant rotators can be divided into two main parts: the generation of the rotation coefficients and the implementation of shift-and-add circuit.

5.2.1 Rotation Coefficients

In hardware implementations, it is not only desirable to reduce the computation error but also the area on the circuit. As mentioned earlier: $\cos\alpha \in [-1, 1]$ and $\sin\alpha \in [-1, 1]$. Therefore, in general it is assumed that C and S are also numbers in the range of $[-1, 1]$ with rounding a certain number of fractional bits, b. There is a number of methods to reduce the coefficient error without increasing the addition cost.

A method based on searching the coefficients by allowing word length increase by E fractional bits is introduced in [14]. The approximation error can be guaranteed to meet $\epsilon \leq 2^{-(N+E+1)}$ with a brute force method by searching coefficients, which fulfill this condition such that the word length increase is at most E fractional bits. Another way of using the additional fractional bits is to realize that there are exactly 2^E different representable coefficients for which $\epsilon \leq 2^{-(N+1)}$, including the one obtained by rounding to b bits. The basic idea is to search these 2^E and select the coefficient value that has the smallest approximation error for allowed complexity called as the addition aware quantization [14]. The allowed complexity is typically assumed to be the same number of additions as required by the coefficient rounded to a coefficient with b fractional bits. This is a generic method and it can be applied on rotation coefficients. Another method, which is well suited especially to rotations, is the so-called *combined coefficient selection and shift-and-add implementation* (CCSSI) [11]. This method refers to a set of rotations that must be optimized together. This joint optimization happens when there is a dependency on the scaling of the rotations. Different optimization problems can be defined for SCR and MCR depending on the scaling that is required and on the hardware layout. Thus, the scaling can be fixed, unity, or arbitrary depending on the freedom to choose the scaling factor. Unity scaling is a particular case of fixed scaling, where the rotation has magnitude of one or, in more general terms, $R = 2^b$. This is equivalent to considering that the binary point is in a different position in the binary representation. Conversely, arbitrary scaling means that R can take any value, i.e., no restriction is set to R. For arbitrary scaling the approximation error is equal to the angular error only, since R will always take the optimal value. However, the scaling for multiple angles is classified based on the relation among the scaling factors of the rotations. More details of the design process can be found from [11].

In pipelined FFT architectures, uniform scaling can be applied on sets of rotations, which means that R is the same for all the rotations of each FFT stage. The purpose is to select the best coefficients of rotations. Thus, when fixed or arbitrary scaling factor is applied, the output of FFT is shifted by a certain factor.

5.2.2 Shift-and-Add Circuit Implementation

This section describes the methods to design the constant rotators circuit for FFT, especially for MCR. There are two main methods to design rotation circuits for FFT. One is based on using combinations of rotation coefficients for other rotation with the aid of additional multiplexers. Thus, it is reducing the coefficients of rotations. Other is merging the rotation and sharing the adders among them by using additional multiplexers.

Regarding the first technique, trigonometric identities are used to reduce the number of required coefficients. This techniques is applied on twiddle factors of W_{16} and W_{32} to reduce the required coefficients from three to two and seven to three, respectively [32]. Thus, the equivalent expression for all the coefficients in twiddle factors of W_{16} and W_{32} is tabulated in Tables 1 and 2, respectively [25, 32].

The architecture for W_{16} twiddle factor is shown in Fig. 14, where a single input is multiplied with any of the coefficient pairs $\{(1, 0), (\cos\frac{\pi}{8}, \sin\frac{\pi}{8}), (\cos\frac{\pi}{4}, \sin\frac{\pi}{4})\}$.

Another implementation technique is based on CCSSI. The coefficient selection has been explained in Sect. 5.2.1. The implementation consists of two steps: first,

Table 1 Trigonometric identities used for W_{16} twiddle factors

Coefficient	Used expression
$\sin\frac{\pi}{4}$	$2\sin\frac{\pi}{8}\cos\frac{\pi}{8}$
$\sin\frac{\pi}{8}$	$\sin\frac{\pi}{8}$
$\cos\frac{\pi}{8}$	$\cos\frac{\pi}{8}$

Table 2 Trigonometric identities used for W_{32} twiddle factor

Coefficient	Used expression
$\sin\frac{\pi}{4}$	$4\cos\frac{\pi}{8}\cos\frac{\pi}{16}\sin\frac{\pi}{16}$
$\sin\frac{\pi}{8}$	$2\cos\frac{\pi}{16}\sin\frac{\pi}{16}$
$\cos\frac{\pi}{8}$	$\cos\frac{\pi}{8}$
$\sin\frac{3\pi}{16}$	$\sin\frac{\pi}{16}\left(2\cos\frac{\pi}{8}+1\right)$
$\cos\frac{3\pi}{16}$	$\cos\frac{\pi}{16}\left(2\cos\frac{\pi}{8}-1\right)$
$\sin\frac{\pi}{16}$	$\sin\frac{\pi}{16}$
$\cos\frac{\pi}{16}$	$\cos\frac{\pi}{16}$

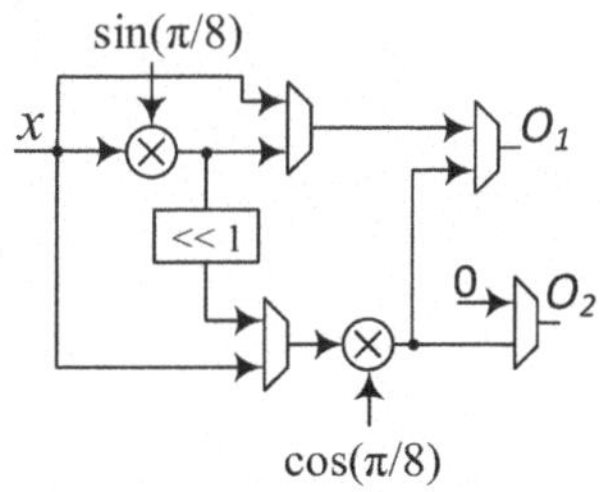

Fig. 14 Architecture for W_{16} twiddle factors [25]

obtain the implementation of a rotation by each SCR and then merge together all the rotations that are carried out by the MCR.

The rotation by an SCR includes the multiplication of the input by C and S, and addition of the products. This corresponds to Eq. (1). The multiplication by C and S is carried out by means of shifts and additions according to the MCM representation of the numbers. Similarly, layout the architecture of all the rotations that are carried by the MCR. Finally, the rotations carried out by the same rotator must be merged together. This is done by adding multiplexers to the inputs of the adders. Figure 15 illustrates an architecture for the computing W_8 twiddle factor and Fig. 15a and b shows the multiplication by 1 and $\frac{\pi}{4}$ rotation, respectively.

The architecture in Fig. 15c is a result of merging together the architectures in Fig. 15a and b. The control signal of the multiplexer controls the multiplication of $\{(1,0), (\cos\frac{\pi}{4}, \sin\frac{\pi}{4})$ [11]. In order to obtain an efficient realization of the rotator, reconfigurable single [13, 35] and multiple constant multiplication [4, 12] techniques can be used. Alternatively, when the number of coefficients is small, which is true in most of the practical cases, the selection of an efficient implementation can be found manually.

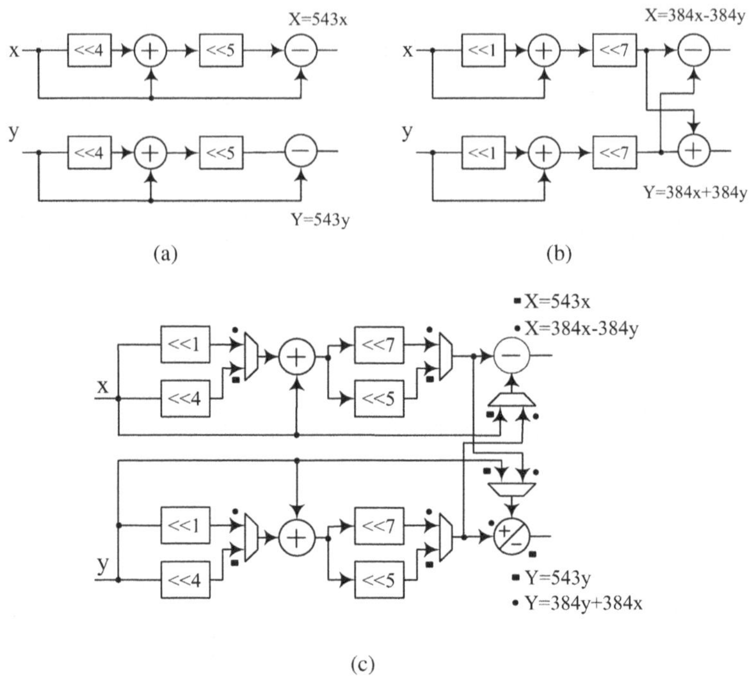

Fig. 15 W_8 twiddle factor: (**a**) multiplication by 1, (**b**) rotation by $\frac{\pi}{4}$, and (**c**) $\{(1), (\cos\frac{\pi}{4}, \sin\frac{\pi}{4})\}$

5.3 CORDIC Rotator

COordinate Rotation DIgital Computer (CORDIC) is one of the most popular algorithms for implementing multiplier-less rotations [1, 7, 8, 36]. It realizes rotation by means of a series of shifts and additions, which reduces the amount of hardware. It is also suitable for cases, where multipliers are not available. However, it may affect the accuracy since it is based on an approximation. The CORDIC algorithm decomposes the angle that has to be rotated, θ, into a sum of M predefined angles, α_i, according to:

$$\theta = \sum_{i=0}^{M-1} \delta_i \alpha_i + \epsilon, \tag{14}$$

where ϵ is the error of the approximation, δ_i indicates the direction of the so-called micro-rotation and

$$\alpha_i = \tan^{-1}(2^{-i}). \tag{15}$$

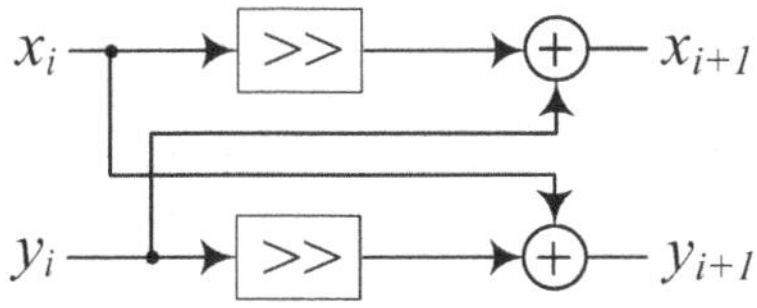

Fig. 16 CORDIC micro-rotation

These angles that define the micro-rotations have the property that they can be rotated by shifts and additions, which reduces significantly the hardware resources. These micro-rotations are carried out as follows:

$$\begin{aligned} x_{i+1} &= x_i - y_i\ \delta_i\ 2^{-i}; \\ y_{i+1} &= y_i + x_i\ \delta_i\ 2^{-i}. \end{aligned} \tag{16}$$

The hardware circuit for calculating the case of $\delta_i = 1$ is depicted in Fig. 16; input samples are rotated with an angle α_i, which is chosen by setting the number of bits that are shifted before the additions and subtractions are carried out.

Usually $\delta \in \{-1, 1\}$. This forces all the micro-rotations to be computed either clockwise or counterclockwise and assures a constant gain for the CORDIC computations, which can be compensated by multiplying the outputs by:

$$K = \prod_{i=0}^{M} \cos(\alpha_i) = \prod_{i=0}^{M} \cos(\tan^{-1}(2^{-i})) = 0.6073. \tag{17}$$

This option is preferable when the circuit is used for rotating several different angles and a constant gain for all of them is required, as happens in the rotators for the FFT. However, in a constant rotator, only a single angle θ can be rotated. In this case, it is better to consider $\delta_i \in \{-1, 0, 1\}$. This approach is called as redundant CORDIC [19], which allows certain micro-rotations to be removed, reducing the number of adders.

There are multiple variations of the CORDIC algorithm. Some of the main modifications are introduced in the following. Surveys on CORDIC techniques can be found, e.g., in [1, 24]. For some of the approaches, it is not straightforward to determine the rotation parameters at run time. Hence, for these methods the design is carried out offline and the control signals are stored in memory rather than the angles. This approach is naturally possible for all techniques and, as the sequence of angles is often known beforehand, most likely advantageous compared to storing the angle values.

The redundant CORDIC considers that $\delta_i \in \{-1, 0, 1\}$ [34] or even $\delta_i \in \{-2, -1, 0, 1, 2\}$ [20]. This allows several rotation angles at each CORDIC stage. However, the scaling for different angles is different, which implies a need for a specific circuit for scaling compensation. The *extended elementary angle set* (EEAS) CORDIC [38] and *mixed-scaling-rotation* (MSR) CORDIC [21, 29] also follow the idea of increasing the number of rotation angles per rotation stage.

The memoryless CORDIC [10] removes the need for rotation memory to store the FFT rotation angles. Instead, the control signals δ_i are generated from a counter. This is advantageous for large FFTs, where the butterfly stages have a large number of rotations. The *modified vector rotational* (MRV) RORDIC [37] allows skipping and repeating CORDIC stages, whereas the hybrid CORDIC [17, 33] divides the rotations into a coarse and a fine rotations. These techniques reduce the number of stages and, therefore, the latency of the CORDIC. The CORDIC II [10] proposes new types of rotation stages: friend angles, uniformly scaled redundant (USR) CORDIC, and nano-rotations. These result in both a low latency and a small number of adders. Finally, the base-3 rotators [18] consider an elementary angle set that is different to that of the CORDIC. All the rotations are generated by combining a small set of FFT angles. This set fits better the rotation angles of the FFT than that of the CORDIC, which results in a reduction in the rotation error, the number of adders, and latency of the circuit.

6 Conclusions

Rotation architecture has an important role in the design of an FFT architecture and has a large effect on the cost of the architecture. This chapter provided an overview over different existing architectures of the rotations especially for FFT. These can be implemented using complex-valued multipliers, constant multipliers, and the CORDIC. Architecture based on the CORDIC and constant multiplication uses shift-and-add circuit, whereas the complex multiplication generally uses complex multiplier and memory to store the coefficients of rotation.

References

1. Andraka, R.: A survey of CORDIC algorithms for FPGA based computers. In: Proc. ACM/SIGDA Int. Symp. FPGAs, pp. 191–200 (1998)
2. Chan, S.C., Yiu, P.M.: An efficient multiplierless approximation of the fast Fourier transform using sum-of-powers-of-two (SOPOT) coefficients. IEEE Signal Process. Lett. **9**(10), 322–325 (2002)
3. Cooley, J.W., Tukey, J.W.: An algorithm for the machine calculation of complex Fourier series. Math. Comput. **19**, 297–301 (1965)
4. Dempster, A.G., Macleod, M.D.: Multiplication by two integers using the minimum number of adders. In: Proc. IEEE Int. Symp. Circuits Syst., vol. 2, pp. 1814–1817 (2005). https://doi.org/10.1109/ISCAS.2005.1464962
5. Finley, T.: Two's complement. Cornell University lecture notes (2000)
6. Garrido, M.: A new representation of FFT algorithms using triangular matrices. IEEE Trans. Circuits Syst. I **63**(10), 1737–1745 (2016)
7. Garrido, M., Andersson, R., Qureshi, F., Gustafsson, O.: Multiplierless unity-gain SDF FFTs. IEEE Trans. VLSI Syst. **24**(9), 3003–3007 (2016)
8. Garrido, M., Grajal, J.: Efficient memoryless CORDIC for FFT computation. In: Proc. IEEE Int. Conf. Acoust. Speech Signal Process., vol. 2, pp. 113–116 (2007)

9. Garrido, M., Gustafsson, O., Grajal, J.: Accurate rotations based on coefficient scaling. IEEE Trans. Circuits Syst. II **58**(10), 662–666 (2011)
10. Garrido, M., Källström, P., Kumm, M., Gustafsson, O.: CORDIC II: A new improved CORDIC algorithm. IEEE Trans. Circuits Syst. II **63**(2), 186–190 (2016). https://doi.org/10.1109/TCSII.2015.2483422
11. Garrido, M., Qureshi, F., Gustafsson, O.: Low-complexity multiplierless constant rotators based on combined coefficient selection and shift-and-add implementation (CCSSI). IEEE Trans. Circuits Syst. I **61**(7), 2002–2012 (2014)
12. Gustafsson, O.: A difference based adder graph heuristic for multiple constant multiplication problems. In: Proc. IEEE Int. Symp. Circuits Syst., pp. 1097–1100 (2007). https://doi.org/10.1109/ISCAS.2007.378201
13. Gustafsson, O., Dempster, A.G., Johansson, K., Macleod, M.D., Wanhammar, L.: Simplified design of constant coefficient multipliers. Circuits Syst. Signal Process. **25**(4), 225–251 (2006)
14. Gustafsson, O., Qureshi, F.: Addition aware quantization for low complexity and high precision constant multiplication. IEEE Signal Process. Lett. **17**(2), 173–176 (2010)
15. Han, W., Erdogan, A.T., Arslan, T., Hasan, M.: High-performance low-power FFT cores. ETRI J. **30**(3), 451–460 (2008). https://doi.org/10.4218/etrij.08.0107.0189
16. Hasan, M., Arslan, T.: Scheme for reducing size of coefficient memory in FFT processor. Elect. Letters **38**(4), 163–164 (2002)
17. Hsiao, S.F., Lee, C.H., Cheng, Y.C., Lee, A.: Designs of angle-rotation in digital frequency synthesizer/mixer using multi-stage architectures. In: Proc. Asilomar Conf. Signals Syst. Comput., pp. 2181–2185 (2011). https://doi.org/10.1109/ACSSC.2011.6190418
18. Källström, P., Garrido, M., Gustafsson, O.: Low-complexity rotators for the FFT using base-3 signed stages. In: Proc. IEEE Asia-Pacific Conf. Circuits Syst., pp. 519–522 (2012)
19. Lee, J.A., Lang, T.: Constant-factor redundant CORDIC for angle calculation and rotation. IEEE Trans. Comput. **41**(8), 1016–1025 (1992). https://doi.org/10.1109/12.156544
20. Li, C.C., Chen, S.G.: A radix-4 redundant CORDIC algorithm with fast on-line variable scale factor compensation. In: Proc. IEEE Int. Conf. Acoust. Speech Signal Process., vol. 1, pp. 639–642, vol. 1 (1997). https://doi.org/10.1109/ICASSP.1997.599849
21. Lin, Z.X., Wu, A.Y.: Mixed-scaling-rotation CORDIC (MSr-CORDIC) algorithm and architecture for scaling-free high-performance rotational operations. In: Proc. IEEE Int. Conf. Acoust. Speech Signal Process., vol. 2 (2003). https://doi.org/10.1109/ICASSP.2003.1202451
22. Liu, H., Lee, H.: A high performance four-parallel 128/64-point radix-2^4 FFT/IFFT processor for MIMO-OFDM systems. In: Proc. IEEE Asia Pacific Conf. Circuits Syst., pp. 834–837 (2008)
23. Loeffler, C., Ligtenberg, A., Moschytz, G.: Practical fast 1-D DCT algorithms with 11 multiplications. In: Proc. IEEE Int. Conf. Acoust. Speech Signal Process., vol. 2, pp. 988–991 (1989). https://doi.org/10.1109/ICASSP.1989.266596
24. Meher, P.K., Valls, J., Juang, T.B., Sridharan, K., Maharatna, K.: 50 years of CORDIC: Algorithms, architectures, and applications. IEEE Trans. Circuits Syst. I **56**(9), 1893–1907 (2009). https://doi.org/10.1109/TCSI.2009.2025803
25. Oh, J.Y., Lim, M.S.: New radix-2 to the 4th power pipeline FFT processor. IEICE Trans. Electron. **E88-C**(8), 1740–1746 (2005)
26. Oppenheim, A., Schafer, R.: Discrete-Time Signal Processing. Prentice Hall (1989)
27. Padgett, W.T., Anderson, D.V.: Fixed-point signal processing. Synthesis Lectures on Signal Processing **4**(1), 1–133 (2009)
28. Parhi, K.K.: VLSI Digital Signal Processing Systems, Design and Implementation. Wiley-Interscience (1999)
29. Park, S.Y., Yu, Y.J.: Fixed-point analysis and parameter selections of MSR-CORDIC with applications to FFT designs. IEEE Trans. Signal Process. **60**(12), 6245–6256 (2012). https://doi.org/10.1109/TSP.2012.2214218
30. Qureshi, F.: Optimization of rotations in FFTs. Ph.D. thesis, Linköping University (2012)
31. Qureshi, F., Gustafsson, O.: Analysis of twiddle factor memory complexity of radix-2^i pipelined FFTs. In: Proc. Asilomar Conf. Signals Syst. Comput., pp. 217–220 (2009). https://doi.org/10.1109/ACSSC.2009.5470121

32. Qureshi, F., Gustafsson, O.: Low-complexity constant multiplication based on trigonometric identities with applications to FFTs. IEICE Trans. Fundamentals **E94-A**(11), 324–326 (2011)
33. Shukla, R., Ray, K.: Low latency hybrid CORDIC algorithm. IEEE Trans. Comput. **63**(12), 3066–3078 (2014). https://doi.org/10.1109/TC.2013.173
34. Takagi, N., Asada, T., Yajima, S.: Redundant CORDIC methods with a constant scale factor for sine and cosine computation. IEEE Trans. Comput. **40**(9), 989–995 (1991). https://doi.org/10.1109/12.83660
35. Thong, J., Nicolici, N.: An optimal and practical approach to single constant multiplication. IEEE Trans. Comput.-Aided Design Integr. Circuits Syst. **30**(9), 1373–1386 (2011). https://doi.org/10.1109/TCAD.2011.2153853
36. Volder, J.E.: The CORDIC trigonometric computing technique. IRE Trans. Electronic Computing **EC-8**, 330–334 (1959)
37. Wu, C.S., Wu, A.Y.: Modified vector rotational CORDIC (MVR-CORDIC) algorithm and architecture. Circuits and Systems II: Analog and Digital Signal Processing, IEEE Transactions on **48**(6), 548–561 (2001). https://doi.org/10.1109/82.943326
38. Wu, C.S., Wu, A.Y., Lin, C.H.: A high-performance/low-latency vector rotational CORDIC architecture based on extended elementary angle set and trellis-based searching schemes. IEEE Trans. Circuits Syst. II **50**(9), 589–601 (2003). https://doi.org/10.1109/TCSII.2003.816923
39. Yang, C.H., Yu, T.H., Markovic, D.: Power and area minimization of reconfigurable FFT processors: A 3GPP-LTE example. IEEE J. Solid-State Circuits **47**(3), 757–768 (2012). https://doi.org/10.1109/JSSC.2011.2176163

Biolabs as Computing Components

Georgi Tanev, Winnie Svendsen, and Jan Madsen

Abstract As a result of biological sciences becoming more quantitative together with the growing economical and societal challenges of improving our health care system, we have witnessed an increasing interest in developing new technologies for the biomedical field. Cyber-medical systems are the fusion of computational and medical technologies aimed to support health management for diagnosis, monitoring, and prevention of diseases. With the advent of miniaturized biochemical laboratories, the classical bench-sized lab robots have transformed into chip-sized complex biochemical laboratories. This development has allowed integration of classical computation with biochemical processes to a degree where computations are moving small amounts of liquids. In this chapter, we survey state-of-the-art microfluidic technologies and argue that even though their applications are still limited to simple passive and fixed structures, they hold the promises of solving many challenging and complex problems related to healthcare. We argue that liquid handling technologies have the potential to scale, but this will require utilization of active components and the ability to abstract their basic operations to a level similar to that of classical computation, i.e., we need to build the equivalent of a *general purpose processor*—a lab-on-chip (or a biochip) which can be programmed and re-programmed.

G. Tanev
Technical University of Denmark, DTU Compute, Lyngby, Denmark

Technical University of Denmark, DTU Bioengineering, Lyngby, Denmark
e-mail: geta@dtu.dk

W. Svendsen
Technical University of Denmark, DTU Bioengineering, Lyngby, Denmark
e-mail: wisv@dtu.dk

J. Madsen (✉)
Technical University of Denmark, DTU Compute, Lyngby, Denmark
e-mail: jama@dtu.dk

S. S. Bhattacharyya et al. (eds.), *Embedded, Cyber-Physical, and IoT Systems*,
https://doi.org/10.1007/978-3-030-16949-7_12

1 Introduction

Lab-on-chip (LoC) technology combines efficient and precise liquid handling techniques together with highly sensitive biosensors in order to miniaturize the traditional wet lab processes and to enable chemical or biological analysis, synthesis, separation, sorting, etc. to be performed on a chip-scale device. Often these devices are called biochips due to their form factor and application mainly positioned in the biomedical field. Carefully engineered biochips, designed to implement biological or medical protocols, are showing advantages over the traditional benchtop laboratory robots by reducing sample and reagent volumes, lowering the reaction time, optimizing materials and operation cost, and maximizing the amount and quality of the acquired data. The substantial contribution of biochips to numerous biological and medical applications is a result of an interdisciplinary endeavor carried out over several decades. This process employed knowledge and expertise from a number of disciplines among which are physics, material science, medical technology, biotechnology, and micro-fabrication processes developed for the electronics industry. Notably, the computer-aided tools developed to design, simulate, and operate these portable miniaturized labs have a central role in the process of LoC miniaturization and automation.

Even though a number of biochips have long passed the proof-of-concept development stage and have already been used to successfully implement commercial protocols addressing real-life problems, biochip utilization still remains somewhat limited. The path to realizing the full potential of these portable labs is a subject to challenges such as efficient design and standardization, lowering the fabrication cost while keeping the yield high, ensuring reliable and fault-tolerant operation, and integration with the already existing lab infrastructure and processes. However, one of the biggest challenges is that biochips are inherently application specific since their design usually closely resembles the structure of the implemented protocol. This has naturally led to a variety of unique biochip designs, most of which have been developed in a process of achieving short-term research goals, without a clear perspective for standardization and reusability on the device or component level. Limited or absent design methodologies and standardized biochip building blocks are considered to be the major roadblocks in the process of realizing the full potential of the LoC technology.

An example of a technological evolution supported by heavy standardization and design automation is the field of electronics and computer architectures. The first computers were designed and built solely as mechanical systems which naturally evolved into being electromechanical systems using components such as electrical switches and relays to program and operate. Later, the slow and unreliable electromechanical components were replaced by the much faster and more reliable vacuum tubes and transistors. These first generations of computers were programmed in a cumbersome way by using mechanical switches, patch cables, and paper cards. High-level programming abstractions were virtually non-existent. Further down the evolution path, supported by the fast development of

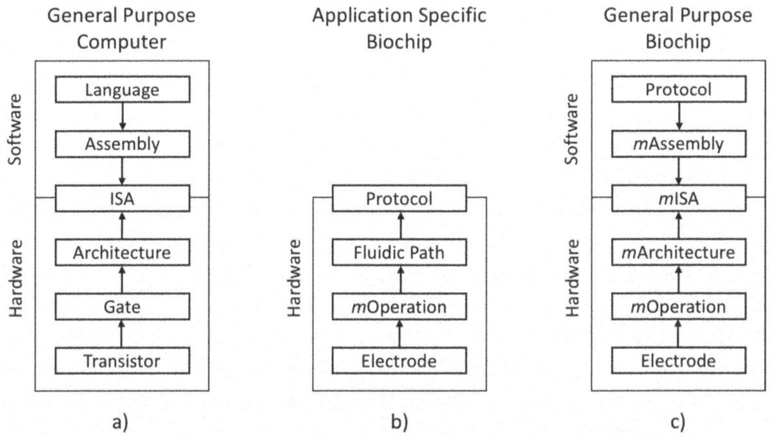

Fig. 1 Design of biochips as compared to modern computers. (**a**) General purpose computer, where the architecture is designed bottom-up to provide a hardware independent programming interface, the ISA, which can be compiled from a high-level programming language in a top-down manner; (**b**) State of the art for biochip design is a bottom-up process providing an application specific architecture; (**c**) The structure of a general purpose biochip structure combining (**a**) and (**b**)

digital electronics, memory became available and computer programs and data started being an essential part of the computer building blocks. This high level of integration between hardware and software naturally increased the level of computational complexity but significantly simplified and enabled a more efficient and abstract programming process as shown in Fig. 1. The full potential of the computer was only realized with the invention of the integrated circuit (IC), which allowed for computers to be built from already well-defined and characterized building blocks and thus next generation of computers required significantly less time to develop due to the high degree of design reusability.

The need for abstracting the programming model from the hardware architecture and implementation-specific details was also identified and addressed in parallel with the hardware development. The model of an instruction set architecture (ISA) has been introduced to abstract from the specifics of the instruction set implementation (ISI). Programming a computer in assembly provides a low level and granular control over computation resources, but certainly, it is a complex, time consuming, and a tedious process. With the standardization of the ISA, a number of high-level programming languages and compilers were introduced with the aim to handle complexity and ensure a consistent computer programming model. As a result, the ISA abstraction allowed for creating reusable and machine independent software libraries to serve as a main component in new designs. Figure 1a illustrates how the ISA works as a platform abstracting the underlying hardware architecture from the application software. The architecture can be built bottom-up from transistors and gates, while the application software is compiled top-down from a high-level computer language to the assembly code initiating the instructions of the platform.

The personal computer as we know it today is an impressive technological accomplishment where a number of subsystems are interconnected through well-defined interfaces to assemble a *general-purpose* computation machine. Modern computers can simultaneously run a number of tasks and programs such as text editor, internet browser, and design tools. The personal computer has evolved to a stage where it is an affordable ubiquitous general-purpose machine even though in its early development a computer used to be an expensive application-specific machine.

In contrast to modern computer systems, the biochip technology still lacks layers of abstraction that would allow for lab protocols to be captured easily and implemented on a biochip. A modern computer can be programmed and used from a user perspective, whereas state-of-the-art biochips lack the level of abstraction needed to bridge the gap between a protocol and low-level implementation details, as illustrated in Fig. 1b. Besides a few widely used manufacturing processes, the search for new materials and operation schemes is still an ongoing process. However, the biochip technology might be in general ready to adapt design methodologies and standardization from well-established and proven fields such as the field of digital electronics and computer science. Just like the early adoption of modular hardware design and a common ISA supported the fast development of new computer systems, biochip scientists and engineers need to focus on developing tools for automated biochip design, formalization, and integration. This will accelerate the development process and it will provide a common standardized architecture as illustrated in Fig. 1c. A few research groups have been developing design automation tools for biochips, e.g., [4, 7, 16]; however, these are still mostly research projects and the results have not yet been widely adopted as compared to design automation tools for microelectronics.

Flexibility and reconfigurability are two very important aspects that need to be considered in the quest of realizing the full potential of the biochip. Just like a field-programmable gate array (FPGA) can be configured to serve a particular application, we envision biochips to allow for a certain degree of reconfiguration and flexibility in the future [15]. Biochips serving the purpose of a general-purpose biochemical computer may appear to be a futuristic vision. Nevertheless, we can argue that the level of miniaturization in the field of digital electronics and computing that we have achieved so far looked as distinct and futuristic to the engineers who developed the first generation of computers in the beginning of twentieth century.

In this chapter, we will focus on outlining the basic characteristics of the biochips and we will discuss their functionality by drawing parallels with concepts borrowed mainly from the fields of digital electronics and computer science. In the following section, we will continue with a review of the two dominant biochip technologies, namely continuous flow microfluidics and digital microfluidics.

2 Digital and Continuous Flow Microfluidics

Microfluidics is about efficient and precise handling of small volumes of liquids in the scale of pico- to microliters. Microfluidics plays a fundamental role in the operation of biochips since it provides the means for liquid manipulations. There are two main classes of active fluid handling technologies—continuous flow microfluidics and digital microfluidics (DMF). Figure 2 shows the main differences between the two classes of biochip technologies. Even though both technologies have advantages and disadvantages, we focus our work mainly on the DMF due to the following three main reasons:

1. DMF are easily programmable and reconfigurable;
2. DMF appear to match the level of abstraction where basic digital electronics can be applied to the DMF operational model; and
3. DMF provide an effective "first order digital" liquid control.

Although we will focus on DMF in the rest of this chapter, for the sake of completeness, we will first briefly discuss the operational principles and instrumentation needs of continuous flow microfluidic chips. This discussion serves the purpose of introducing the reader to the continuous flow microfluidics, as well as it is used as a starting point to compare the two fluid handling technologies and to argue the advantages of DMF in the context of programmable biochips.

2.1 *Continuous Flow Microfluidics*

A traditional class of microfluidic devices is based on continuous flow, where liquid transport is carried out in micrometer-sized channels utilizing air pressure as a driving force. Active control and routing of the liquid streams is achieved by using either mechanical valves or electrokinetic mechanisms. Alternatively, channel sizing and interconnection patterns can be specifically designed to passively navigate and manipulate fluids. Usually, continuous flow devices have fixed functionality and are adequate for simple and well-structured protocols. This is partially dictated

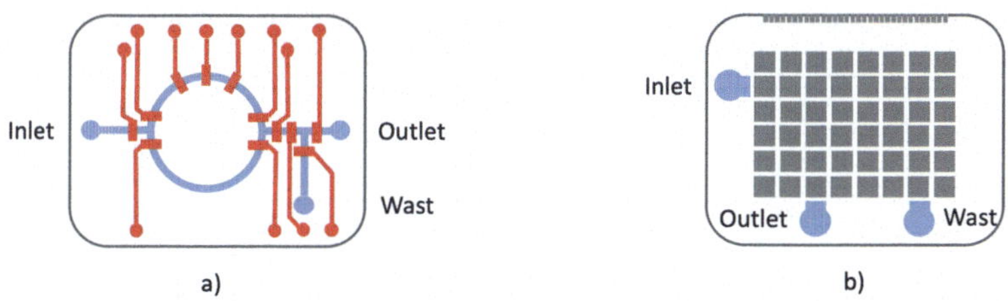

Fig. 2 Microfluidic chip examples. (**a**) Flow-based microfluidic chip (FMF), blue channels are for liquid flow, red are control channels using air pressure. (**b**) Digital microfluidic chip (DMF)

by an on-chip network of channels and control components being fixed once a device is fabricated and partially due to the fact that individually controlling routing components on the chip usually require off-chip equipment, such as pressure pumps. Figure 2a shows a simple example of a mixer component. The blue channels are flow channels carrying the liquids to be handled and the red channels are control channels, controlling air pressure to close or open valves (red boxes). In this case, two different liquid volumes can be placed into the top and bottom parts of the mixing circle through the inlet by setting the valves correctly. Mixing is obtained by fast iterative turning on/off the top three valves of the mixing circle, by which the liquids in the mixing circle will rotate and hence, mix over time.

Continuous flow microfluidics often requires external bulky devices such as pumps and valves to control on-chip processes as shown in Fig. 3. Moreover, these actuators use electric power to operate which means that control signals need to cross a few energy domains before they actually reach the chip. The energy domain crossing is illustrated in Fig. 4 where an automated system is used to control on-chip

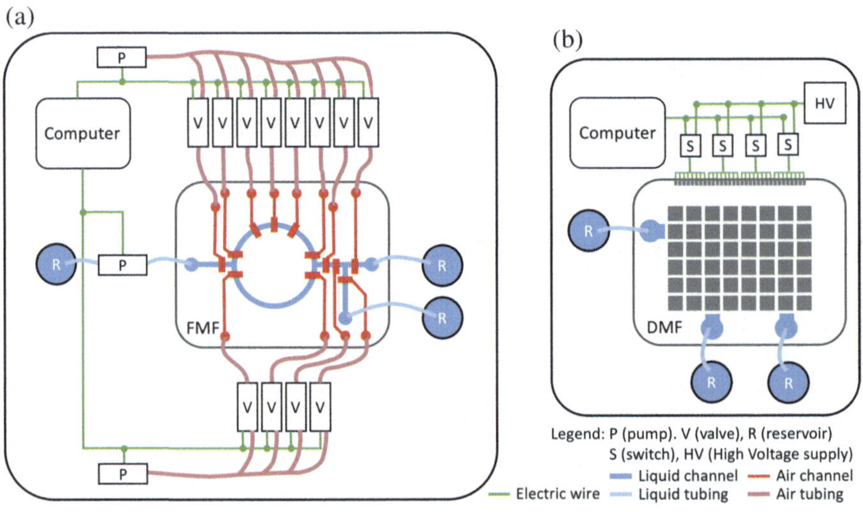

Fig. 3 Microfluidic chip examples with external control functions. (**a**) Continuous flow-based microfluidic chip, on-chip valves are controlled through off-chip pressure valves feed from pumps. Pumps and off-chip valves are controlled by a computer. (**b**) Digital microfluidic chip

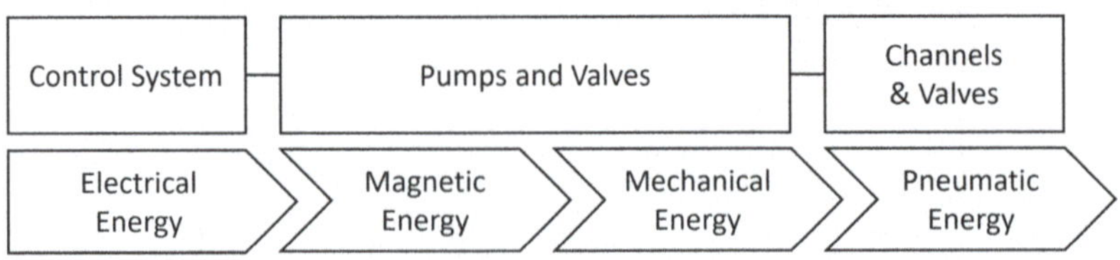

Fig. 4 Continuous flow microfluidics energy domain crossing. *Top:* Control signal chain. *Bottom:* energy domain crossing

valves. To actuate a single vale on the chip, first, an electric control signal needs to be applied to an external electromagnetic valve, which converts electrical current into a magnetic field in order to attract a valve piston, and as a result, a pressure is applied to pneumatically actuate the on-chip valve. From this example, it is evident that the electric control signal needs to cross a few energy domains and the control process involves a number of mechanical moving parts. Continuous flow microfluidics often requires a costly and bulky external instrumentation which combined with the need of precision fabrication tools and a rather fixed operation model makes them a less attractive choice for further exploration in the context of programmable biolabs.

2.2 *Digital Microfluidics*

Digital microfluidics is a technology that allows discrete droplets to be actuated on the top of an insulated array of electrodes. There are two main types of DMF chip constructions—open and closed. The construction of the two types of chips is shown in Fig. 5. An open DMF system is a two-dimensional (2D) array of individually addressable conductive electrodes fabricated on a carrier substrate. Actuation of a droplet on the surface of the chip is achieved by applying a control voltage on a pair of electrodes beneath the droplet as shown in the left side of Fig. 5. Utilizing this control technique, droplet transport, merging and mixing can be achieved. Even though splitting has been reported on an open system [8], it lacks precise control and is unreliable. In a closed DMF system the sample is sandwiched between two parallel plates—an electrode patterned substrate and a top plate. The top plate is usually made of a conductive indium tin oxide coated glass, which is connected to ground potential and thus serving the purpose of a common grounded control electrode. A droplet in a closed chip is controlled in the same way as in the open chip—a control voltage is applied between the top plate and an electrode adjacent to the actuated droplet. The possible fluidic operations on a closed chip overlap with those available on the open chip but with the important addition of a controlled and precise drop splitting [12].

Both the open and the closed DMF chip configurations operate based on a phenomenon known as electrowetting on dielectric (EWOD) which is used to

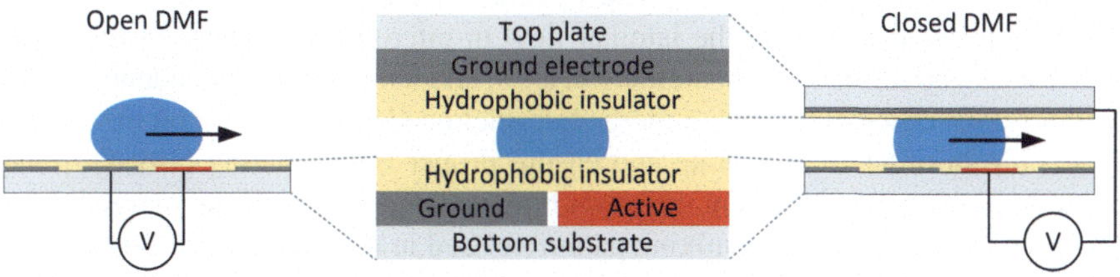

Fig. 5 Digital microfluidics chip construction. *Left:* an open DMF chip. *Center:* chip structure. *Right:* a closed DMF chip

control the solid-to-liquid contact angle as a function of an applied voltage potential [5, 11]. The surface of the electrodes is covered with an isolation hydrophobic layer which does not allow ohmic current to flow through the liquid. Another function of the hydrophobic layer is to provide a low friction interface between the actuated droplets and the chip. Aqueous, chemical, or biological droplets placed on the hydrophobic surface of the biochip take a close to spherical or hemispherical shape to minimize its contact angle with the surface. The qualities of the hydrophobic isolation layer define operation parameters such as control voltage, contact angle, and reliability. For instance, based on the materials and the thickness of the coating, the control voltage varies in the range of a few [9, 10] to hundreds of volts [1]. Applying a control voltage under the hydrophobic layer changes the layer surface properties from hydrophobic to hydrophilic, thus the liquid-to-solid contact angle decreases. This decrease of contact angle causes the droplet to flatten its shape which creates a surface tension gradient that allows a controlled displacement of the droplet along an actuation direction. Pattering a surface with a 2D array of electrodes enables droplets to be transported in arbitrary paths on the surface of the chip.

The EWOD provides a direct mechanism to convert an electrical signal into a physical displacement of a droplet with a crossing of a single energy domain. Furthermore, this elegant operation principle allows for space and instrumentation equipment optimization since it eliminates the need for bulky components with moving mechanical parts. This makes digital microfluidics an attractive choice for further research in the context of programmable biolabs.

3 The Biolab as a Computer

The basic building block of a DMF chip is an electrode which can be connected to either a high potential or to ground. Setting a pair of adjacent electrodes in opposite states allows a droplet to be moved. Hence, the main function of a DMF electrode is to act as a droplet router to one of its adjacent electrodes thus allowing for droplets to be transported, merged, mixed, and split by actuating them over predefined paths as illustrated in Fig. 6. While droplet transport and merging can be considered a discrete operation since they are performed in a single step, droplet mixing and splitting require more than one actuation step. When two droplets are merged together, they do not immediately become a homogeneously mixed fluid due to the predominant effect of the laminar flow in microliter volumes. At this small scale, mixing by diffusion is an inefficient, slow, and temperature dependent process. Therefore, for efficient mixing, a merged droplet needs to be moved on the surface of the chip in order to accelerate and ensure proper liquid mixing [14]. Different actuation paths have a different effect on how fast and how well the merged drops mix and a traditional circular mixing pattern is used in the example shown in Fig. 6.

Controlled splitting of droplets is required for precisely dispensing analytes from on-chip reservoirs, when a single droplet needs to be part of more than one reaction, or in order to provide a required analyte concentration. Splitting is

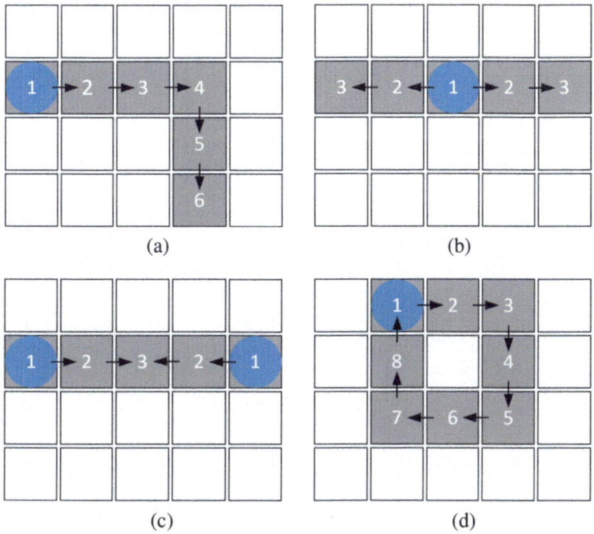

Fig. 6 Electrode operations patterns on a 2D array. (**a**) Transport. (**b**) Splitting. (**c**) Merging. (**d**) Mixing

achieved by simultaneously activating at least three electrodes beneath the droplet, thus stretching it over the surface area of the activated electrodes. Then the middle electrode is deactivated, which recovers the hydrophobicity of the surface and allows the created surface tension gradient to split the droplet in two smaller ones as shown in Fig. 6b.

Besides the plain digital behavior of moving droplets in steps on the chip, droplets can have a unique composition of chemical or biological materials, resembling the *information* contents of a droplet. Hence, a single droplet can serve the function of a fluid vehicle, a reaction chamber, or both. The fact that such a potentially complex fluid composition is naturally encapsulated and appears as a unit makes evident the similarities between a droplet and a traditional data packet used in the computer science field. Comparing a droplet to a digital data packet together with the routing function of the DMF electrodes resembles the network-on-chip (NoC) communication model used in the field of very-large-scale integration (VLSI) design. The composition of a traditional NoC can be described as a graph where each node is composed of a data router and a computational unit (CU) as shown in Fig. 7a. Links between routers establish communication channels that allow data to be transported in both directions and be passed to the CU for processing. A DMF chip maps well to this abstraction where the equivalent of a NoC router is served by a DMF digital electrode, and the NoC CU is compared to a functional element (FE) associated with each electrode. The base case of FE is a digitally controlled switch, which allows an electrode to be connected or disconnected to a control signal. Moreover, a FE can be associated with sensing or control functionalities such as temperature monitoring, biosensing, or heating.

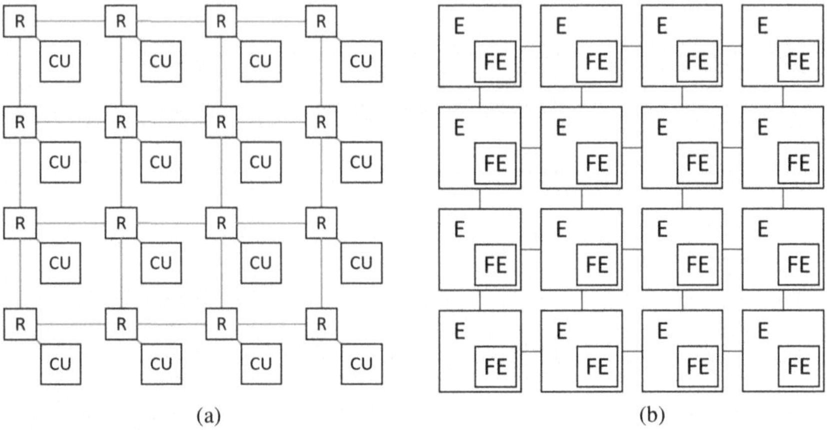

Fig. 7 NoC architecture. *Green* lines represent datapath. *Blue* lines show fluidic path. (**a**) Traditional NoC. (**b**) DMF as a NoC

Similar to the traditional NoC where electrical wires are used to carry data between routers, DMF uses electrodes and the EWOD principle to pass droplets between neighboring electrodes. Another parallel between the NoC and DMF is the special functionality associated with each node of the network—a NoC can be composed of homogeneous or heterogeneous nodes. A homogeneous NoC is a network where all CU are identical, e.g., a CU is a single processor in a multiprocessor NoC architecture. A heterogeneous NoC is composed of CU with different functions such as a general purpose processor, a memory, or an accelerator. Likewise, a DMF chip that only consists of digital electrodes is considered to be a homogeneous design and although droplet actuation has a major role in implementing the biochip functionality, often heating and sensing are also required in order to implement a complete protocol. To accommodate for that, DMF can be heterogeneous systems where FEs implement control or sensing functions.

A NoC router can implement a data packet buffering. Likewise, a DMF electrode can also mimic storage of a single droplet by keeping it on its surface. Fluidic operations such as transport, merging, mixing, and splitting are performed based on this pass-and-store mechanism. To control the array of routing electrodes and functional elements, an external biochip controller is needed. In order to abstract the biochip programming from the implementation details of the DMF chip, this controller needs to implement a biochip independent instruction set. To address this, in the next section we will borrow a well-established control abstraction from the field of digital electronics and apply it to model a biochip controller architecture.

3.1 A Finite State Machine with a Microfluidic Path

A finite state machine (FSM) is an abstract computational model, which can only be in one of its designed states. The traditional FSM is a regular sequential circuit and it consists of a *state register*, *next state logic*, and *output logic* as shown in Fig. 8. However, in contrast to the traditional sequential circuits, the FSM might not exhibit a repetitive state pattern, but rather the next state can be a function of the current state and an external input signal. This allows a FSM to be used to decode an external input signal, match it to one of its defined states and, as a result, generate the next state of the FSM and output control signals. However, the FSM does not perform any data operations but rather serves as a control unit and is thus often referred as a *control path*. To allow for data manipulation, a regular sequential digital circuit consisting of data routing and functional components is needed and is often called a *datapath*. Complex logical operations can be performed by using the control signals of a FSM to control the dataflow in a datapath. This computational model is known as a finite state machine with datapath (FSMD).

The FSMD model can be used to model one of the fundamental components of any computer architecture, namely the central processing unit (CPU). An ISA defines how the decoded instructions will control the functional units and data operations of the datapath. A traditional CPU is composed of a number of functional units such as registers, buses, multiplexers, adders, and multipliers, which are designed to manipulate data. The computational flexibility of the traditional CPU comes mainly from the fact that simple atomic operations can be translated into

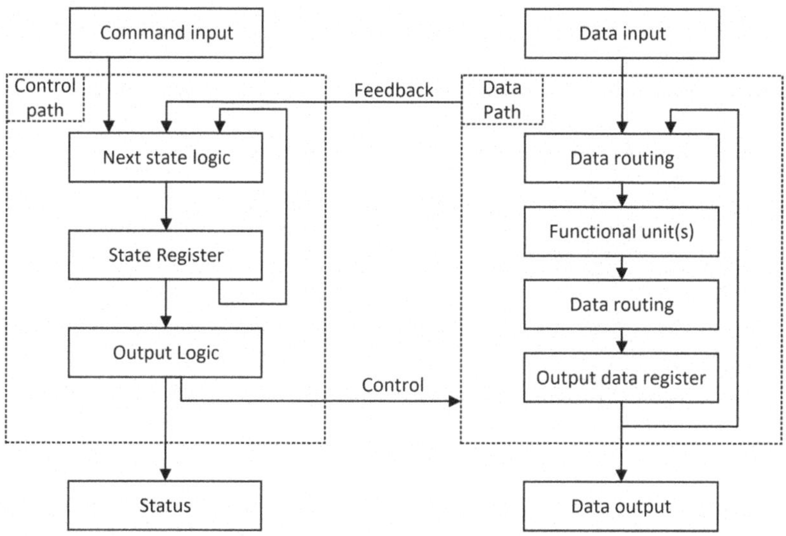

Fig. 8 Traditional FSMD

instructions, logically grouped in a complex program, and then executed on an instance of an ISI.

Similar to the datapath of a CPU, which is designed to operate with digital signals, a DMF chip operates with liquid droplets. Hence, if we substitute the traditional datapath in the FSMD shown in Fig. 8 with a DMF chip and use the FSM control signals to drive the electrodes, the resulted architecture inherits all of the characteristics of a FSMD and in addition, is able to operate with fluids instead of digital data. This digital-to-liquid interface allows commands to be passed to the FSM and executed on the biochip as shown in Fig. 9.

Keeping the biochip control in the digital domain enables for a hardware abstracted programmatic control of electrodes and FEs on a DMF-based biochip. Concepts for programming, standardizing, and automation of biological protocols have previously been studied [2, 3, 13]. Designing a formal vocabulary and a set of grammatical rules for a functional, expressive, and easy to use microfluidics programming language is not in the scope of this chapter. However, in order to demonstrate the concept of a biochip as a computer, we will use a minimalistic set of high-level fluidic commands which are shown in Table 1. These commands can be used to capture the behavior of a simple lab protocol.

A FSM is a computer science model connected to the idea of grammar and language that allows for structure and rules. This allows for a compiler that can process a high-level protocol description and compile it into a sequence of biochip commands. These commands can be further decomposed into low-level instructions supported by the target fluidic ISA. Specifying a fluidic ISA allows protocols to be expressed as a computer program and their execution to be automated and monitored

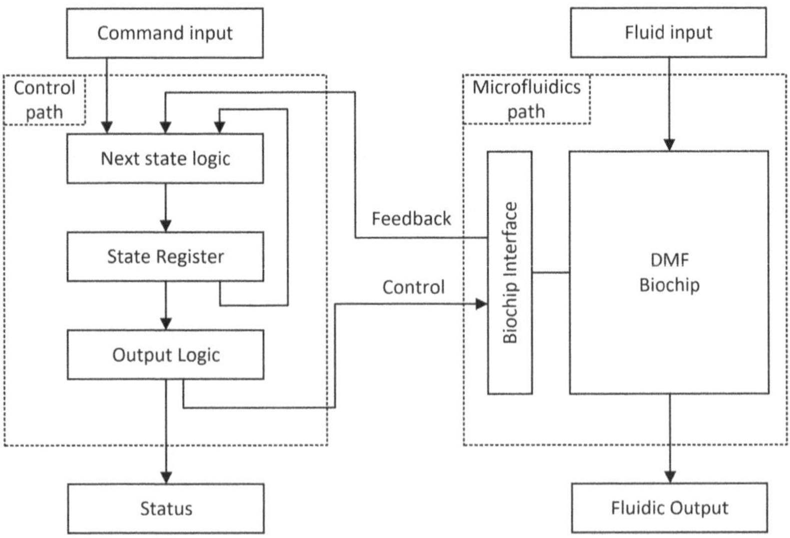

Fig. 9 FSM with a fluidic path

Table 1 A subset of biochip commands

Biochip commands	Description
dispense(r, p)	Dispense from reservoir *d* with parameters *p*
move(d, l1, l2, p)	Move droplet *d* from location *l1* to *l2* with process parameters specified by *p*
join(d1, d2, l)	Join droplet *d1* with *d2* at location *l*
mix(d, p)	Mix droplet *d* with mixing parameters *p*
split(d, p)	Split a droplet with splinting parameters *p*
incubate(d, p)	Incubate droplet *d* with incubation profile *p*
sense(d, s, p)	Sense the content of droplet *d* with sensor *s* and sensing parameters *p*
dispose(d)	Send droplet *d* to waste

in real time. For instance, a move operation on the biochip will be compiled to an electrode activation sequence, or an incubation operation will generate all the required commands for operating the heating functional element and the required timing to be used in the incubation process. We can also draw parallels between the types of instructions such as a dispense from a reservoir or dispose to waste, being equivalent to input/output (IO) operations, or incubation being similar to storing data into a register.

To demonstrate the concept of mapping biochip commands into a microfluidics chip-specific language we will use the fluidic operation transport which was shown earlier in Fig. 6. Droplet transport is achieved by sequentially turning on and off electrodes on the chip. Turning an electrode *on* means connecting the electrode to a high-voltage potential and turning an electrode *off* means connecting it to a low potential. This is equivalent to the traditional data handling memory operations where a single bit, register, or a memory location can be set or cleared. Two low-level instructions can be used to change the state on an electrode: set an electrode which connects the addressed electrode to a high potential, and clear an electrode which sets the addressed electrode to a low potential. For clarity we will call these instructions correspondingly *setel* and *clrel*. The instructions are limited to a single operand which is the manipulated electrode number. Figure 10a shows an example of a droplet transport operation which can be captured by the command *move(droplet<d>, electrode<10>, electrode<15>, 0)* and compiled to the instruction sequence shown in Fig. 10b.

Control flow instructions can be implemented as well by using qualitative or quantitative feedback data extracted from FEs. An example of such a control flow command would be to use a FE with heating and temperature sensing capabilities in order to ensure a certain temperature profile applied to a droplet before the protocol continues.

Text-based programming is a straightforward process for experienced users, nevertheless it has a steep learning curve. An alternative way to map a lab protocol to a structured executable code is to provide a graphical programming environment. A lab protocol can be drawn as a flow diagram with a unique start and end state, and any sequence of states in between these two states, to encapsulate the steps (or

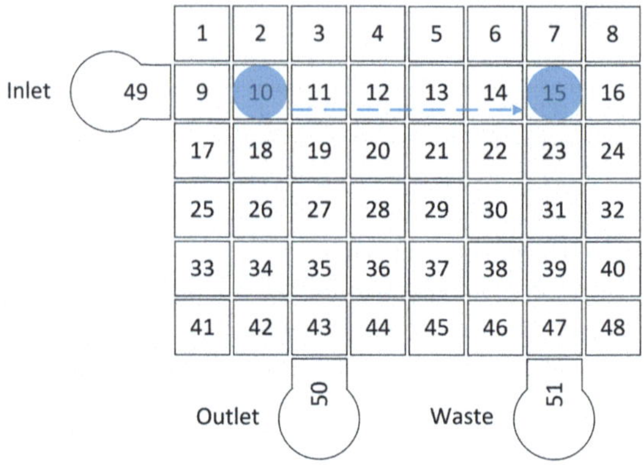

(a)

Step	Command
1	*setel 11*
2	*clrel 11*
3	*setel 12*
4	*clrel 12*
5	*setel 13*
6	*clrel 13*
7	*setel 14*
8	*clrel 14*
9	*setel 15*
10	*clrel 15*

(b)

Fig. 10 Transport of a droplet from chip electrode 10 to electrode 15 and electrode actuation sequence. Droplet actuation path is shown with the blue arrow. (**a**) Chip example. (**b**) Transport commands sequence

possible steps) of the protocol. Each node of the flow diagram defines an operation and its associated parameters. The connection branches define the sequence of the operations.

4 Proof of Concept Through Implementation

In this section, we will identify and discuss some of the challenges of instrumenting biochips and interfacing to them as if they were computing components. The aim is to develop a flexible and reconfigurable control and instrumentation platform that can provide a standardized interface to a biochip viewed as an independent fluidic path. Based on platform-based design [17] and design for changeability principles [6], we can design a microfluidics instrumentation platform that can accommodate for simultaneous evolution of both biochip and instrumentation [19].

Modularity is a proven design technique that allows the functionality of a system to be partitioned into smaller self-contained and decoupled subsystems. Subsystems communicate with each other over well-defined interfaces which inevitably increases the implementation complexity, but allows for different modules to be added, removed, or upgraded without modifying the whole system. An example of such a modular platform is the PC where different input/output cards, peripheral devices, memories, etc. can be attached to a motherboard. The architecture is not strictly defined but rather relies on a set of design guidelines, rules, and constraints in order to ensure interoperability between the attached components.

A block diagram of a modular microfluidics platform is shown in Fig. 11. It consists of a *user interface*, a *control path*, and a *fluidic path*. The user interface

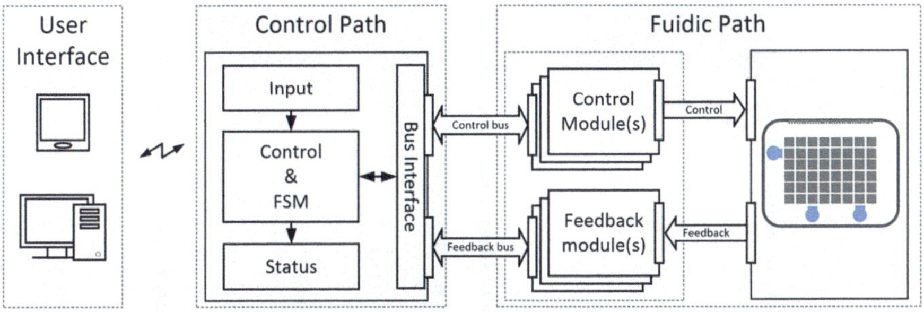

Fig. 11 Implementation of the FSM with a fluidic path, including control and feedback modules, and a wirelessly connected user interface

is a program or a collection of software tools that run on an external device such as a PC or a smartphone. The user interface is an easy to use front-end with the main purpose to capture lab protocols, serve as a compiler, communicate and manage the control path, and provide general control and monitoring functionality. The second part of the system is the actual FSM, which controls the fluidic path. The fluidic path itself consists of a biochip and a set of interface modules required to make the link between the digital electronics and the fluidic FEs. Keeping the instrumentation modules as a part of the fluidic path allows the control path to remain independent from the biochip instrumentation needs.

The user interface of the platform is implemented on a smartphone or a PC due to their ubiquity, great computational power, and familiarity to a broad range of users. Translating a lab protocol into a sequential program can borrow design and implementation elements from the modern text- and graphical-based programming languages. Compiling a protocol into the ISA a given biochip can utilize conventional compilation processes and techniques, making it virtually identical to traditional computer programming. Programming and compilation are well studied and understood and they can be as simple as directly mapped scripting or as complex as multistage and parametric compilation and optimization process. Nevertheless, in a prototyping phase of a hardware/software platform, the first goal is to specify, implement, and verify the main functionality and leave the optimization for later stages in the development process. Therefore, a simple smartphone application and a PC-based tool were developed and their interfaces are shown in Figs. 12 and 13. Both tools support a scripting language used to describe a control sequence, which is then parsed by the application and sent to the controller over a wireless link. Additionally, the smart phone application implements a touch screen interface for manual control of the digital electrodes by simply clicking on them.

The controller is implemented as a hierarchical software state machine running on a stand-alone embedded system. We call this module the main board as shown in Fig. 14. The main board can receive commands from the user interface, decode them, and control the fluidic path attached to the bus interface. The bus interface consists of two communication buses, namely the *control* and the *feedback bus*,

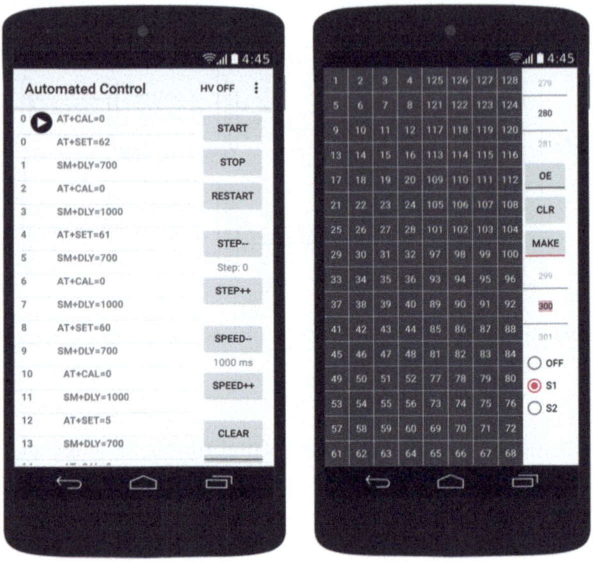

Fig. 12 User interface. *Left:* automated script controller interface. *Right:* manual controller interface

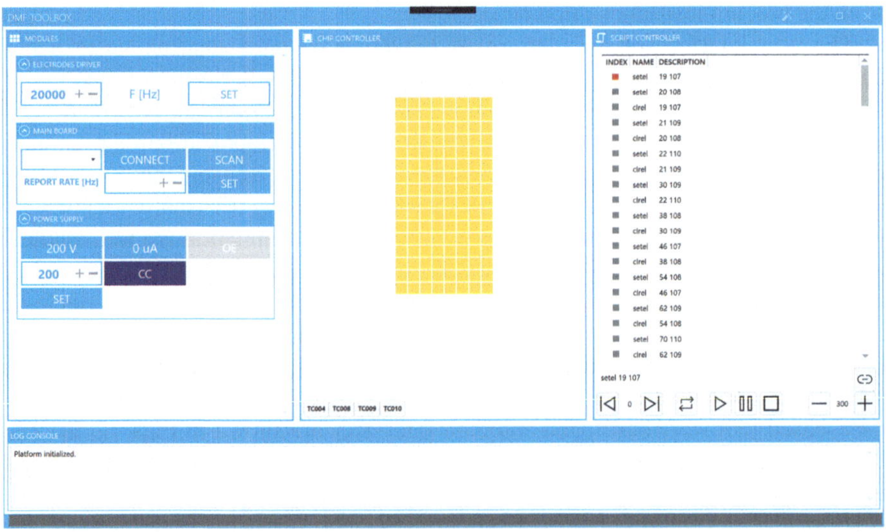

Fig. 13 User interface of the PC-based tool

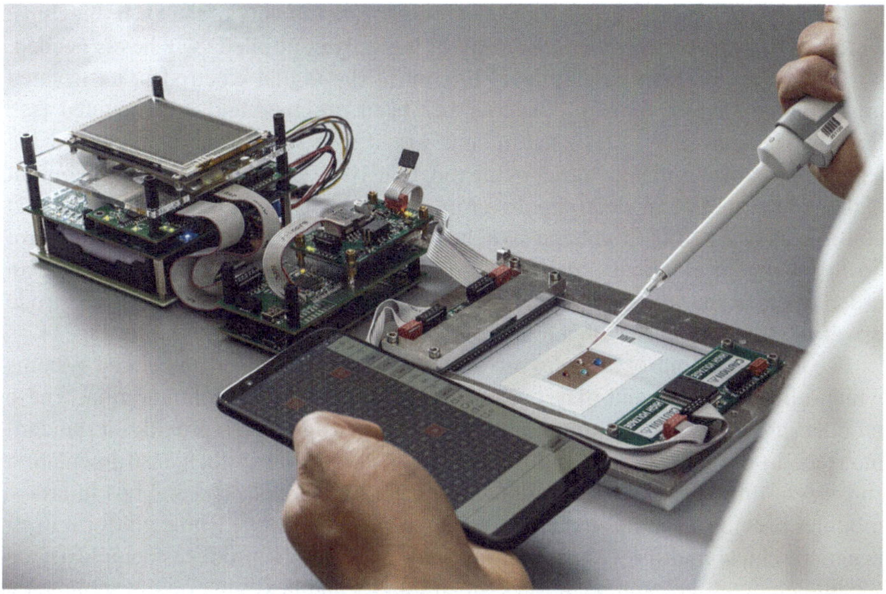

Fig. 14 DMF prototype. *Left:* main board, including battery power and a display. *Center:* instrumentation subsystem. High-voltage power converter and electrodes drivers interface. *Right:* biochip including electrode driver modules

which connect to the instrumentation modules needed by the specific instance of the biochip. The bus interface consists of both data and power lines and supports detection and identification of the connected instrumentation modules. This allows the external controller to read the configuration of the fluidic path and verify if the resources for a particular experiment are available.

A digital biochip is usually composed of a number of FEs as discussed in Sect. 3 (see Fig. 7b), which are the main building blocks of the fluidic path of the biochip. They not only determine the processing capabilities of a particular biochip but also specify what type of interface circuitry is needed in order to link them with the controller. Therefore, it is essential for an instrumentation platform to support a unified interface model to connect any FE configuration to the controller. The basic form of a FE is a digital electrode which is controlled by an electronic switch. Depending on the thickness and the qualities of the isolation layer covering the digital electrodes, the associated electronics switch needs to be able to operate with a potential of a few tens of volts to a few hundreds volts. The switch also acts as a voltage level shifter from a standard digital voltage level to the chip-specific electrode activation level. A FE can also implement functions such as heaters or biosensors and they naturally require instrumentation that is more advanced. Moreover, to facilitate the needs of different lab protocols, biochips are often constructed as a heterogeneous network of FEs.

To accommodate for the basic droplet routing functionality of the DMF biochips, a step-up voltage converter and an array of digitally controlled switches is needed. To provide a sufficient voltage level to control the digital electrodes, an isolated high-voltage step-up flyback converter has been designed and implemented (see Fig. 14. The converter is designed to be powered by 5 V as an input and to provide a programmable regulated output in the range of 10–300 V. The high-voltage output is then connected to an electrode driver module, which has a digital level input that controls an array of high-voltage switches. An electrode driver module can control up to 64 digital electrodes and the modules are designed to be daisy chained in order to allow an arbitrary number of electrodes to be controlled. Two electrode driver modules together with a high-voltage power supply can be seen in Fig. 14.

Feedback and control modules are designed as self-contained embedded systems which allow for plug-and-play operation and easy system reconfiguration. These modules are based on a common microcontroller design which provides a standard interface to the control path and thus modules only differ in their instrumentation subsystem implementation. The standardization of the controller and bus interface modules provide a solid foundation for design reusability. The whole platform itself is a hierarchical multiprocessor distributed system, which introduces complexity on the implementation level, but allows for simplicity and flexibility on the application level.

5 Biochip Perspectives

Microfluidic biochips integrate different biochemical preparation and analysis functions on a single miniaturized chip, effectively handling biological processes at the submillimeter scale. Besides the small scale which makes these biolabs portable or handheld, the miniaturization has several advantages as compared to conventional biochemical analyzers at the macro-scale. They consume reduced sample and reagent volumes, provide faster biochemical reactions, ultra-sensitive detection, higher system throughput, and have the capability to integrate and process several assays on the same biochip.

Furthermore, biochip solutions avoid the use of pipetting every time liquids have to be moved. In classical biolabs, whether operated by humans or lab robots, each pipetting creates a waste of a pipetting plastic tip. Although such a plastic tip is no more than 30–70 mm in size, the fact that a new tip is needed for every operation results in a huge amount of plastic waste. In contrast, for the operation of biochips, pipetting may only be needed for loading the sample liquids onto the chip, which results in highly sustainable lab solutions.

In this chapter, we have focused on the liquid handling, i.e., moving, mixing, and splitting droplets, which is an essential part of sample preparation. Our aim is to show that it is possible to develop fully programmable and reconfigurable general purpose biochips, which can provide new possibilities for miniaturized systems for chemistry and life sciences. As our DMF platform supports instrumentation through the control and feedback modules, we are able to interfere with a large range of sensors, such as colorimetric and electrochemical biosensors, as well as actuators, such as heaters. Sensing and actuation components can be integrated into the electrodes or added as add-on modules.

The presented platform allows for the development of a wide range of applications, such as in vitro diagnostics (point-of-care or self-testing for disease diagnostic and monitoring), drug discovery (high throughput screening), biotech (process monitoring, cell cloning for cell fabrics, including PCR for DNA amplification [18]), and ecology (agriculture and environmental monitoring). The platform is currently being tested and further developed in projects on diagnostics (metabolic disorders, mastitis, and multi-bacterial screening) and biotech (replacing lab robots for cell fabric cloning and validation).

Acknowledgements The work presented in this chapter is supported by the Copenhagen Center for Health Techology, CACHET.

References

1. Mohamed Abdelgawad and Aaron R. Wheeler. Low-cost, rapid-prototyping of digital microfluidics devices. *Microfluidics and Nanofluidics*, 4(4):349–355, Apr 2008.
2. Am Amin, Mithuna Thottethodi, Tn Vijaykumar, Steven Wereley, and Stephen C. Jacobson. Aquacore: a programmable architecture for microfluidics. *Proceedings of the 34th annual international symposium on Computer architecture*, pages 254–265, 2007.
3. Vaishnavi Ananthanarayanan and William Thies. Biocoder: A programming language for standardizing and automating biology protocols. *Journal of biological engineering*, 4(1):13, 2010.
4. Krishnendu Chakrabarty and Yang Zhao. Digital Microfluidic Biochips: A Vision for Functional Diversity and More than Moore. *Lecture Notes in Electrical Engineering*, 105 LNEE:263–285, 2011.
5. Longquan Chen and Elmar Bonaccurso. Electrowetting - From statics to dynamics. *Advances in Colloid and Interface Science*, 210:2–12, 2014.
6. Ernst Fricke and Armin P. Schulz. Design for changeability (DfC): Principles to enable changes in systems throughout their entire lifecycle. *Systems Engineering*, 8(4):342–359, 2005.
7. Daniel Grissom, Christopher Curtis, Skyler Windh, Calvin Phung, Navin Kumar, Zachary Zimmerman, Kenneth O'Neal, Jeffrey McDaniel, Nick Liao, and Philip Brisk. An open-source compiler and PCB synthesis tool for digital microfluidic biochips. *Integration, the VLSI Journal*, 51:169–193, 2015.
8. Yiyan Li, Roger Chen, and R. Jacob Baker. A fast fabricating electro-wetting platform to implement large droplet manipulation. *Midwest Symposium on Circuits and Systems*, pages 326–329, 2014.
9. Yan You Lin, Randall D. Evans, Erin Welch, Bang Ning Hsu, Andrew C. Madison, and Richard B. Fair. Low voltage electrowetting-on-dielectric platform using multi-layer insulators. *Sensors and Actuators, B: Chemical*, 150(1):465–470, 2010.
10. Hyejin Moon, Sung Kwon Cho, Robin L. Garrell, and Chang Jin Kim. Low voltage electrowetting-on-dielectric. *Journal of Applied Physics*, 92(7):4080–4087, 2002.
11. Frieder Mugele and Jean-Christophe Baret. Electrowetting: from basics to applications. *Journal of Physics: Condensed Matter*, 17(28):R705–R774, 2005.

12. N Y Jagath B Nikapitiya, Mun Mun Nahar, and Hyejin Moon. Accurate, consistent, and fast droplet splitting and dispensing in electrowetting on dielectric digital microfluidics. *Micro and Nano Systems Letters*, i, 2017.
13. Jason Ott, Tyson Loveless, Chris Curtis, Mohsen Lesani, and Philip Brisk. BioScript: Programming Safe Chemistry on. 2(November), 2018.
14. Phil Paik, Vamsee K. Pamula, and Richard B. Fair. Rapid droplet mixers for digital microfluidic systems. *Lab on a Chip*, 3(4):253–259, 2003.
15. Francois Patou, Maria Dimaki, Winnie E. Svendsen, Klaus Kjaegaard, and Jan Madsen. A smart mobile lab-on-chip-based medical diagnostics system architecture designed for evolvability. *Proceedings - 18th Euromicro Conference on Digital System Design, DSD 2015*, pages 390–398, 2015.
16. Paul Pop, Mirela Alistar, Elena Stuart, and Jan Madsen. *Fault-Tolerant Digital Microfluidic Biochips: Compilation and Synthesis*. 2015.
17. Alberto Sangiovanni-Vincentelli and G. Martin. Platform-based design and software design methodology for embedded systems. *IEEE Design & Test of Computers*, 18(6):23–33, 2001.
18. Vijay Srinivasan, Vamsee K. Pamula, and Richard B. Fair. An integrated digital microfluidic lab-on-a-chip for clinical diagnostics on human physiological fluids. *Lab on a Chip*, 4(4):310–315, 2004.
19. Georgi Tanev, Winnie Svendsen, and Jan Madsen. A modular reconfigurable digital microfluidics platform. *Symposium on Design, Test, Integration and Packaging of MEMS/MOEMS, DTIP 2018*, pages 1–6, 2018.

Teaching Embedded System Design Among K-12 Students Based on Trees-Forest-View Methodology

Shengqi Yang, Kainian Xie, and Mansun Chan

Abstract Embedded system design is a complex engineering process and embedded system and product is becoming a common and critical part in industry and everyday life. Learning embedded computing is recognized as an essential skill for all students at an early age to develop in order to be competitive in an increasing digital world. Teaching embedded system design among students at an early age is getting more and more important. However, because of the complexity of embedded system, teaching how to design it is a challenging topic for undergraduate and graduate students in college. Apparently it is even more challenging and interesting how to introduce this topic with hands-on projects among K-12 students. In this research, a trees-forest-view methodology is introduced and embedded into two newly developed courses as a step-by-step way to get K-12 students to enter embedded system design world. Research and experiment results show that this methodology inspires big interest and achieves great feedback among a group of K-12 students.

1 Introduction

In China, there is a story almost knew to everybody: the blind men and the elephant. A group of blind men touch the elephant and try to figure out the elephant shape. One man who touched the elephant tooth said that the elephant is like a big carrot; the other one who touched the elephant ear said that the elephant is like a big fan. This story tells that people cannot make an overall judgment of something on the basis of one-sided viewpoint or draw a conclusion on the basis of partial understanding. There is another story to describe people who can see, however, only the trees but not the forest. It has one's view of the important overshadowed

S. Yang (✉) · K. Xie
Shanghai Moocxing Technology Co. Ltd, Shanghai, China

M. Chan
Hongkong University of Science and Technology, Clear Water Bay, Hong Kong

S. S. Bhattacharyya et al. (eds.), *Embedded, Cyber-Physical, and IoT Systems*,
https://doi.org/10.1007/978-3-030-16949-7_13

Fig. 1 (**a**) The blind men and the elephant (**b**) only see the trees but not the forest

by the trivia. Figure 1 shows these two stories. Embedded education is facing the similar situation for students at different levels.

Due to the common availability and popularity of embedded system and product, more and more parents, students, and teachers realize that learning embedded computing is an essential skill for students at different ages to stay competitive in an increasing digital world. At college level, a lot of instructors use ARM based boards plus C/C++ type of programming, or FPGA boards with some soft core to implement lab projects. For both ARM and FPGA platforms and other kinds of platforms, the design process is quite complex and involves a lot of knowledge from different domains, such as programming languages (C/C++, VHDL/Verilog), computer architecture and instruction set, board input/output/peripherals, Linux driver/OS, compiling/synthesis/optimization, interface/function/module, etc. If each piece/component of the whole system is treated as a tree, the whole system looks like a forest. Quite often with just a few projects in one class, student may get chance to know how the tree looks like while lost in the forest or how the forest looks like but miss the details of each tree. This problem can happen similarly during introducing embedded system design among K-12 students. How to solve this problem is the topic of this research.

In this study, two different levels of course are designed for the purpose of introducing embedded system design among K-12 students. Level 1 introduces electronics and circuit design to the student. With the basic understanding of electronic world, level 2 builds the knowledge of a mini tiny embedded system design (a self-driving smart car) for student. On one side, we expose all the individual units or trees view to students. On the other side, students have the chance to use these individual units to build their forest view of the system step by step. Research results show that our two level courses achieved the desired experience among K-12 students and build their confidence to learn more complex embedded system design.

2 Level 1 Course: Build Electronic Circuits

2.1 Motivation

Electronic circuit is the basic building block of the embedded system. Advancements of electron devices have revolutionized our modern world [1], both in terms of our lifestyle and industrial manufacturing activities. While electronic technologies are indispensable in our everyday life, students and youngsters do not have many chances to acquire the skill of circuit construction until a very late stage of their education. With the belief that interesting electronic projects can be introduced at a much earlier stage in our education system, we initiated a number of workshops and exploration camps to let students at the age of around 10 years start to explore the fascination of electronic circuits by constructing some interesting electronic gadgets [2] such as a running light indicator, an electronic piano, an infrared detector, etc.

Delivering such electronic circuit exploration workshops can achieve multiple goals. The hands-on reverse engineering approach is expected to arouse students' interest in electronics without in-depth understanding that usually takes much longer time to acquire. Once the students develop their interest, they will pursue the required knowledge without much guidance in this information-rich generation [3]. Secondly, this level 1 course is designed to help students to bridge the gap between understanding of abstract concept (circuit diagram) and implementing the abstract idea in a real environment with constrained rules.

2.1.1 Building in Mind

It is obviously true to us that building something, which seems very simple, in mind is very different from building it in reality, especially in constrained reality. The following figure shows a simple circuit (Fig. 2). It has four components: electrical wire, battery, resistor, and an LED light. Each component has two ends. By doing the cascading connection of each component's end, student can build this simple circuit. However this circuit building process is done in student's mind. As an experiment, we as instructor ask one student who is a third grade student that if we give him the components and explain how they are connected, can he get the connection done? His answer is very quick and straightforward: of course YES.

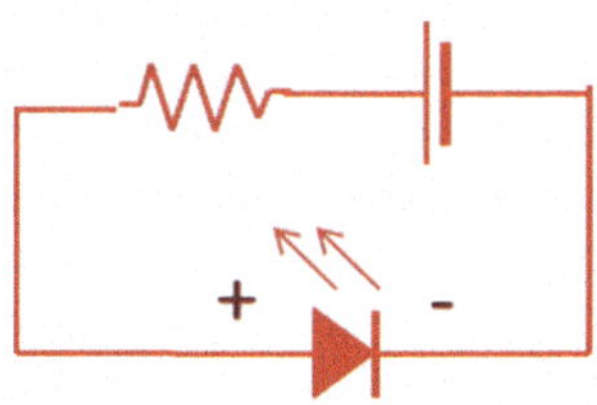

Fig. 2 A simple LED light circuit

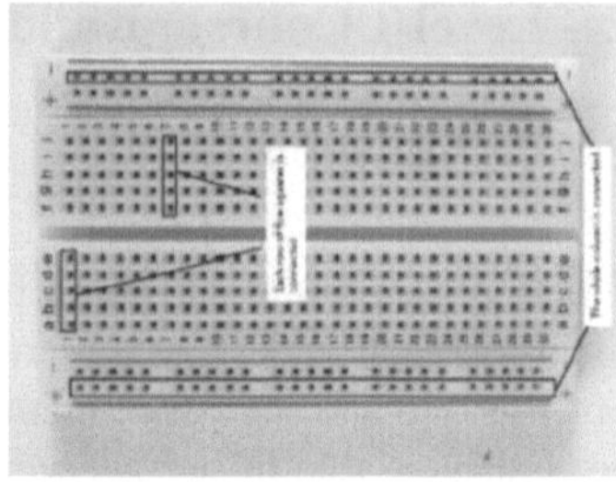

Fig. 3 Circuit building breadboard

2.1.2 Building in Reality

Then we showed the building block to this student: a breadboard as shown below. We told him the connection rules: all the holes in one row are electrically connected, while different rows are not connected. As an example, all holes in row 1 (a b c d e) are connected, all holes in row 2 (a b c d e) are connected, but row 1 and row 1 are isolated. The student told us that he can get the circuit diagram in mind quickly mapped to the board. However, he spent at least 30 min and still could not light up the LED light. He felt frustrated. This is what we expected because the circuit diagram is one dimensional in his mind, while it is two dimensional once being mapped to board. The transition is not easy (Fig. 3).

As the third goal for this level 1 course, it illustrates the trees-forest-view idea to the students. This course uses six individual projects to expose some tiny systems (such as digital piano, digital display) to students with forest view and work with them on the details of each component with tree view. Students have the full opportunity to build the small forest piece by piece and gain the depth knowledge of each piece while enjoying the final working system.

2.2 Delivery of the Level 1 Course

2.2.1 Build Fundamental

In the actual delivery of the introductory session, the workshop can start with the demonstration of how to use a wire stripper (Fig. 4 middle) as the basic tool.Since the hands-on approach is the essence of the workshop, participants should be given the first task of cutting and stripping wires in plastic coatings (Fig. 5 bottom-left). It is then natural to explain about electrical wires. A daily life analogy of water pipes can be used. When explaining about electrical wires, the two important quantities in electronic engineering, namely current and voltage, should be introduced. The corresponding units of current and voltage need to be included as well. With the concept of voltage introduced, a battery (Fig. 5 bottom-middle) can be depicted as a voltage source or generator. Based on the concepts of voltage and current, the concept of resistance can then be explained. The daily life example of water flow

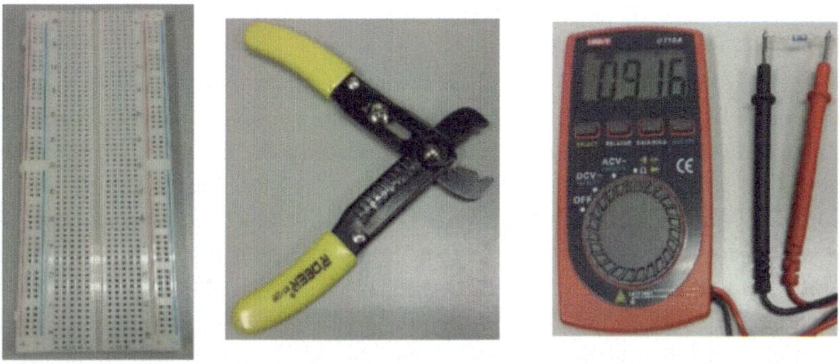

Fig. 4 Very basic tools required for electronic circuit construction: (left) a breadboard, (middle) a wire stripper, and (right) a digital multimeter

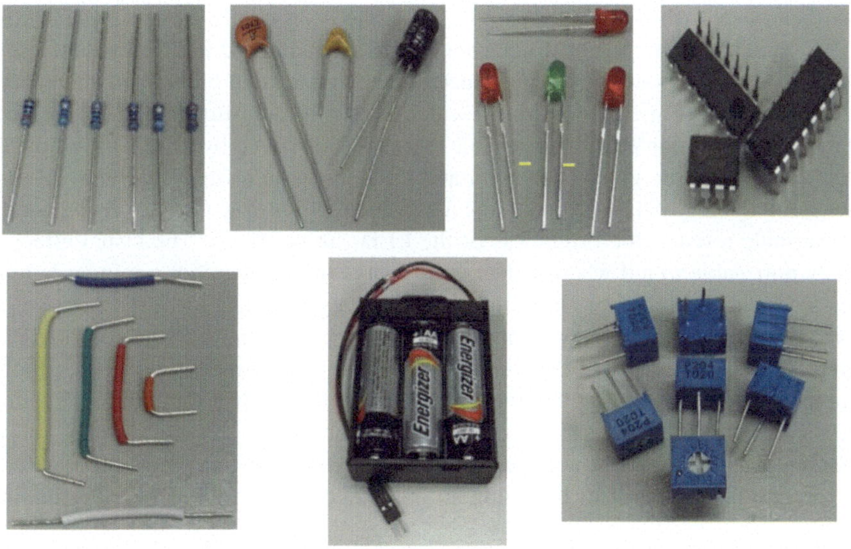

Fig. 5 Very inexpensive components readily available for electronic circuit construction: (top-left) resistors, (top-middle-left) capacitors, (top-middle-right) LEDs, (top-right) chipsets, (bottom-left) wires, (bottom-middle) batteries inside a battery box, and (bottom-right) variable resistors

can be used to illustrate the idea of resistance. After explaining about resistance, it is natural to proceed to the introduction of resistors together with the color code of discrete component resistors (Fig. 5 top-left). The second hands-on task would be reading the color code of a few resistors to figure out the resistance. Afterwards, the use of a multimeter (Fig. 4 right) for measurements of current, voltage, and resistance can be taught. With basic demonstrations, students should be able to learn easily how to use the multimeter (Fig. 4 right). The third hands-on task would be the measurements of the resistors and checking the resistance against the values read from the color code.

The next key apparatus to introduce is the breadboard (Fig. 4 left) as a handy circuit board for quick prototyping of electronic circuits. The properties of the breadboard should be carefully explained, especially the electrical connections of the pin holes along different rows and columns. As the fourth hands-on task, a multimeter (Fig. 4 right) can be used to verify the electrical connections along a row and a column and the electrical isolation between different rows. Without explaining the theory behind, the light-emitting diode (LED) (Fig. 5 top-middle-right) can be introduced as a light indicator which works only when current flows in one direction but not the other. Students should be taught how to identify the two different electrodes of an LED (Fig. 5 top-middle-right). With the positive and negative electrodes of the LED explained, the use of a battery box (Fig. 5 top-right), which has a positive and negative terminals as well as a switch to turn on the voltage, can then be described. This leads to the fifth hands-on task in using a battery box, an LED, and a resistor to construct a simple circuit of an LED light indicator (Fig. 6 left). The last component to be introduced in the first session of the workshop is the variable resistor (Fig. 5 bottom-right) which allows the adjustment of the resistance. Quick demonstration about the variable resistance measured by a digital multimeter (Fig. 4 right) can be given. The sixth and final hands-on task is simply modifying the circuit in the fifth task by including a variable resistor (Fig. 6 right). By adjusting the variable resistor, the brightness of the LED can be varied. The visual effect is easily noticeable to tell whether the circuit works or not. A minor reminder here is that some variable resistors have three pins. Wrong connections of the pins would give no change of the resistance and thus having no brightness change of the LED. Table 1 gives a summary of the components, electronic engineering concepts, tools, and the hands-on tasks taught in the first session of the electronic circuit exploration workshop. It can be seen that the delivery of such electronic circuit exploration workshop is both highly educational and of much fun. With this preparation, we can move onto the first project involving an integrated circuit to generate the running light.

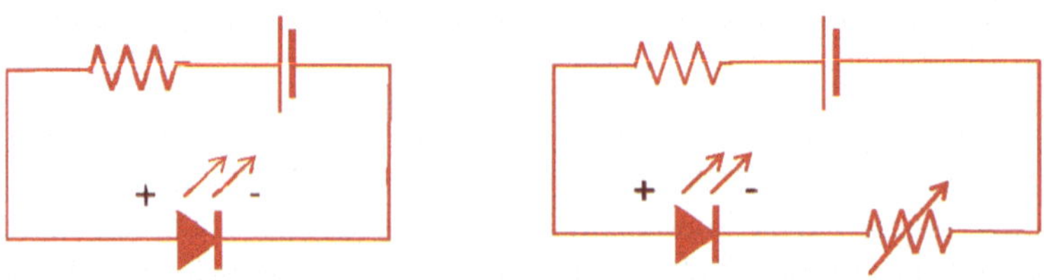

Fig. 6 Simple basic LED indicator circuit for hands-on circuit construction tasks with noticeable visual effects when working (left) with fixed brightness; (right) with a variable resistor for adjusting the resistance and hence the brightness

Table 1 List of components, electronic engineering concepts, tools, and hands-on tasks for an introductory session on electronic circuit exploration

Component	Engineering concept	Tool	Hands-on task
Electrical wires	Current and voltage and the corresponding units (ampere and volt)	Wire stripper	Cutting and stripping wires
Resistors	Resistance	n/a	Reading the resistance values from the color codes of resistors
Multimeter	DC measurement	Multimeter	Measurements of the resistance of resistors
Breadboard	Circuit building platform, connection	Multimeter	Verification of the electrical connections of a breadboard
Light-emitting diode (LED)	Unidirectional current flow	n/a	Light up an LED
Battery box	DC power source; distinguishing positive and negative terminals	Screw driver to put batteries inside	Circuit construction of a simple light indicator
Variable resistor	Variable resistance	Multimeter	Circuit construction of a simple light indicator with adjustable brightness

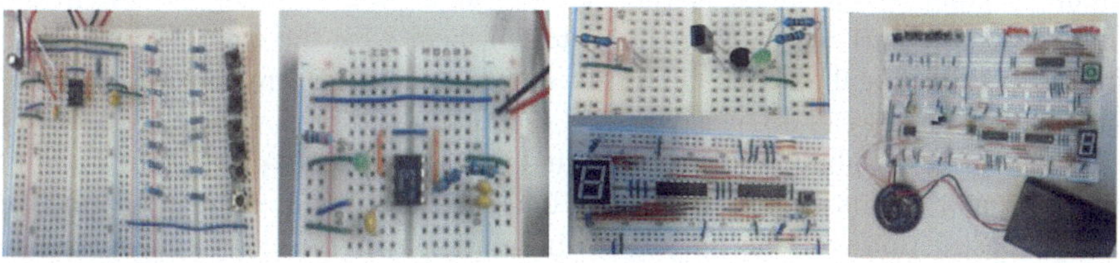

Fig. 7 Hands-on circuit building projects

2.2.2 Start the Trees-Forest-View Process

In this level 1 course, we designed six individual projects as illustrated below (Fig. 7). Here we use one project to illustrate the trees-forest-view process, a running light LED indicator circuit.

Showing the Full Forest View to the Students

A running light indicator circuit is a good choice for this purpose because it uses essentially one IC, several LEDs and resistors as well as a switch. The students by now should be familiar with LEDs and their use with a resistor connected in series

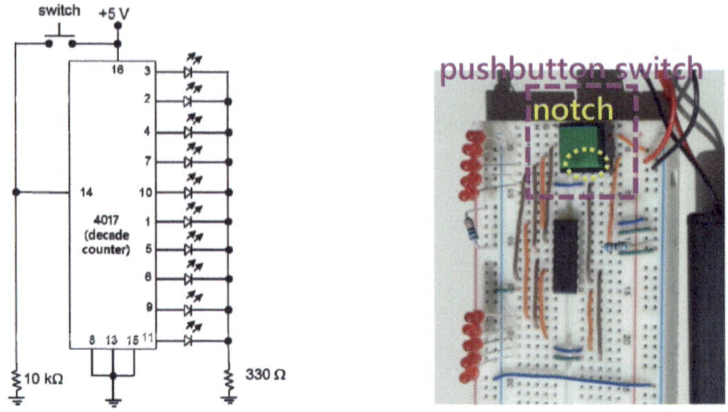

Fig. 8 LED running light indicator circuit using a CMOS 4017 decade counter IC: (left) schematic diagram; (right) circuit constructed on a breadboard

to control the electric current and hence the brightness. The IC is the only new component in this circuit.

In the actual delivery of the circuit construction session, the circuit schematic diagram (Fig. 8 left) can be first shown to tell the students what circuit is to be built. The students may not even understand what the circuit is by looking at the circuit diagram. To ease their possible confusion, a photo of the neatly assembled circuit (Fig. 8 right) can then be shown to tell them what the circuit is like. With the photo shown or the actual circuit for demonstration, the instructor can briefly tell what the circuit can do by saying that each LED will be turned on sequentially by pressing the pushbutton switch once and then so on and so forth. Such a brief explanation or demonstration should give the kids encouragement in attempting the circuit construction with the forest view in mind. Having visual effects controlled by a switch is part of the reason for choosing such a circuit for the workshop. It is possible that the students may not be impressed by the demonstration of the circuit and they have little idea of what the circuit is for. To arouse their interest, the instructor can tell examples of application of such LED display circuits. One common example for students living in large cities is the route map panel with LED indicators showing the metro stations (Fig. 9). The circuit (Fig. 8 left) can also be adapted to make traffic lights. If the workshop time allows, it can also bring up the engineering advantages of using LEDs for displaying information related to fixed positions and information that remains unchanged for months or even a few years. By showing this full system or the forest view, students get a high level picture of what it looks like and how it works and are prepared to dive into details of each piece of this whole system.

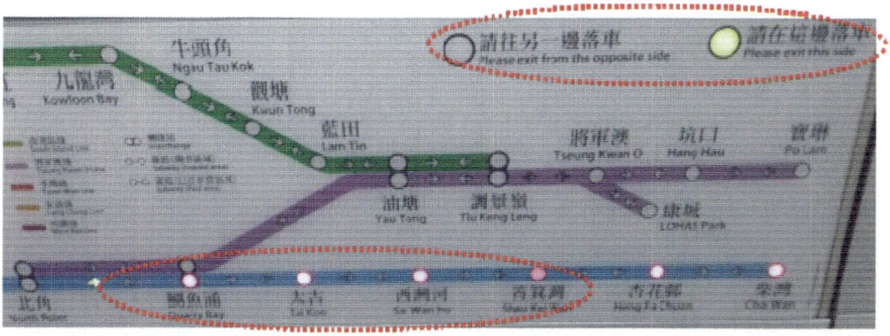

Fig. 9 Example of application of LED display circuits: the route map panel with LED indicators showing the metro stations

Building Forest with All the Trees

Reading Circuit Schematic Diagrams

After giving the students an overall idea about the LED light indicator circuit and its extended applications in daily life, the instructor can move onto explain a general idea of a circuit schematic diagram or simply schematic for short. A few key points can be covered quickly in a concise way. Firstly, the circuit schematic is the language of representing the actual electronic circuits in written form showing the connections of various devices and components. Secondly, it is common that the schematics may not show the devices and components as they are in the actual circuits. Since the students are not expected to learn the theory or complicated concepts, in-depth details should be avoided in teaching the ideas above. This also helps students to prevent losing forest view of the mini system.

IC Chip: Mounting onto Breadboard

On reading the circuit schematic, students should notice a rectangular box with 16 connections that represents the IC chip of the CMOS 4017 decade counter [4] to be used. Although it can be explained that a decade counter is a digital circuit which counts from zero to nine, there is no need to explain the operation principles. As the first hands-on task, the student is asked to mount the IC chip onto the breadboard, with the two columns of pins straddling the middle channel/track of the breadboard.

Circuit Construction Divided into Three Parts

A typical engineering technique, namely "divide and conquer," can be shared with the students. The construction of the apparently difficult circuit would be much easier by dividing the circuit connections into simpler parts then building one by one (Fig. 10). All electronic circuits need electrical power to function. It is sensible

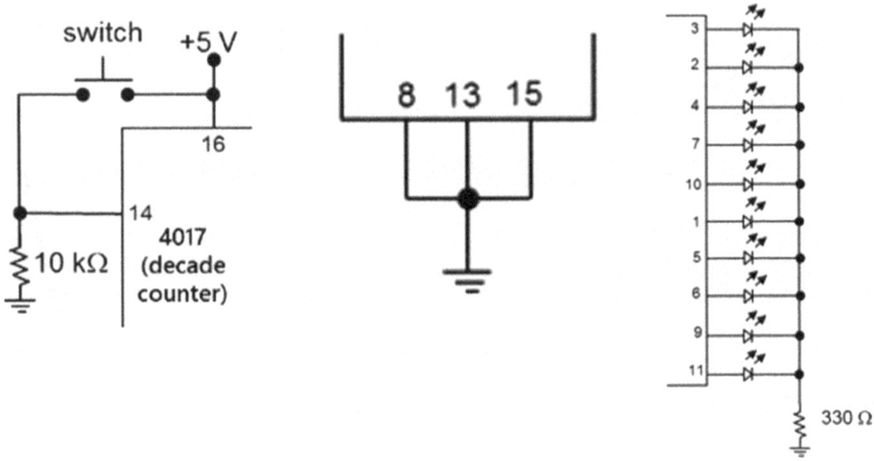

Fig. 10 Construction of the LED running light indicator circuit divided into three parts: (left) power supply part, (middle) pin connections directly to ground, and (right) decade counter output pins connected to ten LEDs in sequence

Table 2 Key steps in the construction of LED running light indicator circuit with the underlying engineering concepts and the components or tools used

Step or task	Component	Engineering concept	Tool
Reading circuit schematic diagram		Positive power supply and ground	
Mounting and dismounting IC chip	4017 decade counter IC chip	Standards of IC pin spacing and position	A pair of forceps and a breadboard
Pin identification of packaged ICs	16-pin decade counter IC	Index marks (notch and dimple) and pin numbering	
Circuit construction part 1	IC, pushbutton switch, resistor, hook-up wires	Divide and conquer; pin connections one by one	A wire stripper
Circuit construction part 2	Hook-up wires	Neat wiring for easy troubleshooting later	
Circuit construction part 3	Ten LEDs, resistor	Common rail tied together	

to build the circuit starting from the power supply connections (Fig. 10 left) and the ground counterpart too (Fig. 10 middle).

Table 2 lists the key steps during the construction process and all the underlying engineering concepts student will learn by composing each of the units or trees into the working running light LED indictor as the final forest.

Fig. 11 Working but not well built circuits (wiring is messy, not neat)

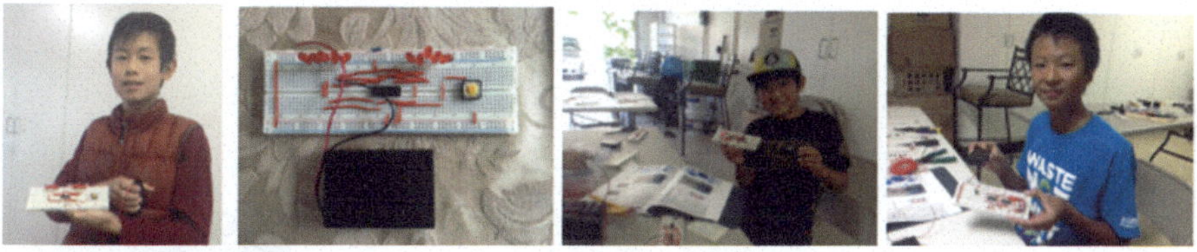

Fig. 12 Working and well laid out circuits

Build Circuit with Better Layout

At the beginning of the class, the easy mistake that students make is bad (messy, not neat) layout of the wires during the connection process as shown in Fig. 11. This messy layout will increase the debugging effort if the finished project does not work. In many cases, a lot of wire sections need to be removed in order to check if one piece of particular wire is connected well or not. The rule is that all wires need to be flatted down on the board without crossing each other and they cannot jump or dance around the breadboard. Should we teach students to avoid this kind of bad layout or not at the beginning of the class? The answer is NO. Within all the six projects in this level class, the first one or two projects are used to teach student lessons on how important it is to build the circuit with good layout in order to avoid expensive debugging time. This making mistake and improving working habit process is also a great way to teach students engineering process. After this building skill ramping period, students gradually build the circuit more and more neatly. Figure 12 shows some finished projects neatly built by students.

2.3 Summary of the Level 1 Course

In this level 1 course, we first show the students how the whole mini system or the mini forest looks like and how it works. Using this forest view, students' interest is inspired and their desire to know the inside details or how each tree is composing the forest is becoming strong. Moving forward we explain each piece of the system without too much details to students and avoid losing forest view in students' mind. By constructing all the pieces together, students get knowledge of each individual

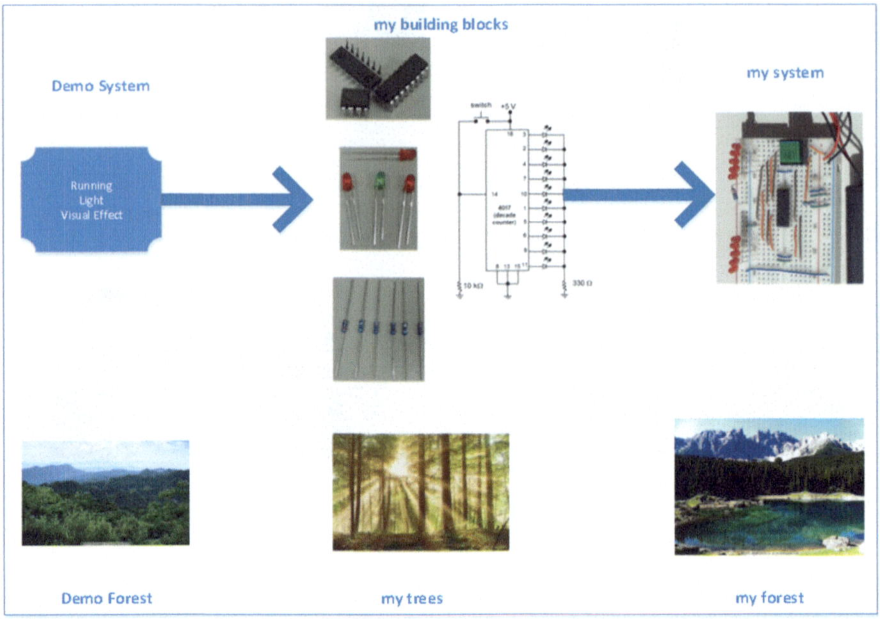

Fig. 13 Trees-forest based teaching methodology

electronic component and how it works and how to integrate them together into the mini whole system. With this process as shown in Fig. 13, students not only learn in depth of the component but also always keep the whole mini system in mind and never lose the view of the forest. This trees-forest-view process helps the students to learn with interest and a full picture.

3 Level 2 Course: Build a Mini Embedded System

3.1 Motivation

Electronic circuit is the important part of a typical embedded system. Level 1 course is designed to help students understand basic concepts and master hands-on experience on how to build electronic circuits. After electronic circuit is introduced to students, a mini embedded system, a smart self-driving car, is developed as the level 2 course.

It is well known that most electronic systems are embedded system. Due to the popularity and common availability, learning embedded system design is an important and critical skill set in digital world. However, due to the design complexity of an embedded system, introducing this topic among K-12 students is very challenging. A suitable, simple enough mini system is desired to deliver this topic among young students. Further embedded system design is a hardware and software co-design process. Although it is very difficult to introduce this concept in depth to students, the course has to touch some of the basics of hardware and software. For software, choosing a simple programming language is required because for most of the students, it is the first time they do programming.

For designing the hardware platform, there are three rules to follow. First, it needs to be simple enough; second, the mini system must be interesting to students; third, the system can be partitioned and assembled piece by piece by a K-12 student (age 8 above). If the mini system is highly integrated, it only gives the forest view to students and hides the details or blocks the tree view of it. Based on the above analysis and rules, we have designed a self-driving smart car which meets the above requirements. This system is very simple and the list of components is shown in Table 3. Figure 14 shows the pictures of the components. As this level 2 course was taught in several technology camps along with other level courses, it was shown to be the most attractive project for students. This confirmed that this mini system meets the above three design rules. Further, this mini system supports our trees-forest-view teaching methodology which allows students to have a full forest view while getting the details of each of the components or trees view.

For choosing the software programming, most K-12 students do not have programming experience and this course may be the first time they touch programming. As a result, the programming language must be simple enough for them to understand condition, control, loop concepts. In this project, we chose Arduino based C-type programming language which meets our requirements. There are some literatures which discussed and compared the advantages and disadvantages for block-based and text-based programming. Arduino does support both text-based and block-based programming. Traditional computer science programming environment is normally text-based programming. Quite a few IDE environments have adopted

Table 3 List of the components inside the mini embedded system (a self-driving smart car)

Components	Number of pieces	Functions
Wheel servo	4	Driving the wheels
Wheels	4	Car wheels
Car body	1	Car assembly platform
Arduino UNO	2	Control signal
Mini board (custom designed)	1	Voltage output to drive the wheel servo
Ultrasonic sensor	1	Sensing the distance between car and obstacles
Mini servo	1	Rotate the sensor
Mini breadboard	1	Connector/bridge
Bluetooth transceiver	1	Sending/receiving Bluetooth signals
Bluetooth control board	1	Handle control
Battery	1	One set with 9 V
Screws	*n*	Support, stabilization, etc.

Fig. 14 Components of the mini embedded system, a self-driving smart car; top-left (the IDE programming environment), top-middle (mini board and Arduino UNO), top-right (ultrasonic sensor, mini servo, Bluetooth handle); bottom-left (wheels with servo), bottom-middle (car body and wheel servo), bottom-right (the mini board for voltage supply to the servo)

the block-based programming approach to lower the programming barrier across a variety of different domains. The representative block-based approaches of visual programming include Scratch [5], Snap! [6], and Blockly [7]. Arduino provided a block-based IDE programming environment as shown in Fig. 13 (top-left). Weintrop [8] concluded that students in both conditions improved their scores between pre- and post-assessments; however, students in the blocks condition showed greater learning gains and a higher level of interest in future computing courses. Students in the text condition viewed their programming experience as more similar to what professional programmers do and as more effective at improving their programming ability.

Our experiments during a few camps did not agree with the finding result in [8]. Here two camps' data is used. There were totally 22 students in the NO 1 camp and 25 students in the NO 2 camp with different age profile as shown in Table 4. For the NO 1 camp during most of the time, students used block-based programming and only one project with text-based programming was introduced in the middle as a comparative study. In contrary, for the NO 2 camp during most of the time, students used text-based programming and only one project with block-based programming was adopted in the middle. The results are shown in Table 4.

As seen from the table, for both camps, block-based programming seems more interesting to most of the students and this is especially true for younger group of students (age 7–10). For this special age group, almost all of them (except one 10 year old student) think block-based programing is more of interest. For camp NO

Table 4 Comparative study of block-based programming and text-based programming

Camp NO	Age	Number of students	Block based is more interesting (Y)
1	7–10	5	5/5
	11–14	10	10/10
	15–16	7	7/7
			New project to use block based (PR)
	7–10	5	2/5
	11–14	10	8/10
	15–16	7	6/7
2			Text based is more interesting (Y)
	7–10	8	1/7
	11–14	12	2/12
	15–16	5	3/5
			New project to use text based (PR)
	7–10	8	4/8
	11–14	12	12/12
	15–16	5	5/51

1, at the end of the class, we tested students on a problem solving using block-based programming. The test results revealed that although block-based programming is more interesting to young students, they seem not fully understand the block relation and sequence order and only two out of five students can pass the test. For camp NO 2, by forcing the students to do project more frequently using text-based programming, young students with age 7–10 has improved performance for the programming test and four out of eight can pass. This improvement suggests that text-based programming is more efficient and direct to deliver the programming concepts such as condition, loop, etc. Further for old student age 11–14 and age 15–16, their test passing ratio (PR) is also improved.

As a result of the above observation, in our camp, for most of the time, we used text-based programming and we adopted block-based programming for some simple projects. The goal of this level 2 course is to build students' knowledge on embedded system design based on the trees-forest-view methodology as shown in Fig. 15.

3.2 *Delivery of the Level 2 Course*

3.2.1 Build Fundamental

For K-12 students, the fundamental concepts of a mini embedded system include two parts: hardware and software. When delivering these concepts, instructors should not dive into a lot of details which can cause students' confidence loss and forest view loss of the whole system. For the hardware part, the instructor focuses on showing the important components on the board and explaining a little bit of their

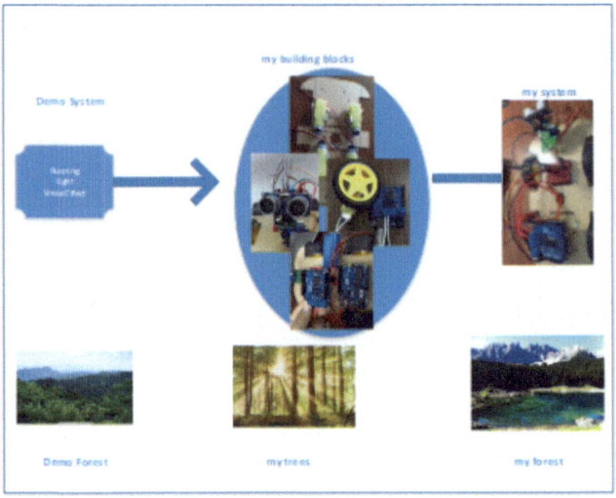

Fig. 15 Trees-forest-view to teach mini embedded system design

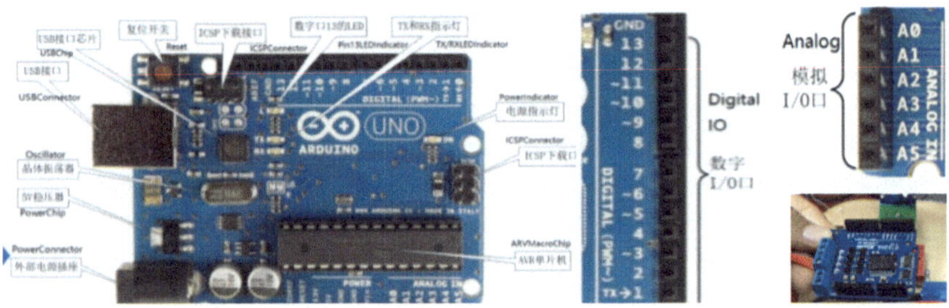

Fig. 16 Introducing the components of an Arduino UNO board; (left) the Arduino board, (middle) the digital pins, (right) the analog pins

functionalities, pointing to the digital and analog voltage input and output pins, and showing the customized board which can output the right voltage level and drive the wheel's servo. The important point for this hardware view is to try to keep them interested while not trying to let them understand all the details about how these components work and this is outside the scope of this level 2 course (Fig. 16).

For the software part, simple Arduino programming guide is reviewed with students within the Arduino IDE environment. For the IDE environment as shown in Fig. 17 (left), students learn how to compile and verify their code and how to identify the possible bug and compiling issues. For the programming language, students are illustrated how to use Arduino C type of language to code the two bodies, void setup() and void loop(). Particularly, students need to understand that loop() body will be executed forever after the board is powered on.

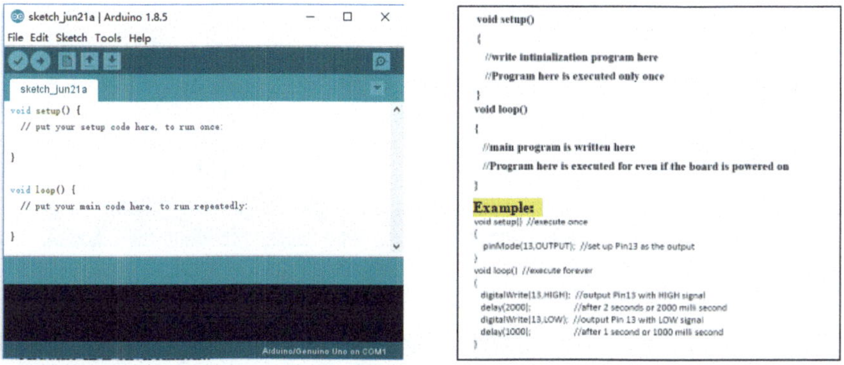

Fig. 17 Simple and intuitive introduction of software program

3.2.2 Start the Trees-Forest-View Process

In this level 2 course, we designed seven individual projects as shown in the following list. Through all these projects, students go through the trees-forest-view learning process to learn how to build a mini embedded system step by step, easy to difficult and trees focus with forest view in mind.

- Step-by-step programming environment setup
- Code and test the Arduino board with running lights
- Code to drive the servo motors
- Hands-on assembling the racing car with motors, sensors, etc.
- Coding sensor for object detection
- Integrate the car with a Bluetooth handle control
- Self-driving design and test of the smart car

Showing the Forest View to the Students

This self-driving smart car shown in Fig. 18 constitutes of three major hardware parts: car body (servo, wheels, and body), programming boards, and ultrasonic sensor. Here divide and conquer method is reinforced and used again to build this mini system from scratch step by step.

Test the Board

In the level 1 course, students already get familiar with LED light and running LED indicator. Here the circuit is used again, but built on Arduino board. To light up one LED light as shown in Fig. 19 (left), the PIN 13 on the Arduino board is used as voltage source and PIN GND is used as voltage sink to close the circuit loop. Figure 20 shows the programming code. Here DELAY concept is introduced to

Fig. 18 The finished self-driving smart car: (left) top view, (right) bottom view

Fig. 19 Lighting up one LED and running LED light

```
void setup()
{
  pinMode(13, OUTPUT);
}
```

```
void loop()
{
  digitalWrite(13, HIGH);  // LED ON
  delay(1000);             // Last for 1000ms
  digitalWrite(13, LOW);   // LED OFF
  delay(1000);             // Last for 1000ms
}
```

```
void setup()
{
  pinMode( 4 , OUTPUT);
  pinMode( 7 , OUTPUT);
  pinMode( 10 , OUTPUT);
}
```

```
void loop()
{
  digitalWrite( 4 , HIGH );
  digitalWrite( 7 , LOW );
  digitalWrite( 10 , LOW );
  delay( 1000 );
  digitalWrite( 4 , LOW );
  digitalWrite( 7 , HIGH );
  digitalWrite( 10 , LOW );
  delay( 1000 );
  digitalWrite( 4 , LOW );
  digitalWrite( 7 , LOW );
  digitalWrite( 10 , HIGH );
  delay( 1000 );
}
```

Fig. 20 Coding the LED light and running LED light

Fig. 21 Use Arduino board to drive a single wheel. (left): wheel with servo; (middle): custom board MOTOR shield; (right): how to drive the servo with Arduino pins

students for how long it is desired to keep the LED ON or OFF. As horizon extension projects, students are asked to program the LED light with running speed faster and faster. In this project, language LOOP is introduced.

Driving One Wheel with Servo

In this step, students learn how to drive one wheel forward and backward with the servo and Arduino board. Figure 21 illustrates the setup. In the previous section, students learn how to use the pins on the Arduino board to output voltage as driving source. The same design approach is applied here as well. The custom designed board, MOTOR SHIELD, is used to drive the wheel servo. Here the instructor does not need to go through details on why the motor shield board connected with Arduino board works. However students need to know that the pin 4, 5, 6, and 7 from Arduino board is connected to motor shield board pin A+/A−/B+/B− which can be used to connect to the servo two input ports and drive the servo. In this section, students will not be given the sample code. They need to program Arduino board based on the illustration or pseudo code as shown in Fig. 20 (right).

Driving the Car

In this step, students learn how to drive the car with hands-on assembly. Figure 22 (left) shows the top view of the car bottom side which has four servo motors. It is not an easy task for young students to assemble the servo motors onto the car body. The assembly process contains a lot of screws which need to be carefully faced to one side and the servos also need to position themselves with one particular side internally or externally based on the screw siding. Normally this process need to be repeated for a few times and students try out and then understand the correct way to assemble all the servo motors correctly. The average assembly time for a 10-year old student is around 45 min with help from instructor. Figure 22 (right) shows the finished assembly. There is another trick during the assembly and testing: how to make sure that the two wheels on one side of the car are synchronized. For the two

Fig. 22 (Left): top view of the car bottom side; (right) top view of the car top side

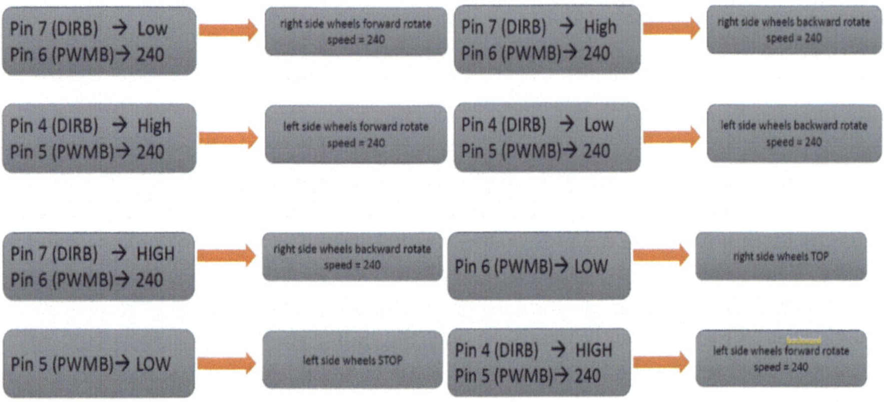

Fig. 23 (Top-left): drive the car forward; (top-right): drive the car backward; (bottom-left): turn the car right; (bottom-right): should the left side wheels drive backward or forward?

wheels on one side, they are controlled by one pair of output ports from the motor shield board, either pair of A+/A− or pair of B+/B−. Because they share the same control, if not connected correctly, there are chances that one wheel drives forward and the other wheel drives backward after power on. Students need to figure out the reason and fix the issue. As shown by all our camps, this process makes students to feel that they are very hands-on and know how to find, try out, and fix the problems. This is a very important component of this course: giving the chance to students to create problem, try out solution, and fix the problem.

After the car assembly is done, it comes to the programming part. Here it is assumed that Arduino board pin 6/7 is used to control the right side wheels, while pin 4/5 is used to control the left side of the wheels. Figure 23 shows the logic to set the pins and drive the car forward, backward, turning right, and turning left. In this section some open questions can be given to students as well to test their understanding, such as the open question shown in Fig. 22 (bottom-right).

Fig. 24 (Left): all the components to mount the ultrasonic sensor on board; (middle) the assembled sensor and the micro servo to rotate the sensor board; (right): mounting the sensor to the car body and connecting the pins to the Arduino board

Again, in this section, student will not get the sample code. Instead, they need to figure out the programming based on the logic shown in Fig. 23.

Ultrasonic Sensor

For this mini embedded system, ultrasonic sensor is used to detect object/obstacle around the car and how far the distance is between the sensor/car body and the object/obstacle. Figure 24 (left) shows all the components to mount the sensor and the micro servo to the car body. The micro servo is used to rotate the sensor to 0°, 90°, and 180°. When the ultrasonic sensor is positioned at 90°, the detection is for the distance between the sensor and the object in front of the car. When it is positioned at 0°, the detection is between the sensor and the object on the left side of the car. For 180°, the sensor measures the distance between itself and the object on the right side. With this detection, the program will make decision whether the car should continue driving forward, or stop, or turn left or turn right. Here again, a lot of screws are needed for the installation. Students achieve great experience for hands-on system assembly.

After the assembly, students are asked to program the board implementing the task functions as listed in Table 5. The purpose of these tasks is to help student understand how a simple self-driving is implemented. The basic idea is to use ultrasonic sensor detecting objects around the car and to find a direction where the car should drive toward. Also this section prepares students with all the needed details before final integration of the self-driving car.

Final Implementation

The simple algorithm to implement the self-driving is shown in Fig. 25. Students are required to refer to this algorithm technology camp pictures and implement the self-driving car. For this project, normally a 10-year old student takes 60 min to finish (with reference to some existing code in the previous sections), 20–30 min to fix all the compiling issues, and 30 min to test drive the self-driving feature.

Table 5 List of tasks to test sensor and distance measurement

Project	Tasks	Purpose
1	Turn the sensor to 0°	Test sensor position
	Hold the position for 1 s	
	Turn the sensor to 90°	
	Hold the position for 1 s	
	Turn the sensor to 180°	
	Hold the position for 1 s	
	Stop	
2	Measure the distance between sensor and objects	Test the distance measurement
	Use Arduino IDE serial monitor to print the distance	
	Stop	
3	Measure the distance between sensor and objects	Combine the sensor rotation and distance measurement
	If distance is <20 cm	
	Rotate the sensor to 0° and measure the distance	
	Rotate the sensor to 180° and measure the distance	
	Rotate the senor to 90° facing to the front	
	End If and Stop	

```
1. Drive the car forward
2. Detect the distance D1 between the sensor and possible ojbect in front
3. If detected distance D1 is less than or equal to 30Cm
4.    Stop the car
5.    Turn the ultrasonic sensor to left : 0 degree
6.    Wait for 1000 msec
7.    Detect the distance D2 between the sensor and possible object on the left
8.    Wait for 100 msec
9.    Turn the ultrasonic sensor to right: 180 degree
10.   Wait for 1000 msec
11.   Detect the distance D3 between the car and possible object on the right
12.   Wait for 100 msec
13.   Turn the ultrasonic sensor to facing the front: degree 90
14.   Wait for 500 msec
15.   If D2>=D3 turn left
16.   else, turn right
17.   End if
18. Else continue forward
19. End if
```

Fig. 25 Simple self-driving algorithm

3.3 Summary of the Level 2 Course

While keeping the full forest view of the whole mini embedded system in students' mind, this level 2 course goes through each component with the tree view without too many details. Further this project involves a lot of hands-on assembly. Young students, as demonstrated in all our technology camps, have the most sense of achievements after they finish the car assembly step by step.

4 Experiments

In this section, we describe how to measure the success of this trees-forest-view based K-12 teaching methodology. There are two traditional ways to do this kind of measurements: score based and questionnaire based. For this level 1 and level 2 program, it is not appropriate to test the students whether they can memorize some basic knowledge with the score based exams. Their learning experience is very hard to be quantized by score. Instead of exam, here questionnaire based approach is used. Table 6 lists all the past technology camps. Column 1 gives the camp time; column 2 and 3 mean the number of students attending each camp; column 4 indicates how many students were invitation based; column 5 gives the number of students who were recommended to this program. At the beginning of the program, because the program was not known to the public, we have to do some invitation and advertisement. Gradually more and more parents and students got to know the program and were recommended to these camps. This is a good indicator that these two programs achieved recognition among students and parents.

Table 7 shows the result on whether these programs are interesting to the students and whether they are willing to attend the next level course. As results demonstrated, our program is different from the traditional classroom courses and it is very interesting to the students. This confirmed that one of our initiatives, enhancing students with hands-on technology projects through interest driven approach, is successful.

Table 8 lists the critical questions in the questionnaire and answers from all the students who attended the past technology camps. From the results we can see that

Table 6 List of the camps and statistics data

Camp time	Student @ level 1	Student @ level 2	Invitation based	Recommendation based
12-2015	3	0	2/2	0/2
08-2016	12	12	5/12	7/12
12-2016	22	22	10/22	12/22
08-2017	25	25	3/25	22/25
12-2017	25	25	4/25	21/25
06-2018	50	50	7/50	43/50

Table 7 Interesting indicator of all the past camps

Camp time	Student @ level 1/2	This course is interesting (%)	Willing to attend the next level (%)
12-2015	3	100	100
08-2016	12	100	100
12-2016	22	100	100
08-2017	25	100	100
12-2017	25	100	100
06-2018	50	100	100

Table 8 Questionnaire and answers

Questionnaire summary
Q: Which part of the program is the most interesting one to you?
A: >90% of students choose hands-on assembly
Q: Do you have the full picture of each of the mini systems in mind?
A: >83% of students choose YES
Q: Do you know enough details of each component to finish the mini systems?
A: >79% of students choose YES
Q: After the course, do you feel confident to finish a different mini system by yourself?
A: >68% of students choose YES
Q: Are you willing to work in engineering field in the future?
A: >87% of students choose YES
Q: Do you want to recommend the course to your friends?
A: 100% of students are willing to recommend this.

more than 90% of the students enjoy the hands-on assembly process which gives them the most sense of achievement and the feeling of how the whole system works with each component. Through this trees-forest-view teaching methodology, more than 79% of students can get enough knowledge of each of the components while keeping the full picture of the whole system in mind. Some of the young students (8–10 year old), due to the level of understanding capability of a complex system, still have some gap on the trees-forest teaching process. This is expected because the mini system is complex enough and its level of difficulty is above some 8–10 year old students. This is further reflected in the next question about the confidence to finish another mini system by themselves. The level of confidence is around 68% and most young students (<11 year old) still feel they are not fully ready yet. However these two programs gave them enough encouragement to continue the purse of engineering activities in the future. Importantly they strongly recommend this program to their friends.

5 Conclusions

Teaching embedded system design among students at different levels and ages is getting more and more important. However, because of the complexity of embedded system, teaching how to design it is a challenging topic for undergraduate and graduate students in college. Apparently it is even more challenging and interesting how to introduce this topic with hands-on projects among K-12 students. The design process is quite complex and involves a lot of knowledge from different domains. If each piece/component of the whole system is treated as a tree, the whole system looks like a forest. Quite often with just a few projects in one class, student may get chance to know how the tree looks like while lost in the forest or how the forest looks like but losing the detail of each of the trees. In this study, two different levels of courses are designed. Level 1 introduces electronics and circuit to the student. With the basic understanding of electronic world, level 2 builds the knowledge of a mini tiny embedded system (a self-driving smart car) for student. On one side, we exposed all the individual units or trees to students. On the other side, students have the chance to use these individual units or trees to build/grow their forest system or a tiny whole system step by step. At the end we evaluated the success of this program. The results show that most of the students have the most sense of achievements by hands-on finishing a mini system assemble and make it work through programming or connection. This program attracted a lot of both girls and boys to attend. Most of them think this class is quite different from their classroom, but interesting to help them understanding engineering process (Figs. 26 and 27).

Fig. 26 Technology camp pictures

Fig. 27 If you have the elephant in mind, even if you are blind, you can draw the full picture of an elephant (right) the same applies to the trees-forest view (detail view and full view)

References

1. Joachim N. Burghartz (editor), Guide to State-of-the-Art Electron Devices, West Sussex, UK: Wiley-IEEE Press, 2013
2. Mansun Chan, "EDS Launches a Series of Exploration Camps in Region 10 to Cultivate Future Electronic Engineers", IEEE Electron Devices Society (EDS) Newsletter, vol. 24, no. 2, pp. 32-34, April 2017
3. Charles Platt, Make: Electronics: Learning Through Discovery, 2nd Edition; San Francisco, California: Maker Media (2015)
4. Charles Platt, Make: More Electronics: Journey Deep Into the World of Logic Chips, Amplifiers, Sensors, and Randomicity; San Francisco, California: Maker Media (2014)
5. M. Resnick, B. Silverman, Y. Kafai, J. Maloney, A. Monroy-Hernández, N. Rusk, E. Eastmond, K. Brennan, A. Millner, E. Rosenbaum, and J. Silver. 2009. Scratch: Programming for all. Commun. ACM 52, 11 (November 2009), 60
6. B. Harvey and J. Mönig. 2010. Bringing "no ceiling" to Scratch: Can one language serve kids and computer scientists? In J. Clayson & I. Kalas, eds. Proceedings of Constructionism 2010 Conference. 1–10
7. N. Fraser. 2015. Ten things we've learned from Blockly. In Proceedings of the 2015 IEEE Blocks and Beyond Workshop (Blocks and Beyond). 49–50
8. D. Weintrop, Comparing Block-Based and Text-Based Programming in High School Computer Science Classrooms, ACM Transactions on Computing Education, Vol. 18, No. 1, Article 3. Publication date: October 2017

The manufacturer's authorised representative in the EU is Springer Nature Customer Service Centre GmbH, Europaplatz 3, 69115 Heidelberg, Germany. If you have any concerns regarding our products, please contact ProductSafety@springernature.com

Printed and bound by CPI Group (UK) Ltd, Croydon, CR0 4YY
24/07/2026
02175970-0004